ras Oncogenes

NATO ASI Series

Advanced Science Institutes Series

A series presenting the results of activities sponsored by the NATO Science Committee, which aims at the dissemination of advanced scientific and technological knowledge, with a view to strengthening links between scientific communities.

The series is published by an international board of publishers in conjunction with the NATO Scientific Affairs Division

A	**Life Sciences**	Plenum Publishing Corporation
B	**Physics**	New York and London
C	**Mathematical**	Kluwer Academic Publishers
	and Physical Sciences	Dordrecht, Boston, and London
D	**Behavioral and Social Sciences**	
E	**Applied Sciences**	
F	**Computer and Systems Sciences**	Springer-Verlag
G	**Ecological Sciences**	Berlin, Heidelberg, New York, London,
H	**Cell Biology**	Paris, and Tokyo

Recent Volumes in this Series

Volume 166—Vascular Dynamics: Physiological Perspectives
edited by N. Westerhof and D. R. Gross

Volume 167—Human Apolipoprotein Mutants 2: From Gene Structure to
Phenotypic Expression
edited by C. R. Sirtori, G. Franceschini,
H. B. Brewer, Jr., and G. Assmann

Volume 168—Techniques and New Developments in Photosynthesis Research
edited by J. Barber and R. Malkin

Volume 169—Evolutionary Tinkering in Gene Expression
edited by Marianne Grunberg-Manago, Brian F. C. Clark,
and Hans G. Zachau

Volume 170—*ras* Oncogenes
edited by Demetrios Spandidos

Volume 171—Dietary ω3 and ω6 Fatty Acids: Biological Effects
and Nutritional Essentiality
edited by Claudio Galli and Artemis P. Simopoulos

Volume 172—Recent Trends in Regeneration Research
edited by V. Kiortsis, S. Koussoulakos, and H. Wallace

Series A: Life Sciences

ras Oncogenes

Edited by
Demetrios Spandidos
The Beatson Institute for Cancer Research
Glasgow, United Kingdom

Plenum Press
New York and London
Published in cooperation with NATO Scientific Affairs Division

Proceedings of a NATO Advanced Research Workshop
on *ras* Oncogenes,
held November 10–15, 1988,
in Vouliagmeni, Athens, Greece

Library of Congress Cataloging in Publication Data

NATO Advanced Research Workshop on *ras* Oncogenes
   (1988: Athens, Greece)
   ras oncogenes.
   (NATO ASI series. Series A, Life sciences; v. 170)
   "Proceedings of a NATO Advanced Research Workshop on *ras* Oncogenes,
held November 10–15, 1988, in Vouliagmeni, Athens, Greece"—T.p. verso
   "Published in cooperation with NATO Scientific Affairs Division."
   Includes bibliographies and index.
   1. *ras* oncogenes—Congresses. I. Spandidos, Demetrios. II. North Atlantic
Treaty Organization. Scientific Affairs Division. III. Title IV. Series. [DNLM: 1. *ras*
Genes—congresses. QZ 202 N2777r 1988]
RC268.44.R37N37   1988   616.99′4042   89-8598
ISBN 0-306-43228-5

Printed in the United States of America

PREFACE

    In recent years we have witnessed a dramatic increase in our
knowledge of cellular events involved in cancer as a result of the
discovery of oncogenes. For the first time the genes responsible for
the development of cancer in humans have been studied in great detail.
In particular these genes have been cloned in bacteria, their primary
nucleotide sequences determined, the proteins encoded by these genes
have been identified and expressed in bacteria and their properties
studied in detail. The possibilities of using the information in
diagnosis and prognosis and in the development of new therapies is
encouraging.

    Ras genes constitute an important family among the fifty oncogenes
that have been discovered so far. Ras genes contribute significantly
to the human tumor burden since around 70% of human tumors have an
increased level of ras gene product and 40% carry a mutated form of
ras gene.

    The present volume contains the contributions to the NATO Advanced
Research Workshop on "Ras Oncogenes" held at Vouliagmeni, Athens, Greece,
November 10-15, 1988. At this meeting 50 researchers in the frontiers
of cancer research met to present their latest advances in the field of
ras oncogenes. As organizer I feel that the Workshop was a success,
and I would like to thank all the participants for contributing to this
volume. I thank the NATO Scientific Affairs Division, Brussels, the
Greek Ministry of Culture and Industry, Energy and Technology, the
Greek National Tourist Organization, Olympic Airways, Cetus
Corporation, and the International Center for Cancer Research for
financial support of the meeting.

                                        DEMETRIOS SPANDIDOS

CONTENTS

STRUCTURE AND ORGANIZATION OF THE *RAS* GENE FAMILY, IN HUMAN

P. Chardin, N. Touchot, A. Zahraoui, V. Pizon, I. Lerosey, B. Olofsson and A. Tavitian

INSERM U-248, 10, Avenue de Verdun 75010 Paris, France

INTRODUCTION

The H-*ras*, K-*ras* and N-*ras* genes code for 21kd GTP/GDP binding proteins possessing a weak GTPase activity and transiently anchored to the inner face of the plasma membrane by a palmitic acid covalently linked to their C-terminus.

Transforming alleles of the H-*ras*, K-*ras* or N-*ras* genes are frequently found in human tumours, where the acquisition of a transforming potential is usually due to a point mutation resulting in the substitution of amino acids (a.a.) 12, 13 or 61. *Ras* proteins are found in organisms as different as yeast and man, and their high phylogenic conservation indicates that they certainly play an essential role, however the precise biochemical function of the *ras* proteins is not understood (Barbacid, 1987).

The fortuitous discovery of two proteins : *rho* in the marine snail Aplysia (Madaule and Axel, 1985) and YPT in yeast (Gallwitz et al., 1983) sharing ≈ 30% homology with *ras* proteins, suggested the existence of a large family.

A sequence of 6 a.a. : DTAGQE in position 57-62 of p21*ras*, was strictly conserved in the *ras* proteins from various organisms as well as in *rho* and YPT. In the first step of this study, using an original oligonucleotide strategy, we took advantage of this conserved sequence to isolate several new members of the *ras* superfamily.
In a second step, the cDNAs isolated by this first strategy were then used under low stringency conditions to search for related genes. We also searched for human homologs of the *Drosophila Melanogaster* Dras3, and the *Aplysia rho* genes.

RESULTS

1°/ Using the oligonucleotide strategy we first isolated a simian cDNA for the *ral*A protein possessing ≈ 50% a.a. identity with H, K or N-*ras* (Chardin and Tavitian, 1986). We then used this probe to search for human *ral* cDNAs under low stringency conditions and isolated the human *ral*A cDNAs encoding a protein with only one amino acid difference from simian *ral*A, we also isolated cDNAs encoding the human *ral*B protein possessing ≈ 85% a.a. identity with human *ral*A, most of the differences being located in the C-terminal end.

We also searched for human homologs of the drosophila D-*ras3* gene. By use of the D-*ras3* cDNAs we isolated human cDNAs encoding the *rap1A* protein sharing ≈ 70% a.a. identity with D-*ras3* and ≈ 50% a.a. identity with H, K or N-*ras* and the *rap2* protein sharing ≈ 50% a.a. identity with both D-*ras3* and H, K or N-*ras* (Pizon et al., 1988a). The rap1A cDNA, used under low stringency conditions, enabled us to isolated human cDNAs for the rap1B protein, very closely related to rap1A : ≈ 95% a.a. identity (Pizon et al. 1988b).

2°/ In the *rho* branch, we used the *rho6*, *rho9* and *rho12* cDNAs isolated by P. Madaule (1985) to isolate complete cDNAs encoding three closely related proteins named *rhoA* (*rho12*), *rhoB* (*rho6*) and *rhoC* (*rho9*) sharing ≈ 90% a.a. identity.

3°/ The extensive screening of a rat brain cDNA library by the oligonucleotide strategy enabled us to isolate ≈ 50 positive clones, 14 were studied in detail and sequenced in the oligonculeotide hybridizing region : 1 was a K-*ras* cDNA, 3 were H-*ras* cDNAs, 6 did not possess the exact oligonucleotide sequence and were not *ras*-related, 4 encoded new *ras*-related proteins that we named *rab1*, 2, 3 and 4. *Rab1* is a mammalian homolog of the yeast YPT protein (≈ 70% a.a. identity) while *rab2*, 3 and 4 possess ≈ 40% a.a. identity with *rab1* and are clearly in the same branch of the *ras* family.

The rat cDNAs encoding *rab1*, 2, 3 and 4 were then used to search for human cDNAs under low stringency conditions. We isolated human *rab1*, 2, 3 and 4 cDNAs encoding proteins sharing more than 95% a.a. identity with their rat counterparts. A cDNA coding for a protein closely related to the first *rab3A* was also isolated, we named it *rab3B* (≈ 75% a.a. identity with *rab3A*).

Two other cDNAs encoding proteins with ≈ 40% a.a. identity with any of the other *rab* proteins were also isolated, and named *rab5* and *rab6*.

Structure of the *ras* superfamily

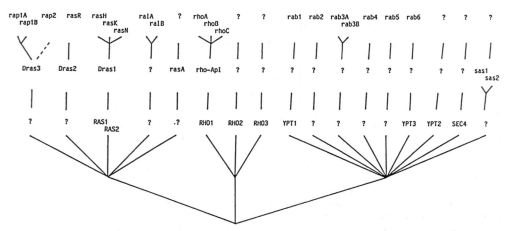

So far, the *ras* superfamily might be subdivided into three main branches represented by *ras*, *rho* and *rab*. On this figure the lower line represents yeast genes (Upper case letters), the middle line : Drosophila, Aplysia or Dictyostelium genes, and the upper line: mammalian genes.

1°/ In human, the *ras* branch includes the three classical *ras* proteins H-*ras*, K-*ras* and N-*ras* : ≈ 85% a.a. identity (reviewed by Barbacid, 1987) ; the *ral A and ral* B proteins : ≈ 85% a.a. identity and ≈ 50% identity with any of the *ras* proteins (Chardin and Tavitian, 1986 and unpublished results) ; the R-*ras* protein : ≈ 55 % identity with *ras* or *ral* and related to drosophila D-*ras*2 (65% a.a. identity) (Lowe et al, 1986), the *rap*1A and *rap*1B proteins 95% identical and sharing ≈ 55% identity with *ras*, the *rap*2 protein ≈ 50% a.a. identity with *ras* (Pizon et al., 1988 a and b) *Rap*1A and *rap*1B are clearly human homologs of the drosophila D-*ras*3 protein, *rap*2 has 60% a.a. identity with *rap*1A or *rap*1B but is not much more closely related to D-*ras*3 than to the other *ras* or *ras*-related proteins, however *rap*2 has a Threonine in position 61, an exception only found in *rap*1A, *rap*1B and D-*ras*3, at present the evolutionary relationships of this protein are not clear. For another gene : Apl-*ras* from Aplysia, no mammalian counterpart as yet been found but it is likely that a mammalian homolog of Apl-*ras* might exist. This suggests that the *ras* branch might include at least a dozen of different proteins.

2°/ In human, the *rho* branch includes the three closely related *rho* proteins : *rho* A, (Yeramian et al., 1987) *rho* B and *rho* C (Chardin et al., 1988) : ≈ 85% a.a. identity and ≈ 30% when compared to *ras*.

The yeast counterpart of these human *rho* proteins is *RHO1*, but a second protein : *RHO*2 possessing ≈ 50% identity with *RHO*1 has also been found and might be considered as a yeast *rho*-like protein (Madaule et al., 1987). Another yeast gene previously known as CDC42 has recently been found to encode a *rho*-like protein (D. Johnson, personal communication). It would be surprising that such *rho*-like proteins do not exist in mammals although they have not been found as yet.

3°/ In human, the third branch is represented by the *rab* proteins : *rab*1, 2, 3A, 3B, 4, 5 and 6 possessing ≈ 35-50% a.a. identity between each other except *rab*3A and *rab*3B that are much more closely related (≈ 78% a.a. identity). *Rab*1 is a mammalian homolog of the S. Cerevisiae YPT1 protein and *rab*6 is a human homolog of an S. Pombe YPT-like protein (D. Gallwitz, personal communication). Three other proteins : one from S. Cerevisiae : SEC4 (Salminen and Novick, 1987) and two from S. Pombe (M. Yamamoto, D. Gallwitz personal communications) as well as two proteins from Dictyostelium: SAS1 and 2 (A. Kimmel personal communication) clearly belong to this *rab* branch.

It is likely that some of the *rab* 2, *rab* 3, *rab* 4 and *rab*5 genes have yeast counterparts that remain to be discovered, while reciprocally, the yeast YPT-like and SEC4 genes and the Dictyostelium SAS genes probably have mammalian counterparts, suggesting that the *rab* branch might include also at least a dozen of different proteins.

In the course of this study it happened to us more and more frequently to re-isolate already known cDNAs, thus we assume that a high proportion of this family has now been discovered and we might estimate the number of different proteins in this family in the range of 30 to 100.

Expression

Most of the *ras* superfamily genes are expressed in all tissues, although several of these genes are expressed at higher levels in one or a few organs (Leon et al., 1987, Olofsson et al., 1988). High levels of expression are frequently observed in brain ; moreover, one of the *rab* genes: *rab*3A appears to be expressed at high levels in brain while it is hardly detected in any other organ (Olofsson et al., 1988).

Each gene displays a specific pattern of expression in the different organs suggesting a complex regulation. A detailed expression study will need the use of *in-situ* hybridization or immunocytochemical staining.

Structural properties of the *ras* superfamily proteins

A three dimensional model of the p21*ras* protein has recently been deduced from X-ray cristallography studies (de Vos et al., 1988) and it is of high interest to locate the major conserved regions on this model.

1./ Four regions : the Glycine rich region in position 10-20, the DTAGQE region in position 57-62 the LVGNK-DL region in position 113-120 the ETSAK region in position 143-147 and a few scattered amino-acids including F28 are involved in GTP binding and are highly conserved in all the proteins of the *ras* family. However significant differences are found at "critical" positions : the *rho* proteins have an Alanine in position corresponding to *ras* Gly 13, the *rab* proteins do not possess a Glycine in position corresponding to *ras* Gly 12 (*rab*3A, *rab*3B and *rab*5 have neither Gly 12 nor Gly 13).

R-*ras* has an Alanine instead of a Threonine in position 144 known to be of high importance since a yeast *RAS*2 protein might be activated by this kind of mutation (Camonis and Jacquet, 1988).

Even region 61 which had been considered as a hall mark of *ras* proteins is not strictly conserved : *rap* proteins possess a Threonine instead of the Glutamine found in *ras* position 61.

2./ Region 32-42 known as the "effector" region is strictly conserved between *ras* and *rap*1 proteins while there is one a.a. change in *rap*2 and R-*ras* and three differences in *ral*.

Nevertheless this region is not highly conserved between the different *rab* proteins, suggesting that they do not follow the same functionnal scheme than the proteins of the *ras* branch.

Only one amino acid : Thr35 is conserved in all the proteins of the *ras* family, suggesting a special role for example as a Ser/Thr Kinase substrate.

3./ Regions 53-56, 63-82 and 97-104 are all located on the same part of the molecule close to loop 32-42 and are ≈ 70% homologous between the different proteins of the *ras* branch. We would predict that these regions are involved in the interaction with a large protein that remains to be characterized. It is noteworthy that small deletions might be introduced in each of these regions taken separately without impairing the transforming potential of a v-H-*ras* p21 (Willumsen et al., 1986) unfortunately the same kind of data is not available for a normal H-*ras* protein. In these regions Tyrosine 71 is conserved in all proteins of the *ras* family, and might be important, for example as a Tyrosine Kinases substrate.

4./ Two small sequences in position 47-49 and 161-164 are located close to the C-terminal tail of p21 *ras*, anchoring it to the plasma membrane, and are frequently conserved. Loop 47-49 is interacting with region 161-164 and is directly linked to the effector loop and might behave as a trigger for

the conformational change induced by the putative "exchange factor" or receptor.
5./ The 20-30 last C-terminal amino acids are highly variable and confer their specificity to the different proteins of a same sub-branch such as H, K and N-ras, several of the proteins of the *ras* branch have many basic amino acids such as Lysines and Arginines in their C-terminal end.
This characteristic is not found in the *rab* branch of the family.
6./ All the *ras* family proteins possess at least one Cysteine close to their C-terminal end, suggesting a membrane localization. However the conservation of specific motifs, from yeast to man, in each sub-branch strongly suggest a role in the targetting of these proteins to different sub-cellular membrane systems.

Biochemical properties of the *ras* superfamily proteins

The high conservation of the GTP binding region strongly suggested that all of these proteins were GTP binding proteins. In fact, the GTP binding ability, and low GTPase activity, have been demonstrated for *ras*, *rho*, YPT, R-*ras*, *ral* and *rab* proteins (Tucker et al., 1986 ; Anderson and Lacal, 1987 ; Wagner et al., 1987 ; Lowe and Goeddel, 1987 ; Chardin et al., Zahraoui et al., in preparation). However, it is noteworthy that many of these proteins have an amino acid difference in positions known to be "critical" in the case of *ras* proteins. The *rho* proteins have an Alanine in position corresponding to *ras* Glycine 13 ; *rab* proteins do not possess a Glycine in position corresponding to *ras* Glycine 12 (*rab* 3 *rab* 5 and *rab* 6 have neither Gly 12 nor Gly 13). Even region 61, which had been considered as a hallmark of *ras* proteins is not strictly conserved : *rap* proteins possess a Threonine instead of the Glutamine found in *ras* position 61. As far as the conclusions drawn from the study of *ras* proteins might be transposed to the other proteins of this family we would predict that *rho*, *rab* and *rap* proteins are constitutively in a slightly more active state than *ras* proteins.
Preliminary results on *ral*, *rap*, *rho* and *rab*1, 2, 3, 4, 5 proteins expressed in E. Coli show that these proteins display very different GTP binding abilities after Western Blot transfer to nitrocellulose, furthermore careful studies on purified proteins show that several biochemical parameters such as the GTP/GDP exchange rates and the GTPase activity vary by at least one order of magnitude from one protein to the other, suggesting that although these proteins all bind GTP they possess significantly different biochemical properties

DISCUSSION

Transforming potential of *ras* related genes

While transforming alleles of the H, K or N-*ras* genes are found in 10 to 70% of human tumors, none of the *ras*-related genes has ever been isolated from the large number of tumors studied by the NIH/3T3 assay in many laboratories, indicating that *ras* related genes are not frequently implicated in tumor formation or can not be detected by this approach. In fact, it seems that neither R-*ras* (Lowe et al. 1987) nor *ral*A (Chardin et al., in preparation) are able to transform NIH/3T3 cells in the usual assay. Therefore the potential involvement of *ras* related genes in transformation remains an open question.

Function of the *ras* superfamily proteins

The first proteins characterized on the basis of their biological activity are H-*ras*, K-*ras* and N-*ras* : the activated proteins transform NIH/3T3 cells, and their most obvious effect, when microinjected are : an increase in membrane ruffling/fluid phase endocytosis/cell motility and the induction of mitosis, that might explain the loss of contact inhibition and the transformed phenotype.

In the fission Yeast Schizosaccharomyces Pombe, disruption of the *ras* 1 gene does not affect growth, but the cell shape is extensively deformed (shortened and swollen), futhermore *ras* 1 is essential for mating and required for efficient sporulation (Fukui et al., 1986).

The second protein isolated on a functional basis in yeast is SEC4 : this mutation results in accumulation of post-golgi vesicles (Salminen and Novick, 1987), and it has been shown that the SEC4 protein associates with secretory vesicles and the plasma membrane (Goud et al., 1988).

In S. Cervisiae, the YPT1 gene is located between the actin (ACT1) and β-tubulin (TUB2) genes. The YPT protein is involved in the organization of the cytoskeleton during vegetative growth. YPT deficiency leads to larger cells, with two or several buds instead of one ; microtubules as well as the actin network are disorganized and nuclear integrity is lost (Schmitt et al., 1987 ; Segev and Botstein, 1987). YPT1 deficiency also causes a build-up of vesicles (Segev and Botstein, 1988). However YPT1 and SEC4 do not complement each other and it is unlikely that SEC4 and YPT1 fulfill similar roles. Therefore, it seems that the *ras* superfamily proteins play a critical role at the membrane/cytoskeleton interface and regulate some of the physical exchanges beween the cell and its environment. However, a direct involvement of the *ras* superfamily proteins in these events remains to be demonstrated.

Ras proteins possess some sequence homologies and some biochemical analogies with the α subunits of G proteins involved in many receptor/effector systems, as well as a common sub-membrane localization ; it was thus postulated that *ras* proteins might transduce signals from growth factor receptors to an effector such as phospholipase C or phospholipase A2. However, recent evidence indicate that the action of *ras* on these phospholipases is an indirect effect (Seuwen et al., 1988; Yu et al., 1988). It is likely that *ras* proteins are transiently activated by growth-factor receptors, but the biochemical basis of this activation is not understood. The careful study of any biochemical parameter in a *ras* transformed cell usually demonstrates a difference when compared to a normal cell, however it is well known that the biochemistry of a transformed cell is deeply altered (Warburg, 1920) and this kind of data provides little information, if any, on *ras* function. What we would like to know is the nature of the proteins directly interacting with p21 *ras*.

At present only one protein has been shown to interact directly with p21 *ras* : the GTPase Activating Protein (GAP). This protein is a 116 K cytoplasmic protein that can give rise to a 55K, that might be a degradation product, still active and probably representing the "*ras* interacting" domain of the protein (Trahey et al., 1987, F. Mc Cormick personal communication).

An elegant genetic approach has localized an essential external loop of the p21ras protein that can not be deleted or mutated without loss of activity, even in a constitutively active protein : v-H *ras* p21 (Willumsen et al., 1986; Sigal et al., 1986)) this region in position 32-42 is known as the "effector" binding site and recent evidence indicate that some mutations in this region impair the interaction with the GAP protein, and abolish the transforming potential, suggesting that GAP might be the effector (Adari et al., 1988; Calès et al., 1988) however for some mutants the correlation between diminished sensitivity to GAP and abolished transforming potential is not clear (D. Lowy, personal communication) and other regions around position 60-70 also seem to be involved in GAP interaction (J.C. Lacal, personal communication).

It is noteworthy that at least two *ras*-related proteins *rap*1A and *rap*1B have exactly the same sequence than *ras* in the "effector" region while *rap*2 and R-*ras* display one difference and *ral* has three. It would of course be of major interest to know whether *rap*1A and *rap*1B really interact with the same "effector" than *ras*, and whether *rap*2, R-*ras* and *ral* also interact with identical or closely related effectors. As *rap* proteins possess in position 61 a Threonine, an a.a. leading to a slight activation of *ras*, it is conceivable that *rap* proteins are in a constitutively active state and interact with the same effector as *ras*, keeping it in a low, basal, activity level, while *ras* would be able to transiently boost this activity in response to an external signal, furthermore it is attractive to speculate that *rap* proteins interact with the same effector than *ras* but have an antagonist effect.

The availability of most of these proteins, expressed in E. Coli should enable us to test these hypothesis.

Acknowledgements

We wish to thank the following colleagues for their gifts of sequences prior to publication : D. Gallwitz and M. Yamamoto for YPT-like sequences from S. Pombe, A. Kimmel for SAS1 and SAS2 from Dictyostelium and D. Johnson for CDC 42 from S. Cerevisiae.

References

Adari, H., Lowy, D.R., Willumsen, B.M., Der, C.J., and Mc Cormick, F. (198). Guanosine Triphosphatase Activating Protein (GAP) interacts with the p21 *ras* effector binding domain. Science, **240**, 518-521.

Anderson, P. and Lacal, J.C. (1987). Expression of the Aplysia californica *rho* gene in Escherichia coli : purification and characterization of its encoded p21 product. Mol. Cell. Biol. **7**, 3620-3628.

Barbacid, M. (1987). *ras* Genes. Ann. Rev. Biochem. **56**, 779-827.

Bar-Sagi, D. and Feramisco, J.R. (1986). Induction of membrane ruffling and fluid-phase pinocytosis in quiescent fibroblasts by *ras* proteins. Science, **233**, 1061-1068.

Calès, C., Hancock, J.F., Marshall, C.J. and Hall A. (1988). The cytoplasmic protein GAP is implicated as the target for regulation by the *ras* gene product. Nature, **332**, 548-551.

Camonis, J. and Jacquet, M. (1988). A new *ras* mutation that supresses the CDC25 gene requirement for growth of Saccharomyces Cerevisiae. Mol. Cell. Biol., **8**, 2980-2983.

Chardin, P. and Tavitian A. (1986). The *ral* gene : a new *ras* related gene isolated by the use of a synthetic probe. EMBO J. **5**, 2203-2208.

Chardin, P., Madaule, P. and Tavitian A. (1988) Coding sequence of human *rho* cDNAs clone 6 and clone 9. Nucleic Acids Res. **16**, 2717.

de Vos, A.M., Tong, L., Milburn, M.V., Mathias, P.M., Jancarik, J., Noguchi, S., Nishimura, S., Miura, K., Ohtsuka, E. and Kim, S.-H. (1988) Science, **239**, 888-893.

Fukui, Y., Kozasa, T., Kaziro, Y., Takeda, T. and Yamamoto, M. (1986). Role of a *ras* homolog in the life cycle of schizo-saccharomyces pombe. Cell, **44**, 329-336.

Gallwitz, D., Donath, C. and Sander C. (1983). A yeast gene encoding a protein homologous to the human c-*has*/*bas* proto-oncogene product. Nature **306**, 704-707.

Goud, B., Salminen, A., Walworth, N.C., and Novick, P.J. (1988). A GTP-binding protein required for secretion rapidly associates with secretory vesicles and the plasma membrane in Yeast. Cell, **53**, 753-768.

Leon, J., Guerrero, I. and Pellicer, A. (1987). Differential expression of the *ras* gene family in mice. Mol. Cell. Biol. **7**, 1535-1540.

Lowe, D., Capon, D., Delwart, E., Sakaguchi, A., Naylor, S. and Goeddel, D. (1987). Structure of the human and murine R-*ras* genes, novel genes closely related to *ras* proto-oncogenes. Cell **48**, 137-146.

Lowe, D. and Goeddel, D. (1987). Heterologous expression and characterization of the human R-*ras* gene product. Mol. Cell. Biol. **7**, 2845-2856.

Madaule, P. and Axel, R. (1985). A novel *ras*-related gene family. Cell **41**, 31-40.

Madaule, P., Axel, R. and Myers, A. (1987). Characterization of two members of the *rho* gene family from the yeast Saccharomyces cerevisiae. Proc. Natl. Acad. Sci. USA **84**, 779-783.

Melançon, P., Glick, B.S., Malhotra, V., Weidman, P.J., Serafini, T., Gleason, M.L., Orci, L. and Rothman J.E. (1987).

Involvement of GTP-binding "G" proteins in Transport through the Golgi Stack. Cell, 51, 1053-1062.

Mozer, B., Marlor, R., Parkhurst, S. and Corces, V. (1985). Characterization and Developmental Expression of a Drosophila *ras* oncogene. Mol. Cell. Biol. 5, 885-889.

Olofsson, B., Chardin, P., Touchot, N., Zahraoui, A. and Tavitian A. (1988). Expression of the *ras*-related *ral*A, *rho*12 and *rab* genes in adult mouse tissues. Oncogene, 2, 231-234.

Pizon, V., Chardin, P., Lerosey, I., Olofsson B. and Tavitian A. (1988a). Human cDNAs *rap*1 and *rap*2 homologous to the *Drosophila* gene Dras3 encode proteins closely related to *ras* in the "effector" region. Oncogene, 2, 201-204.

Pizon, V., Lerosey, I., Chardin, P. and Tavitian A. (1988b). Nucleotide sequence of a human cDNA encoding a *ras*-related protein (*rap*1B). Nucleic. Acids. Res., 16, 7719.

Salminen, A. and Novick, P. (1987). A *ras*-like protein is required for a post-golgi event in yeast secretion. Cell 49, 527-538.

Schejter, E. and Shilo, B.-Z. (1985). Characterization of functional domains of p21 *ras* by use of chimeric genes. EMBO J. 4, 407-412.

Schmitt, A., Wagner, P., Pfaff, E. and Gallwitz, D. (1986). The *ras*-related YPT1 gene product in yeast : a GTP-binding protein that might be involved in microtubule organization. Cell 47, 401-412.

Segev, N. and Botstein, D. (1987). The *ras*-like yeast YPT1 gene is itself essential for growth, sporulation, and starvation response. Mol. Cell. Biol. 7, 2367-2377.

Segev, N., Mulholland, J. and Botstein, D. (1988) The Yeast GTP-binding YPT1 protein and a mammalian counterpart are associated with the secretion machinery. Cell, 52, 915-924.

Seuwen, K., Lagarde, A. and Pouyssegur, J. (1988). Deregulation of hamster fibroblast proliferation by mutated *ras* oncogenes is not mediated by constitutive activation of phosphoinositide-specific phospholipase C. EMBO J. 7, 161-168.

Sigal, I.S., Gibbs, J.B., D'Alonzo, J.S., and Scolnick, E.M. (1986), Proc. Natl. Acad. Sci. USA, 83, 4725-4729.

Swanson, M., Elste, A., Greenberg, S., Schwartz, J., Aldrich, T. and Furth, M. (1986). Abundant expression of *ras* proteins in *Aplysia* neurons. J. Cell Biol. 103, 485-492.

Touchot, N., Chardin P. and Tavitian, A. (1987). Four additional members of the *ras* gene superfamily isolated by an oligonucleotide strategy : molecular cloning of YPT-related cDNAs from a rat brain library. Proc. Natl. Acad. Sci. USA 84, 8210-8214.

Trahey, M. and Mc Cormick, F. (1987). A cytoplasmic protein stimulates Normal N-*ras* p21 GTPase, but does not affect oncogenic mutants. Science, **238**, 542-545.

Tucker, J., Sczakiel, G., Feuerstein, J., John, J., Goody, R. and Wittinghofer, A. (1986). Expression of p21 proteins in Escherichia coli and stereochemistry of the nucleotide-binding site. EMBO J. **5**, 1351-1358.

Wagner, P., Molenaar, C., Rauh, A., Brökel, R., Schmitt, H. and Gallwitz, D. (1987). Biochemical properties of the *ras*-related YPT protein in yeast : a mutational analysis. EMBO J., **6**, 2373-2379.

Warren, G., Davoust, J. and Cockcroft, A. (1984). Recycling of transferrin receptors in A431 cells is inhibited during mitosis. EMBO J., **3**, 2217-2225.

Willumsen, B.M., Papageorge, A.G., Kung, H.F., Bekesi, E., Robins, T., Johnsen, M., Vass, W.C., and Lowy, D.R. (1986). Mutational analysis of a *ras* catalytic domain. Mo. Cell. Biol., **6**, 2646-2654.

Yeramian, P., Chardin P., Madaule, P. and Tavitian, A. (1987). Nucleotide sequence of human *rho* cDNA clone 12. Nucleic, Acids Res. **15**, 1869.

Yu, C.-L., Tsai, M.-H. and Stacey, D. (1988). Cellular *ras* activity and phospholipid metabolism. Cell **52**, 63-71.

Zahraoui, A., Touchot, N., Chardin, P. and Tavitian, A. (1988). Complete coding sequences of the *ras* related *rab*3 and 4 cDNAs. Nucleic. Acids. Res. **16**, 1204.

TRANSCRIPTIONAL REGULATORY SEQUENCES IN THE H-RAS1 GENE

Demetrios A. Spandidos[1,2]

[1]The Beatson Institute for Cancer Research
Garscube Estate
Bearsden
Glasgow G61 1BD
Scotland, UK.

[2]Biological Research Center
National Hellenic Research Foundation
48 vas Constantinou Avenue
Athens 11635
Greece

INTRODUCTION

The H-ras1 gene is a member of the ras family (1,2). It is the smallest of the family occupying approximately 3.5 kb of DNA. It encodes a protein of 21,000 daltons which possesses GDP and GTP binding and GTPase activities. The viral p21 also exhibits autophosphorylating activity. Although the function of the ras p21 proteins is not known they are assumed to be signal transducers because of their structural and functional similarities with G proteins (1,2).

In this paper I discuss the evidence for transcriptional regulation of the H-ras1 gene.

EXPRESSION OF RAS GENES IN NORMAL TISSUES

Ras gene expression in normal tissues has been analyzed at the RNA (3,4) and protein (5,6) levels.

Employing molecular hybridization techniques it has been demonstrated that H-ras1 is expressed in a variety of tissues in the adult and during development (3,4). A particular study (4) has shown that certain tissues like brain, kidney, heart and skeletal muscle contain higher levels of H-ras1 specific RNA transcripts when compared to i.e. lung, spleen or thymus. At day 10 during gestation the levels of H-ras1 transcripts are higher than at days 13, 16 or 19.

Similar results have been obtained by examining ras p21 levels using an immunohistochemical method of monoclonal antibodies (5,6). Ras p21 was detected in every adult or fetal tissue but some tissues, i.e. brain, kidney and heart contained higher level than others, i.e. pancreas or breast (6). Using Western blots ras p21 was also found to be elevated in heart tissue (5).

EXPRESSION OF RAS GENES IN TUMORS

Ras genes have been found to be expressed at higher levels in tumors as compared to adjacent normal tissue (7-14). Thus, analysis at the RNA or protein level in breast (8,12), colon (7,10), cervical (13), head and neck (9) and bladder (11) lesions has provided evidence to support the view that elevated ras p21 levels contribute to the development of cancer. Moreover, in some cases increased ras p21 levels has been shown to correlate with poor prognosis (14). Furthermore using oligonucleotide hybridization analysis structural mutations in the coding sequences of the H-ras1 gene have been detected in human primary breast carcinomas (15).

CELL TRANSFORMATION STUDIES

The effect of over-expressing the normal or the mutant T24 H-ras1 genes has been tested on early passage and immortalized cells (16,17). Linking a strong transcriptional enhancer from a retrovirus LTR or SV40 to the normal H-ras1 and transfecting the resulting recombinant plasmid into immortalized cells resulted in tumorigenic conversion of these cells (16,17). Similar recombinants have been shown to immortalize early passage rodent cells (17). Moreover, when the mutant T24 H-ras1 gene was linked to the Moloney LTR or SV40 enhancer transfection of the resulting recombinant plasmids into early passage cells triggered malignant conversion of these cells (17,18).

Thus, quantitative alterations of H-ras1 gene expression may result in the developments of cancer phenotypes. Therefore, it is suggested that analogous phenomena affecting transcriptional regulation may occur in the development of tumours in vivo.

REGULATORY SEQUENCES AT THE 5' END OF THE H-RAS1 GENE

Nucleotide sequence analysis of the 5' end of the H-ras1 gene has shown that this region has a high GC content (19-22). Several initiation start sites have been found using primer extension (20) or S1 mapping (23). Moreover, no TATA box has been found near the transcription initiation sites and the evidence is consistent with the view that the H-ras1 promoter has a dispersed nature. DNA fragments, varying in length from 0.8 kb to 51 bp have been found to have promoter activity (20-25) which is bidirectional (23,24) and responds to the phorbol ester TPA (22). An enhancer element within the 0.8 kb SstI fragment spanning the promoter region has also been described (24).

ENHANCER ACTIVITY OF THE VTR SEQUENCES

Nucleotide sequence analysis of the 3' end of the H-ras1 gene has shown a variable tandem repeat (VTR) sequence consisting of 28 bp which in the cloned T24 H-ras1 gene is repeated 29 times (19). The VTR sequence has been found to possess a transcriptional enhancer activity assayed in short term (26) and stable (26) transfection assays. Moreover, it has been found to possess an enhancer consensus sequence (15). Since the VTR element is highly polymorphic it has been suggested that the activity of this enhancer is variable within the human population (26).

REGULATORY SEQUENCES IN THE H-RAS1 INTRONS I AND IV

Transfection studies have shown that the first intron contains at least two positive and two negative regulatory elements (23). One of these positive elements is located in a sequence in the middle part of the first intron which is conserved between various mammalian species (23). Another sequence in the first intron which is defined by the 6 bp deletion in the

T24 H-ras1 gene has been found to negatively regulate transcription. This sequence shows homology to the "E" element of the immunoglobulin heavy chain enhancer and the Sp1 binding site (28).

A point mutation in the IV intron of the T24 H-ras1 gene has also been shown to be responsible for increased expression and transforming activity of the gene (29). This finding also suggests the existence of a regulatory sequence in the IV intron.

CONCLUSIONS AND FUTURE PROSPECTS

The transcriptional control of the H-ras1 gene although not completely understood is complex. A number of regulatory sequences within or near the end of the gene are involved. Delineating the role of each of these transcriptional regulatory elements is a task for the near future. In fact future progress on the role of ras oncogenes in human cancer will depend to a great extent on the progress we make to unravel their transcriptional control. The development of new techniques such as the use of oligonucleotide hybridization (30) and RNase mismatch (31) to identify ras mutations in structural and regulatory regions. The study of their effects on ras expression hold great promise in understanding of the molecular basis of carcinogenesis by ras oncogenes and in assessing the clinical significance of ras mutations in human cancers.

REFERENCES

1. M. Barbacid. Ras genes. Ann. Rev. Biochem. 56:779 (1987).
2. D.A. Spandidos. Ras oncogenes in cell transformation. ISI Atlas of Science. Immunology 1:1 (1988).
3. R. Muller, D.J. Slamon, J.M. Tremblay, M.J. Cline and I. Verma. Differential expression of cellular oncogenes during pre- and postnatal development of the mouse. Nature 299:640 (1982).
4. J. Leon, I. Guerrrero and A. Pellicer. Differential expression of the ras gene family in mice. Mol. Cell. Biol. 7:1535 (1987).
5. D.A. Spandidos and T. Dimitrov. High expression levels of ras p21 protein in normal mouse heart tissues. Biosci. Rep. 5:1035 (1985).
6. M.E. Furth, T.H. Aldrich and C. Cordon-Cardo. Expression of ras proto-oncogene proteins in normal human tissues. Oncogene 1:47 (1987).
7. D.A. Spandidos and I.B. Kerr. Elevated expression of the human ras oncogene family in premalignant and malignant tumors of the colorectum. Br. J. Cancer 49:681 (1984).
8. D.A. Spandidos and N.J. Agnantis. Human malignant tumors of the breast as compared to their respective normal tissue have elevated expression of the Harvey ras oncogene. Anticancer Res. 4:269 (1984).
9. D.A. Spandidos, A. Lamothe and J.N. Field. Multiple transcriptional activation of cellular oncogenes in the human head and neck solid tumors. Anticancer Res. 5:221 (1985).
10. A.R.W. Williams, J. Piris, D.A. Spandidos and A.H. Wyllie. Immunohistochemical detection of the ras oncogene p21 product in an experimental tumor and in human colorectal neoplasms. Br. J. Cancer 52:687 (1985).
11. M.V. Viola, F. Fromowitz, S. Oravez, S. Deb and J. Schlom. Ras oncogene p21 expression is increased in premalignant lesions and high grade bladder carcinoma. J. Exp. Med. 161:1213 (1985).
12. N.J. Agnantis, C. Petraki, P. Markoulatops and D.A. Spandidos. Immuno-histochemical study of the ras oncogene expression in human breast lesions. Anticancer Res. 6:1157 (1986).
13. N.J. Agnantis, D.A. Spandidos, H. Mahera, P. Parissi, A. Kakkanas, A.

Pintzas and N.X. Papaharalampous. Immunohistochemical study of ras oncogene expression in endometrial and cervical human lesion. Eur. J. Gyn. Oncol. 7:1 (1988).

14. T. Clair, W.R. Miller and Y.S. Cho-Chung. Prognostic significance of the expression of a ras protein with a molecular weight of 21,000 by human breast cancer. Cancer Res. 47:5290 (1987).

15. D.A. Spandidos. Oncogene activation in malignant transformation: A study of H-Ras in human breast cancer. Anticancer Res. 7:991 (1987).

16. E.H. Chang, M.E. Furth, E.M. Scolnick and D.R. Lowy. Tumorigenic transformation of mammalian cells induced by a normal human gene homologous to the oncogene of Harvey murine sarcoma virus. Nature 297:478 (1982).

17. D.A. Spandidos and N.M. Wilkie. Malignant transformation of early passage rodent cells by a single mutated human oncogene. Nature 310:469 (1984).

18. D.A. Spandidos. Mechanism of carcinogenesis: the role of oncogenes, transcriptional enhancers and growth factors., Anticancer Res. 5:485 (1985).

19. D.J. Capon, E.Y. Chen, A.D. Levinson, P.H. Seeburg and D.V. Goeddel. Complete nucleotide sequences of the T24 human bladder carcinoma oncogene and its normal homologue. Nature 302:33 (1983).

20. S. Ishii, G.T. Merlino and I. Pastan. Promoter region of the human Harvey ras proto-oncogene: similarity to the EFG receptor proto-oncogene promoter. Science 230:1378 (1985).

21. S. Ishii, J.T. Kadonaga, R. Tjian, J.N. Brady, G.T. Merlino and I. Pastan. Binding of the Sp1 transcription factor by the human Harvey ras1 proto-oncogene promoter. Science 232:1410 (1986).

22. D.A. Spandidos, R.A.B. Nicholls, N.M. Wilkie and A. Pintzas. Phorbol ester responsive H-ras1 gene promoter contains multiple TPA-inducible/AP-1 binding consensus sequence elements. FEBS Letters. In Press.

23. H. Honkawa, W. Masahashi, S. Hashimoto and T. Hashimoto-Gotoh. Identification of the principal promoter sequence of the c-H-ras transforming oncogene: Deletion analysis of the 5'-flanking region by focus formation assay. Mol. Cell. Biol. 7:2933 (1987).

24. D.A. Spandidos and M. Riggio. Promoter and enhancer like activity at the 5'-end of normal and T24 Ha-ras1 genes. FEBS Lett. 203-169 (1986).

25. W.S. Trimble and N. Hozumi. Deletion analysis of the c-Ha-ras oncogene promoter. FEBS Lett. 219:70 (1987).

26. D.A. Spandidos and L. Holmes. Transcriptional enhancer activity in the variable tandem repeat DNA sequence downstream of the human Ha-ras1 gene. FEBS Lett. 218:41 (1987).

27. J.B. Cohen, M.V. Walter and A.D. Levinson. A repetitive sequence element 3' of the human c-Ha-ras1 gene has enhancer activity. J. Cell. Phys. Suppl. 5:75 (1987).

28. D.A. Spandidos and A. Pintzas. Differential potency and trans-activation of normal and mutant T24 human H-ras1 gene promoters. FEBS Letters 232:269 (1988).

29. J.B. Cohen and A.D. Levinson. A point mutation in the last intron responsible for increased expression and transforming activity of the c-Ha-ras oncogene. Nature 334:119 (1988).

30. J.L. Bos, E.R. Fearon, S.R. Hamilton, M. Barlaam-de Vries, J.H. van Boom, A. Van der Eb and B. Vogelstein. Prevalence of ras gene mutations in human colorectal cancers. Nature 327:293 (1987).

31. K. Forrester, C. Almoguera, K. Han, W.E. Grizzle and M. Perucho. Detection of high incidence of k-ras oncogenes during human colon tumorigenesis. Nature 327:298 (1987).

MUTATIONAL ACTIVATION OF HUMAN RAS GENES

J.L.Bos

Laboratory for Molecular Carcinogenesis
Sylvius Laboratories
P.O.Box 9503
2300 RA Leiden The Netherlands

The introduction of new assays for point mutations in chromosomal DNA has enabled us to identify activated ras genes in a large number of human tumors (Table I; ref.1). Thus it became clear that certain tumor types never or only occasionally harbor a mutated ras gene, whereas other tumor types have an activated ras gene more or less frequently (20% to 90% of the cases). On the basis of clinical or histopathological features, a tumor that has a mutated ras gene does not differ from a tumor of the same type that does not. This indicates that the activation of ras is not an essential event. However, since activated ras proteins have cell-transforming properties, and (nearly) all malignant cells of a tumor carry a mutated gene, it is likely that the activation of ras does contribute to the development of tumors. Apparently, other genetic events have similar effects as ras activation.

The activation of ras genes is only one step in the process of tumor development. From animal model systems it is known that ras activation can be an early or even the initial event (2,3), as well as a late event (4). In human tumors, the analysis of benign (precursor) tumors has led to the same conclusion, although the activation appears to be mostly an early event. In colon adenomas, which are considered to be the precursor of colon adenocarcinomas, the incidence of mutated ras genes is the same as in the malignant tumors indicating that the mutation is an early event (5). In very small adenomas, however, the incidence is much lower, suggesting that the mutation occurs in the adenoma tissue and is responsible for further progression of the adenoma. Most of the small adenomas, however, regress and it might be that mutated ras genes predispose an adenoma to become a malignant carcinoma. In that case, the mutational event is very early or even initial. In some tumors where both adenoma and carcinoma tissues were analyzed a mutated ras gene was found only in the carcinoma tissue, indicating that for development of colon cancer the timing of ras activation is not invariant.

The conclusion that the mutational activation of ras can be both an early and a late event was also drawn for myeloid leukemia (6-8). In some patients with myelodysplastic syndrome or "preleukemia" it was found that the mutation was already present in nearly all the bone marrow cells very early in the clinical course of the disease. In these cases the mutation was also present in peripheral mature T-lymphocytes indicating that the mutation has occurred in a hematopoietic stem cell (6,8). The mutation remained present during evolution into acute myeloid leukemia (AML) as

Table I. Incidence of Ras mutations

Tumor	percentage tumor samples with an activated RAS gene	preferential activated RAS gene
breast adenocarc.	0	-
ovary carc.	0	-
cervical carc.	0	-
esophageal carc.	0	-
glioblastomas	0	-
neuroblastomas	0	-
stomach carc.	0	-
lung, squamous cell carc.	0	-
lung, large cell carc.	0	-
lung, adenocarc.	30	K-ras
colon, adenocarc.	50	K-ras
colon, adenoma	50	K-ras
pancreas carc.	90	K-ras
seminomas	40	K-, N-ras
melanomas	20	N-ras
bladder carc.	6	H-ras
myeloid disorders:		
-myelodysplastic syndrome	30	N-ras
-acute myeloid leukemia	30	N-ras
-chronic myeloid leukemia	0	-
lymphoid disorders:		
-acute lymphoid leukemia	0	-
-non-Hodgkin lymphomas	0	-
-Hodgkin lymphomas	0	-
pediatric leukemias:		
-myeloid	30	N-ras
-lymphoid	10	N-ras

For details and references see ref.1

well as during complete clinical remission and after the subsequent relapse (8). In other patients the mutation was not present in the first "preleukemic" bone marrow sample but occurred during the progression into AML (7,8). In these patients cells with a mutated ras gene disappeared during clinical remission and did not reappear in the following relapse (8). Apparently, in the development of AML ras activation can both be an early and a late event.

It is still unclear whether in some tumors ras activation is involved in the metastatic progression. Initially, it has been suggested that ras mutations might be involved in the metastatic progression of melanomas (9). However, we have found in five cases that a mutation was present both in the metastasis and in the primary tumor (10), indicating that in these cases the mutational activation was most likely an earlier event.

With animal model systems it has been shown that ras mutations can be induced by a variety of chemical mutagens. This prompted us to investigate whether a correlation exists between the presence of ras gene mutations and mutagenic agents like, for instance, tabacco smoke. Of 15 adenocarcinomas of the lung which were found to harbor a mutated ras gene (45 tumors tested) 14 were isolated from 32 patients with a recent smoking history, 1 from 4 patients that had not smoked for 5 years and

Table II KRAS codon 12/13 mutations in tumors of pancreas, colon and lung

		pancreas n=28	colon n=60	lung n=14
K12	cys G-T	36*	12	43
	ser G-A	0	12	0
	arg G-C	4	0	0
	val G-C	28	16	21
	asp G-A	32	32	29
	ala G-C	0	7	7
K13	asp G-A	0	21	0

* indicated are percentages of the total number of KRAS 12/13 mutations

none in patients that have never smoked (6 patients) or whose smoking history was unknown (3 patients) (13). These data strongly suggest the involvement of a mutagenic agent in the activation of ras genes in lung tumors. Whether mutagenic agents are involved in the induction of ras mutations in other tumors is still unknown. The comparison of the types of K-ras mutation found in tumors from pancreas, colon and lung (Table II) revealed however a clear difference in the mutation spectra (12). In colon tumors the majority of the mutations are G-T transversions, whereas in pancreas and lung tumors the mutations are mainly G-A transitions. These differences may indicate the involvement of different mutagenic agents in the induction of the ras gene mutation, although tissue-specific effects, like metabolic conversion of mutagens and DNA repair cannot be excluded.

In conclusion, the analysis of mutated ras genes in human tumors has revealed that mutational activation of ras genes is a rather common phenomenon. It is, however, not an obligatory event for the development of a tumor, and, when it occurs, it is not restricted to a certain stage in tumor development. The analysis has also revealed the possible involvement of chemical mutagens in the mutational event and, as a consequence, in the induction of the tumor. However, the analysis of mutated ras genes has not yet provided a clear contribution in the treatment of tumors.

ACKNOWLEDGEMENT

This work was supported by a grant from the Dutch Cancer Foundation (K.W.F.)

REFERENCES

1. J. L.Bos, Ras gene mutations and human cancer, in: Molecular genetics and the diagnosis of cancer, Elsevier (1989) (Ed. J. Cossman). in press.
2. S. Sukumar, N. Notorio, D. Martin-Zanca, and M. Barbacid, Induction of mammary carcinomas in rats by nitrosomethylurea involves malignant activation of H-ras-1 locus by single point mutations, Nature 306:658 (1983).
3. M. Quintanilla, R. Brown, M. Ramsden, and A. Balmain, Carcinogen-specific mutation and amplification of Ha-ras during mouse skin carcinogenesis, Nature 322:78 (1986).

4. K. H. Vousden, and C.J. Marshall, Three different activated ras genes in mouse tumors; evidence for oncogene activation during progression of a mouse lymphoma, EMBO J. 3:913 (1984).

5. B. Vogelstein, E.R. Fearon, S.R. Hamilton, S.E. Kern, A.C. Preisinger M. Leppert, Y. Nakamura, R. White, A.M.M. Smits, and J.L. Bos, Genetic alterations during colorectal-tumor development. N.Engl.J Med. 319:525 (1988).

6. J. W. G. Janssen, A.C.M. Steenvoorden, J. Lyons, B. Anger, J.U. Böhlke, J.L. Bos, H. Seliger, and C.R. Bartram, Ras gene mutations in acute and chronic myelocytic leukemias, chronic myeloproliferative disorders, and myelodysplastic syndromes, Proc.Natl.Acad.Sci. USA 84:9228 (1987).

7. H. Hirai, M. Okada, H. Mizoguchi, H. Mano, Y. Kobayski, J. Nishida, and F. Takaku, Relationship between activated N-ras oncogene and chromosomal abnormality during leukemic progression from myelodysplastic syndrome, Blood 71:256 (1988).

8. J. J. Yunis, A.J.M. Boot, M.G. Mayer, and J.L. Bos, Preponderance of N-ras mutation in myelodisplastic syndromes with monocytic features and poor prognosis, submitted.

9. A. P. Albino, R. Le Strange, A.I. Oliff, M.E. Furth, and L.J. Old, Transforming ras genes from human melanoma: a manifestation of tumour hetergeneity? Nature 308:69 (1984).

10. L. J. van 't Veer, B.M. Burgering, R. Versteeg,, A.J.M. Boot, D.J. Ruiter, S. Osanto, P.I. Schrier, and J.L. Bos, N-ras mutations in human melanoma, submitted.

11. S. Rodenhuis, R.J.C. Slebos, A.J.M. Boot, S.G. Evers, W.J. Mooi, S.C. Wagenaar, P.Ch. van Bodegom and J.L. Bos, Incidence and possible clinical significance of K-ras oncogene activation in adenocarcinoma of the human lung, Cancer Res. 48:5738 (1988).

12. V. T.H.B.M. Smit, A.J.M. Boot, A.M.M. Smits, G.J. Fleuren, C.J. Cornelisse, and J.L. Bos, KRAS codon 12 mutations occur very frequently in pancreatic adenocarcinomas, Nucl.Acids Res. 16:7773 (1988).

EARLY DETECTION OF ras ONCOGENES ACTIVATED BY CHEMICAL CARCINOGEN-
TREATMENT

R. Kumar, S. Sukumar, and M. Barbacid

BRI-Basic Research Program
Frederick Cancer Research Facility
P.O. Box B
Frederick, MD   21701

INTRODUCTION

Molecular analysis of H-ras oncogenes induced by a single dose of
nitrosomethylurea (NMU) in pubescent rats revealed their consistent
activation by G→A transitions, the type of mutation preferentially
induced by NMU. These results suggested that NMU is directly responsible
for the malignant activation of ras oncogenes during initiation of
carcinogenesis. We have now modified this experimental system to examine
the precise mechanisms by which ras oncogenes contribute to carcinogenesis. Our results suggest that although ras oncogenes may become
activated during the carcinogenic insult, they require specific cooperation with normal developmental programs to exert their carcinogenic
properties. Definitive demonstration of this hypothesis requires
identification of activated ras oncogenes in the putative target cells
prior to the manifestation of the neoplastic phenotype. In an attempt to
reach the sensitivity levels necessary to verify this hypothesis, we have
optimized the detection of ras DNA sequences amplified by the polymerase
chain reaction (PCR) technique and have succeeded in identifying ras
oncogenes at the single-cell level.

CONTRIBUTION OF ras ONCOGENES TO NEOPLASIA

H-ras oncogenes have been identified in most breast carcinomas of
rats exposed to a single intravenous dose of NMU during puberty. We
reasoned that the specific development of mammary tumors in these
animals, and perhaps the reproducible activation of H-ras oncogenes,
might be a direct consequence of the active developmental stage of the
mammary gland during the carcinogenic insult. In order to separate the
activation of ras oncogenes from sexual development and to examine their
individual contribution to carcinogenesis, we modified the basic protocol·
of this animal model system. Rats were exposed to a single intravenous
dose of NMU 2 days after birth and then either allowed to develop

normally or treated with the antiestrogen drug Tamoxifen and subsequently ovariectomized.

Neonatal exposure of rats to NMU did not significantly reduce the incidence of mammary carcinoma in those animals allowed to proceed through sexual development. Thirty-five of 44 animals (80% incidence) injected with a single dose of NMU 48 hours after birth developed mammary carcinomas with a mean latency period of 3.7 months. The histopatholog- ical properties of these mammary carcinomas were indistinguishable from those of tumors obtained by exposing pubescent animals to this carcino- gen. The presence of H-ras oncogenes in these tumors was established by direct identification of the diagnostic 12th codon G→A transitions using the PCR technique followed by selective oligonucleotide hybridization (see below). Seventeen of 35 mammary carcinomas (49%) induced by neonatal exposure to NMU contained G→A activated H-ras oncogenes. These results demonstrate that NMU-induced activation of H-ras oncogenes in mammary carcinomas of rats is independent of the timing of the carcino- genic insult. The reason for this lower frequency of H-ras activation in neonatal-treated animals (48%) compared to pubescent animals (86%) remains to be determined. However, it possibly reflects the smaller number of cells that can be targeted by the initiating carcinogen in the neonatal animals.

The neonatal NMU treatment protocol allowed us to separate the cancer initiating event from the promotional role played by sexual hormones. Mammary carcinogenesis was drastically reduced in the 2-day-old NMU treated rats that were subsequently ovariectomized. Whereas 80% of the NMU-tested control animals developed mammary carcinomas (see above), only 2 out of 27 (7%) ovariectomized rats exhibited tumored breasts. These results suggest that if NMU activates H-ras oncogenes during its brief period of mutagenic activity in the exposed animals, the nontumored ovariectomized rats should harbor H-ras oncogenes in their mammary epithelium. Therefore, reconstitution of sexual differentiation in these ovariectomized animals should induce tumor growth and at least a fraction of these tumors should carry G→A activated H-ras oncogenes.

To test this hypothesis, 12 rats were exposed after birth to NMU and ovariectomized before reaching sexual development. After 4 months, none exhibited any palpable breast tumors. In contrast, more than 60% of the NMU-treated control animals had already developed sizable mammary carcinomas. Injection of a single dose of estrogen to the ovariectomized tumor-free animals elicited the appearance of mammary carcinomas in six of the animals in less than 6 weeks. When submitted to PCR analysis, three (50%) of these six mammary tumors contained H-ras oncogenes activated by G→A transitions. These results illustrate the requirement of estrogen-induced differentiation of mammary epithelial cells to trigger neoplastic development. Moreover, the presence of H-ras onco- genes activated by the NMU-induced G→A mutations suggests that these oncogenes became activated in the neonatal animal but could not exert their tumorigenic properties until the harboring cells became engaged in hormone-induced differentiation processes. These results raise the exciting possibility that H-ras oncogenes reside in normal cells without eliciting any neoplastic phenotype. If so, neoplasia will result from the combined action of a resident oncogene and cellular pathways involved in developmental and/or proliferative programs.

DETECTION OF RAS ONCOGENES AT THE SINGLE CELL LEVEL

Molecular analysis of the precise timing of ras oncogene activation requires detection of their activating mutations in a small number of cells prior to the appearance of neoplastic, or even pre-neoplastic, phenotypes. For this purpose, we have developed a novel strategy that improves the detection of sequences amplified by the PCR technique. This strategy utilizes liquid hybridization combined with a gel retardation assay. Reactions in liquid media allow a better control of the stringency of hybridization by adjusting either the temperature or the ionic strength. More importantly, liquid hybridization makes it possible to use restriction fragment length polymorphisms (RFLPs) to identify closely related allelic variants such as _ras_ oncogenes. Hybridization of the radiolabeled probe to the amplified DNA results in the formation of hemiduplexes that can be separated from the free probe by electrophoresis in nondenaturing polyacrylamide gels. The extent of retardation of the annealed radioactive probe reflects the size of the amplified product. Since the mobilities of the amplified sequences depend on the selection of the amplimers, our strategy makes it possible to detect different amplified targets in a single reaction.

Individual normal rat kidney (NRK) cells were obtained by micromanipulation of a cell suspension. One, 2, 5 or 10 NRK cells were placed in reaction vials and lysed prior to PCR amplification. Following electrophoresis, radioactive hybrids representing 1/20 of the amplified sample could be easily detected in less than 4 hours. Detection of H-_ras_ sequences in single cells was not limited to those derived from established cell lines. Kidney and mammary glands were obtained from a fresh rat cadaver. Cells were prepared by enzymatic digestion of finely minced tissues, washed repeatedly in PBS, counted and distributed into amplification vials by appropriate dilution. Amplification of H-_ras_ codon 12 sequences in tissue-derived cells was as efficient as in cells cultured in vitro. Comparable hybridization intensities could be routinely observed when 1 to 50 cells were used to initiate PCR amplification. A 2- to 10-fold stronger signal was usually detected when 100 to 10,000 cells were amplified. However, amplification of more than 50,000 cells under our standard conditions hampered the efficiency of polymerization (PCR) and yielded weaker hybridization signals.

A dramatic demonstration of the use of PCR in the analysis of single-copy genes in isolated cells is the amplification of H-_ras_ sequences in one-cell mouse embryos. For this purpose, fertilized mouse ova (kindly provided by Michael Rosenberg, Molecular Genetics of Oncogenesis Section, Mammalian Genetics Laboratory) were individually collected by micromanipulation and disrupted, and their DNAs were amplified in the region corresponding to codon 61 of the resident mouse H-_ras_ proto-oncogene. The amplified H-_ras_ sequences were efficiently detected in each of the samples, including the one containing 1/20 of the DNA resulting from the amplification of a single mouse embryo.

The liquid hybridization/gel retardation assay allows the use of RFLPs as an alternative or complementary approach to detect single point mutations. In rat H-_ras_ oncogenes, the NMU-induced GGA→GAA mutation in the 12th codon eliminates the GAGG sequence recognized by MnlI, where the underlined G is the mutated base. This restriction endonuclease cuts the DNA seven bases 5′ of the GAGG recognition sequence, generating a 40-base pair fragment that can be readily identified after electrophoresis in polyacrylamide gels. Resistance of the amplified 62-base pair DNA fragment to MnlI digestion is diagnostic of the activating GGA→GAA mutation.

To determine the extent to which such an allelic imbalance in the PCR substrate would interfere with the amplification and detection of H-_ras_ oncogenes, we designed a simple reconstitution experiment. A segment of the H-ras gene including 12th-codon sequences was amplified in a series of cell mixtures containing different amounts of NRK and NMU58 cells. The amplified sequences were subsequently identified by the liquid hybridization/gel retardation technique using probes specific for the normal (GGA) and mutated (GAA) 12th codons at their respective discriminating temperatures. Detection of the H-ras oncogene harbored in a single heterozygous NMU58 cell was not affected by the presence of up to 10,000 normal NRK cells. These results indicate that the PCR technique allows representative amplification of individual alleles, even when they are highly underrepresented.

EARLY DETECTION OF ras ONCOGENES BY A POSITIVE IDENTIFICATION STRATEGY

We have employed a new experimental approach for the detection of ras gene mutations in tissues soon after carcinogen treatment of the animals. This method exploits the PCR procedure and utilizes designed mismatched amplification primers. The combination of the introduced mismatch in the PCR primer and the mutated 12th codon sequences generate a novel diagnostic restriction site. Tissue samples (mammary glands are individually isolated and analyzed) are subjected to PCR amplification. An aliquot of the PCR product is cleaved with the diagnostic restriction enzyme followed by analysis of the products by the gel-retardation assay.

Such RFLP analysis have been developed for the detection of G→A mutation in the 12th codon of H-ras and K-ras genes of the rat. Neonatal rats treated with NMU on day 2 post-partum were sacrificed two weeks later. Tissue samples were analyzed for the presence of specifically mutated ras oncogenes. We have now identified both H- and K-ras oncogenes in these samples. In the seven animals tested so far, one or two mammary glands in each animal contained ras oncogenes. This represents a major improvement in the detection technology. We are presently extending these experiments in order to address several fundamental questions in carcinogenesis.

ACKNOWLEDGEMENT

Research supported by the National Cancer Institute, DHHS, under contract No. NO1-CO-74101 with Bionetics Research, Inc.

RAS MUTATIONS IN PRIMARY MYELODYSPLASIA AND IN PATIENTS FOLLOWING CYTOTOXIC THERAPY

A. Jacobs, R.A. Padua, G. Carter, D. Hughes, R.E. Clark,
D. Oscier, D. Bowen, and J.F. McCormick

Cardiff, London, Bournemouth, U.K.; San Francisco, USA

The ras gene family, H-ras, K-ras and N-ras, code for 21kD proteins that have GTPase activity and have been implicated in the control of cell proliferation (McKay et al. 1986; Trahey et al. 1987). Mutations in these genes give rise to abnormal protein products that have the capacity to transform certain cells to a malignant phenotype. Ras mutations have been found in a wide range of human malignancies and N-ras has been particularly implicated in AML, chronic myeloid leukaemia (CML), and acute lymphoblastic leukaemia (ALL) (Janssen et al.1985; Hirai et al. 1985). Activation of these genes has been associated with mutations in codons 12/13 or codon 61 and lesions of this type have recently been described in MDS. Hirai et al. (1987) described three MDS patients with codon 13 mutations in N-ras and Liu et al. (1987) two patients with a similar mutation in K-ras. We have assessed the mutational status of members of the ras gene family by polymerase chain reaction and hybridisation with synthetic oligonucleotide probes in 50 patients with MDS together with the use of a nude mouse tumorigenicity assay in some cases (Padua et al. 1988). We have also examined material from 70 haematologically normal patients in complete remission following cytotoxic therapy for lymphoma, but in whom a higher than normal incidence of MDS and leukaemia might be expected in subsequent years.

Methods

High molecular weight DNA was extracted from blood or bone marrow cells obtained from 50 patients with MDS. Sequences of about 100bp were amplified by the polymerase chain reaction using primers flanking codons 12-13 and 61 of Ha, K and N-ras. The primer sequences and probes have been described previously (Padua et al. 1988).

Results

Activated ras genes were found in 21 cases, including three with primary acquired sideroblastic anaemia (PASA). N-ras mutations were found in 14 cases together with 6 K-ras and 2 H-ras mutations (Table 1). Double mutations were found in different ras genes in one patient with PASA and one patient with chronic myelomonocytic leukaemia (CMML). Although the results show that ras mutations may occur in the more benign forms of MDS, the highest frequency was found in CMML. DNA from 17 patients with mutations detected by hybridisation with the synthetic probes were all positive when tested in NIH 3T3 transformation assays (Padua et al. 1988). DNA from one patient in whom no mutation was found by direct hybridisation was positive in nude mouse tumorigenicity assay and the tumours yielded a mutant human N-ras gene on further analysis. During the period of this study 8 of the 21 patients with mutant ras genes transformed to AML compared with 4 of the 29 patients with no mutation. Transfection of DNA from the four patients without ras mutations detected by hybridisation with mutant specific probes but who subsequently developed AML, showed that two of these gave rise to transformation of NIH3T3 cells and tumour formation in nude mice. In neither of these cases were the tumours found to contain an activated ras gene. Our own work and that of Takaku et al. (1987) suggests that tumorigencitiy assays offer a more sensitive method of detecting activated oncogenes than DNA hybridisation.

More recent studies using probes directed at possible mutations in codon 117 of N-ras, revealed one case of sideroblastic anaemia with an Asp substitution at this position. We have found no correlation between the presence of any of the mutant ras genes and clonal cytogenetic abnormalities.

Of the 70 patients studied following cytotoxic therapy and screened for N, K and H-ras mutations by DNA hybridisation, preliminary data show one patient with an N12 valine substitution. This patient has been in complete remission from Hodgkins disease for 9 years. The N12 Val mutation is not detectable in paraffin-block sections of the original tumour tissue, or in unaffected tissues from the same patient. These findings suggest that the N12 Val positive cells are not relapsed Hodgkins disease, and confirm the somatic origin of this mutation.

Discussion

Presumably, ras activation plays a part in malignant progression by conferring selective growth advantage on the cells in which the mutations

24

occur and, possibly, by interfering with their differentiation programme. The basis for preferential activation of a particular ras gene is not clearly understood. In haemopoietic cells, N-ras and K-ras p21 proteins are expressed at higher levels than H-ras p21 (Shen et al. 1987) and this is associated with a relatively high frequency of activated N-ras, and a somewhat lower frequency of activated K-ras, in haemopoietic malignancies. This might suggest a specific role for the N-ras and K-ras proteins in the growth control of these tissues with its consequent disruption by the production of an abnormal protein. Preferential activation of H-ras is found in urinary tract cancer (Tabin et al. 1982) and K-ras activation is commonly found in lung and colon cancer (McCoy et al. 1983). In rodents, H-ras mutations are not only found with a high frequency in chemically induced epithelial tumours (Sukumar et al. 1983; Balmain et al. 1984) but normal mouse skin epithelium shows a high expression of the normal H-ras p21. While there is now a substantial body of evidence implicating the ras genes in the pathogenesis of a wide range of malignancies, their precise role is obscured by ignorance of their normal cellular function. There are differences between normal and abnormal p21 ras proteins. Mutant proteins have been shown to have reduced GTPase activity, though reduced GTPase activity does not necessarily correlate with transforming ability (Trahey et al. 1987).

Evidence from chemically induced skin tumours in mice suggests that H-ras mutation is an early event in the malignant process and may be found in premalignant papillomas before they undergo carcinomatous transformation (Balmain et al. 1984). Similar experiments in chemically induced rat mammary tumours suggest an early occurrence of H-ras mutations (Sukumar et al. 1983). The presence of mutations in patients with PASA, the most benign form of MDS, and possibly an early stage in the preleukaemic process, suggests that it may also be an early event in this condition. However, the apparently high rate of leukaemic transformation in patients with mutant genes makes it unlikely that this is a common early event. Hirai et al (1988) have reported two cases of MDS in which the amount of mutated DNA was related to the number of blast cells in the bone marrow as the disease progressed to leukaemic transformation, suggesting that the mutation is a late event characterising the fully leukaemic phenotype.

Current findings indicate that N-ras mutations may be found both in early and in late stages of haemopoietic malignant progression. Mutations appear to have a frequency of about 30-40% both in MDS and in AML. Shen et al (1987) have suggested that there is a heterogeneity of leukaemic cells with respect to the presence of an activated ras oncogene and that in many

Table 1. RAS mutations in MDS. Square brackets indicate mutations of the two patients with double mutations. RARS = refractory anaemia with ringed sideroblasts; RA = refractory anaemia; RAEB = refractory anaemia with excess blasts; CMML = chronic myelomonocytic leukaemia.

	No. RAS mutants/ no. samples	Substitution
Controls	0/10	
Myelodysplasia		
RARS	2/13	1 x H61 Arg (A → G) ⌉ 1 x N12 Val (G → T) ⌋ 1 x K12 Asp (G → A)
RA	4/13	1 x N12 Asp (G → A) 1 x N12 Val (G → T) 1 x N61 Leu (A → T) 1 x K12 Asp (G → A)
RAEB	3/8	1 x N12 Ala (G → C) 1 x N12 Asp (G → A) 1 x K12 Arg (G → C)
CMML	11/16	1 x N12 Ala (G → C) 1 x N12 Asp (G → A) 3 x N13 Ala (G → C) 3 x K12 Asp (G → A) ⌉ 1 x H12 Val (G → T) ⌋ 3 x N61 His (A → C)
TOTAL	20/50 (40%)	

26

patients only a fraction of the malignant cells carry the mutant gene. One possible explanation of this is that the mutations occurs late, after a preleukaemic clone has already emerged, and simply gives the premalignant clone an additional growth advantage. An alternative explanation is that the ras mutation occurs early in the preleukaemic process and there is later evolution with emergence of a clone in which the transforming gene has been deleted. The observation of mutant ras genes in 21 preleukaemic patients, many of whom have no increase in the number of blast cells, argues for the latter explanation, though there is no evidence that it is an initiating event in leukaemogenesis. Further support for this view is the occurrence of patients with acute leukaemia in whom a mutant ras gene is demonstrated on initial diagnosis but in whom it cannot be detected at a later stage when the disease relapses (Farr et al. 1988).

While ras mutations appear to be an important factor in the evolution of malignancy, the fully malignant phenotype requires the interaction of a number of different lesions, not necessarily occurring in a constant sequence. The data suggest that N-ras mutations in an abnormal clone of haemopoietic cells may confer a growth advantage at an early stage in the leukaemogenic process and may predispose to further evolution. Later genetic events, including karyotype abnormalities, probably determine the morphological and functional phenotype of the clinical disorder.

ACKNOWLEDGEMENTS

This work was supported by the Leukaemia Research Fund. The Tables are taken from Padua et al. (1988).

REFERENCES

Balmain A, Ramsden M, Bowden GT, Smith J. Activation of the mouse cellular Harvey-ras gene in chemically induced benign skin papillomas. Nature 1984;307:658-660.

Hirai H, Tanaka S, Azuma M, Anraku Y, Kobayashi Y, Fujisawa M, Okabe T, Urabe A. Transforming genes in human leukaemia cells. Blood 1985; 66: 1371-1378.

Hirai H, Kobayshi Y, Mano H, Hagiwara K, Maru Y, Omine M, Mizgouchi H et al. A point mutation at codon 13 of the N-ras oncogene in myelodysplastic syndrome. Nature 1987;327:430-432.

Hirai H, Okada M, Muzgouchi H, Mano H, Kobayashi Y, Nishida J, Takaku F. Relationship between an activated N-ras oncogene and chromosomal abnormality during leukaemic progression from myelodysplastic syndrome. Blood 1988; 71: 256-258.

Janssen JWG, Steenvoorden ACM, Collard JG, Nusse R. Oncogene activation in human myeloid leukaemia. Cancer Research 1985;45:3262-3267.

Liu E, Hjelle B, Morgan R, Hecht F, Bishop JM. Mutations of the Kirsten-ras proto-oncogene in human preleukaemia. Nature 1987;330:186-188.

McCoy MS, Toole JJ, Cunningham JM, Chang EH, Lowy DR, Weinberg RA. Characterization of a human colon/lung carcinoma oncogene. Nature 1983;302: 79-81.

McKay IA, Paterson H, Brown R, Toksoz D, Marshall CJ, Hall A. N-ras and human cancer. Anticancer Research 1986;6:483-490.

Padua RA, Carter G, Hughes D, Gow J, Farr C, Oscier D, McCormick F, Jacobs A. Ras mutations in myelodysplasia detected by amplification, oligo-nucleotide hybridisation and transformation. Leukaemia - in the press.

Shen WP, Aldrich TH, Venta-Perez G, Franza BR, jr, Furth ME. Expression of normal and mutant ras proteins in human acute leukaemia. Oncogene 1987; 1: 156-165.

Sukumar S, Notario V, Martin-Zanca D, Barbacid M. Induction of mammary carcinomas in rats by nitroso-methylurea involves malignant activation of H-Harvey-ras gene in chemically induced benign skin papillomas . Nature 1984;307:658-660.

Tabin CJ, Bradley SM, Bargmann CI, Weinbergh RA, Papageorge AG, Scolnick EM, Dhar R, Lowy DR. Mechanism of activation of human oncogene. Nature 1982; 300:143-147.

Trahey M, Milley RJ, Cole GE, Innis M, Paterson H, Marshall CJ, Hall A, McCormick F, Biochemical and biological properties of the human N-ras p21 protein. Molecular and Cellular Biology 1987; 7:541-544.

HIGH FREQUENCY OF ras ONCOGENE ACTIVATION

IN BENIGN AND MALIGNANT HUMAN THYROID TUMOURS

Nick R. Lemoine, Edward S. Mayall, Fiona S. Wyllie, and David Wynford-Thomas

CRC Thyroid Tumour Biology Research Group, Department of Pathology, University of Wales College of Medicine, Heath Park, Cardiff, CF4 4XN

INTRODUCTION

Tumours of the follicular epithelium of the human thyroid gland represent a multi-stage model of epithelial tumorigenesis with a low frequency of malignant progression (Williams 1979). Human thyroid follicular adenomas are genuine benign neoplasms which occur much more frequently and affect a younger age group than follicular carcinomas (Sommers 1982), while at the other end of the spectrum, undifferentiated thyroid carcinomas are rare tumours which occur predominantly in the elderly and are thought to arise as a progression from differentiated carcinoma.

We recently demonstrated the presence of activated ras oncogenes in a small series of differentiated thyroid follicular and papillary carcinomas analyzed by transfection (Lemoine et al 1988a). This stimulated us to examine a larger series encompassing the whole spectrum of thyroid neoplasia from benign adenomas through to undifferentiated carcinomas, using polymerase chain reaction amplification and sequence-specific oligonucleotide hybridization to search for ras oncogene mutations.

MATERIALS & METHODS

Thyroid adenomas can be divided on the basis of their microscopic appearance into two major groups: microfollicular (also called fetal) and macrofollicular (also called colloid) types (Hedinger 1974). Our series comprised 8 macrofollicular (cases 1-8 in Figure 1) and 16 microfollicular adenomas (cases 9-24 in Figure 1). The malignant tumours examined comprised 10 follicular carcinomas (cases 1-10 in Figure 2) and 10 undifferentiated carcinomas (cases 11-20 in Figure 2). The majority of cases were in females which reflects the well-documented epidemiology of these tumours (Sommers 1982).
Selective amplification of regions around codons 12/13 and 61

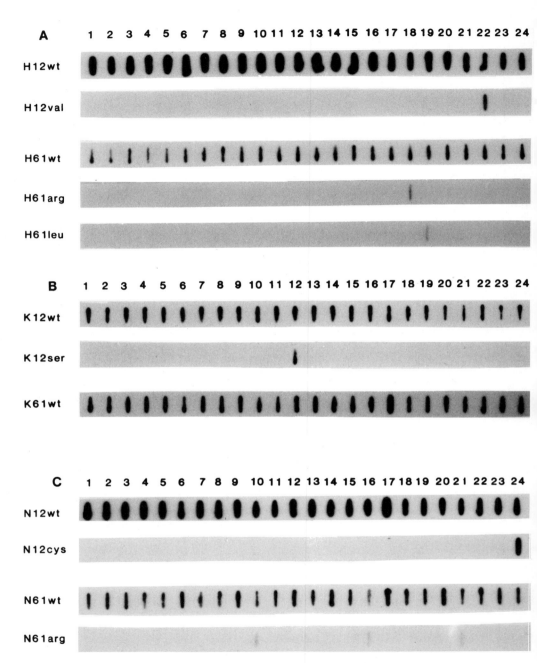

Figure 1. Slot-blot analysis of benign thyroid lesions for
mutations in A) Ha-ras, B) Ki-ras and C) N-ras oncogenes. Only
those filters that showed specific hybridisation at the
discriminating temperature are shown

of each ras oncogene by the PCR technique (Saiki et al 1985) was
performed directly on thin sections from formalin-fixed,
paraffin-embedded tissue blocks, up to 10 years old, selected
from the archives of the Department of Pathology, University of
Wales College of Medicine by a method similar to Shibata et al
(1988). Heat-stable Taq polymerase (Cetus Corporation) was used
to amplify DNA in the presence of specific paired primers. Probes
and primers correspond to those listed in Padua et al., (1988), and
probe labelling and hybridization conditions were those described
by Farr et al (1988).

Transfection analysis using the nude mouse tumorigenicity
assay (Fasano et al 1984) was performed where fresh tissue was
available for the preparation of high molecular weight DNA.

RESULTS

Thyroid adenomas

Mutated ras oncogenes were identified in 8 cases (Figure 1),
all of which were microfollicular adenomas. Codon 61 of N-ras (3
cases with N61 gln ->arg substitutions) and Ha-ras (one case with
H61 gln ->arg and one case with H61 gln ->leu) were the most
common targets for mutation, although single examples of mutation
at N12, H12 and K61 were found. 4/8 cases involved A->G
transition at position 2 of codon 61.

6 of the 10 cases (all microfollicular adenomas) subjected to
transfection analysis showed transforming activity which was due
to activated N-ras in 3 cases, activated Ha-ras in 2 cases, and
activated Ki-ras in 1 case (detected by Southern blot analysis of
first round NIH3T3 transformants, data not shown). Corresponding
mutated ras oncogenes were found in the genomic DNA of each of
these cases by PCR and oligonucleotide probing.

Differentiated thyroid carcinomas

Mutated ras oncogenes were identified in four of the ten
differentiated follicular carcinomas (cases 4, 6, 7 & 9 in Figure
2). It is notable that H61 gln->arg mutation, A-> G transition at
position 2, occurred in two of these four cases.

Undifferentiated thyroid carcinomas

Six of the ten undifferentiated carcinomas were found to
contain mutated ras oncogenes by oligonucleotide probing (cases
11, 12, 13, 14 (weak signal only), 19 & 20 in Figure 2). In these
cases the most common mutation was G-> T transversion at codon 12
of Ha-ras or Ki-ras, which contrasts with the transitions more
commonly seen in the differentiated tumours (both benign and
malignant).

Only four DNAs from undifferentiated carcinomas (cases 12,
13, 14 and 15) were subjected to transfection assay: in each case
there was concordance of the results of oligonucleotide
hybridisation and the identity of transforming gene detected in
the nude mouse tumour DNA.

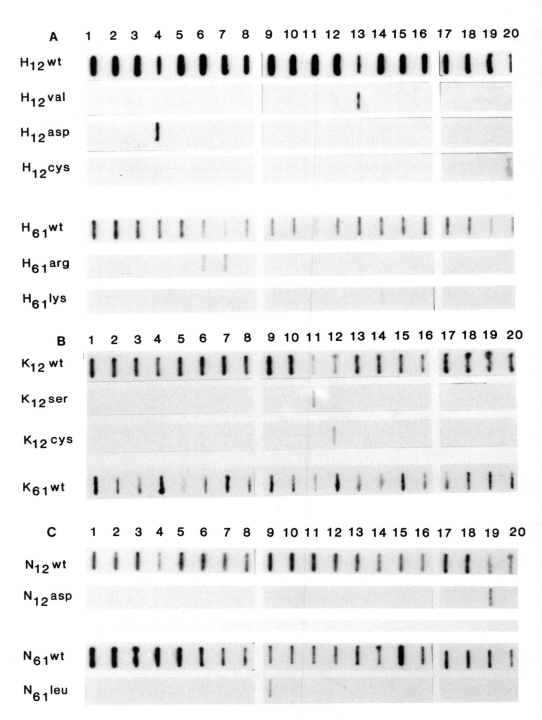

Figure 2. Slot-blot analysis of malignant thyroid tumours for mutations in A) Ha-ras, B) Ki-ras and C) N-ras oncogenes. Only those filters that showed specific hybridisation at the discriminating temperature are shown.

DISCUSSION

It appears that activation of ras oncogenes by mutation
occurs at a relatively early stage of thyroid tumorigenesis,
being present in 33% of follicular adenomas overall. This is one
of the first reports of ras activation in unequivocally benign
human tumours. The first example of ras oncogene activation in a
genuinely benign tumour was reported by Leon et al (1988) who
found that one out of ten human keratoacanthomas contained
activated Ha-ras. More recently, Vogelstein et al (1988) have
performed a detailed analysis of 80 colonic adenomas and 92
colonic carcinomas (which constitute an analogous multi-stage
model of tumorigenesis to that of thyroid neoplasms) and found
mutant ras alleles in 9% of adenomas less than 1 cm diameter and
in 58% of adenomas larger than 1 cm (as well as in 47% of
carcinomas which confirmed the results of the previous studies of
these tumours, Bos et al 1987a, Forrester et al 1987).

The frequency of detection of ras oncogene activation is
similar at all stages of tumorigenesis in the human thyroid: 33%
of adenomas (50% of microfollicular), 40% of follicular
carcinomas and 60% of undifferentiated carcinomas. This would
implicate ras oncogene mutation among the early events of
neoplastic progression in this system. A similar situation is
postulated for human myeloid neoplasia where ras mutations are
observed at similar frequency in the premalignant condition of
myelodysplasia (Hirai et al 1987, Liu et al 1987, Janssen et al
1987, Padua et al 1988) as in acute myeloid leukaemia (Bos et al
1987b, Farr et al 1988), and also for epithelial tumours of the
human colon (Bos et al 1987a, Forrester et al 1987, Vogelstein et
al 1988). However, the fact that small adenomas (less than 1 cm)
contained mutant ras oncogenes relatively infrequently
(significantly less often than larger adenomas) suggests that ras
activation is not usually the very first genetic alteration, and
hence cannot be responsible for the initiation of neoplasia in
these tumours.

The evidence from animal models is strongly in favour of ras
oncogene activation as an early event. There a high frequency of
ras oncogene activation in carcinogen-induced tumours, including
those of the rat thyroid gland (Lemoine et al 1988b), and the
activating mutations are those expected from the known mutagenic
action of the initiating carcinogen (Zarbl et al 1985, McMahon et
al 1987, Bizub et al 1986, Quintanilla et al 1986, Yamasaki et al
1987, Hochwalt et al 1988, Wiseman et al 1986). In the mouse skin
tumour model, the ras activation has been shown to be present in
benign tumours induced by carcinogen (Balmain et al 1984) as well
as the subsequent carcinomas (Balmain & Pragnell 1983).

Activated ras oncogenes were found in 50% of microfollicular
adenomas, but were found in none of the macrofollicular lesions.
It is of interest that while microfollicular areas are commonly
seen also in follicular carcinomas (which are presumed to arise
from pre-existing follicular adenomas), macrofollicular
architecture in these malignant tumours is very rare. In fact,
macrofollicular lesions appear to have such a low potential for
neoplastic progression, it has been postulated that they might
alternatively represent localised foci of nodular hyperplasia
(Rosai 1981).

The most frequently detected amino acid substitution amongst the differentiated tumours (benign and malignant) was that of glutamine to arginine at position 61 of Ha-ras or N-ras, which results from an A to G transition. This mutation did not appear at all amongst the undifferentiated cancers. Transition mutations in general far outnumber transversions in the differentiated tumours (6 transitions to 2 transversions in follicular adenomas, 3 transitions to 1 transversion in follicular carcinomas) while the situation is completely reversed for the undifferentiated carcinomas, where transversions are more common (4 transversions to 2 transitions). It is possible that this difference in mutation pattern is the result of a difference in aetiological (initiating) agents for differentiated and undifferentiated tumours. Alternatively, it may reflect a difference in biological activity conferred by the various mutant ras oncogenes, with position 61 arginine mutants being less likely to progress to the undifferentiated phenotype.

REFERENCES

Balmain A., Pragnell I.B. (1983) Mouse skin carcinomas induced in vivo by chemical carcinogens have a transforming Harvey-ras oncogene. Nature, 303: 72-74.

Balmain A., Ramsden M., Bowden G.T., Smith J. (1984) Activation of the mouse cellular Harvey-ras gene in chemically induced benign skin papillomas. Nature, 307: 658-660.

Bizub D., Wood A.W., Skalka A.M. (1986) Mutagenesis of the Ha-ras oncogene in mouse skin tumors induced by polycyclic aromatic hydrocarbons. Proc. Natl. Acad. Sci. USA, 83: 6048-6052.

Bos J.L., Fearon E.R., Hamilton S.R., & 4 others (1987a) Prevalence of ras gene mutations in human colorectal cancers. Nature, 327: 293-297.

Bos J.L., Verlaan-de Vries M., van der Eb A.J. & 4 others (1987b) Mutations in N-ras predominate in acute myeloid leukemia. Blood, 69: 1237-1241.

Farr C.J., Saiki R.K., Erlich H.A., McCormick F., Marshall C.J. (1988) Analysis of RAS gene mutations in acute myeloid leukemia by polymerase chain reaction and oligonucleotide probes. Proc. Natl. Acad. Sci. USA, 85: 1629-1633.

Fasano O., Birnbaum D., Edlund L., Fogh J., Wigler M. (1984) New human transforming genes detected by a tumorigenicity assay. Mol. Cell. Biol., 4: 1695-1705.

Forrester K., Almoguera C., Han K., Grizzle W.E., Perucho M. (1987) Detection of high incidence of K-ras oncogenes during human colon tumorigenesis. Nature, 327: 298-303.

Hedinger C. in Histological Typing of Thyroid Tumours. World Health Organization: Geneva (1974).

Hirai H., Kobayashi Y., Mano H., & 6 others (1987) A point mutation at codon 13 of the N-ras oncogene in myelodysplastic syndrome. Nature, 327: 430-432.

Hochwalt A.E., Solomon J.J., Garte S.J.(1988)Mechanism of H-ras oncogene activation in mouse squamous carcinoma induced by an alkylating agent. Cancer Res. 48: 556-558.

Janssen J.W.G., Steenvoorden A.C.M., Lyons J. & 5 others (1987) RAS gene mutations in acute and chronic myelocytic leukemias, chronic myeloproliferative disorders, and myelodysplastic syndromes. Proc. Natl. Acad. Sci. USA, 84: 9228-9232.

Lemoine N.R., Mayall E.S., Wyllie F.S., & 6 others (1988a) Activated ras oncogenes in human thyroid cancers. Cancer Res., 48: 4459-4463.

Lemoine N.R., Mayall E.S., Williams E.D., Thurston V., Wynford-Thomas D. (1988b) Agent-specific ras oncogene activation in rat thyroid tumors. Oncogene, in press.

Leon J., Kamino H., Steinberg J.J., Pellicer A. (1988) H-ras activation in benign and self-regressing skin tumors (keratoacanthomas) in both humans and an animal model system. Mol. Cell. Biol., 8: 786-793.

Liu E., Hjelle B., Morgan R., Hecht F., Bishop J.M. (1987) Mutations of the Kirsten-ras proto-oncogene in human preleukaemia. Nature, 300: 186-188.

McMahon G., Davis E., Wogan G.N. (1987) Characterization of c-Ki-ras oncogene alleles by direct sequencing of enzymatically amplified DNA from carcinogen-induced tumours. Proc. Natl. Acad. Sci. USA, 84: 4974-4978.

Padua R.A., Carter G.A., Hughes D., & 5 others (1988) RAS mutations in myelodysplasia detected by amplification, oligonucleotide hybridization, and transformation. Leukemia 2: 503-510.

Quintanilla M., Brown K., Ramsden M., Balmain A. (1986) Carcinogen-specific mutation and amplification of Ha-ras during mouse skin carcinogenesis. Nature, 322: 78-80.

Rosai J. (1981) in Ackerman's Surgical Pathology (ed. J.Rosai), Mosby: St Louis.

Saiki R.K., Scharf S., Faloona F., & 4 others (1985) Enzymatic amplification of beta-globin genomic sequences and restriction site analysis for diagnosis of sickle cell anemia. Science 230: 1350-1354.

Shibata D.K., Arnheim N., Martin W.J. (1988) Detection of human papilloma virus in paraffin-embedded tissue using the polymerase chain reaction. J. Exp. Med., 167: 225-230.

Sommers S.C. (1982) in Endocrine Pathology, General and Surgical. (ed. J.M.B. Bloodworth), Williams & Wilkins: Baltimore.

Vogelstein B., Fearon E.R., Hamilton S.R., & 7 others (1988) Genetic alterations during colorectal-tumor development. N. Engl. J. Med, 319: 525-532.

Williams E.D. (1979) The aetiology of thyroid tumours. Clinics in Endocrinology and Metabolism, 8: 193-207.

Wiseman R.W., Stowers S.J., Miller E.C., Anderson M.W., Miller J.A. (1986) Activating mutations of the c-Ha-ras protoncogene in chemically induced hepatomas of the male B6C3 F1 mouse. Proc. Natl. Acad. Sci. USA, 83: 5825-5829.

Yamasaki H., Hollstein M., Martel N., Cabral J.R.P., Galendo D., Tomatis L. (1987) Transplacental induction of a specific mutation in fetal Ha-ras and its critical role in postnatal carcinogenesis. Int. J. Cancer, 40: 818-822.

Zarbl H., Sukumar S., Arthur A.V., Martin-Zanca D., Barbacid M. (1985) Direct mutagenesis of Ha-ras-1 oncogenes by N-nitroso-N-methylurea during initiation of mammary carcinogenesis in rats. Nature, 315: 382-385.

ACKNOWLEDGEMENTS: This work was supported by grants from the Cancer Research Campaign of Great Britain. We thank Dr C.J. Farr and Dr R.A. Padua for providing some of the oligonucleotides.

ALTERATIONS OF C-Ha-*ras* GENE ARE ASSOCIATED WITH c-*myc* ACTIVATION IN HUMAN CERVICAL CANCERS

Guy Riou, Michel Barrois, Zong-Mei Sheng, and Danyi Zhou

Laboratoire de Pharmacologie Clinique et Moléculaire, Institut Gustave Roussy 94800 Villejuif, Cedex, France

INTRODUCTION

In underdeveloped countries, mortality by cervical cancer is the first cause of mortality by cancer in women (Parkin *et al.*, 1984), and continues to increase, in contrast with the situation found in developed countries where the general trend is towards a reduction in both incidence rates and mortality. However clinical observations of the last decade have suggested that there has been an increase in the number of cervical cancer patients aged under 35 years (Hall and Monaghan, 1983). Furthermore those cancers seem to be more aggressive. The biological behaviour of invasive carcinoma of the uterine cervix is not always predictable. Lymph node involvement is the most important prognostic factor, however the absence of invaded lymph nodes does not always guarantee recovery. Therefore it is quite important to find out new biological markers whose presence in early stages of cancers could predict the risk of recurrences. We have previously shown that amplification and/or overexpression of c-*myc* gene were associated with tumoral progression of invasive cervical cancers (Riou *et al.*, 1985; Riou *et al.*, 1987; Riou, 1988a,b) and that elevated levels of c-*myc* RNA in cancers at early stage were correlated with a high risk of relapse for patients (Riou *et al.*, 1987). Among other proto-oncogenes involved in the genesis of human cancers, the *ras* family genes are thought to play an important role (for review, see Barbacid, 1987 and Bos, 1988). Somatic mutations (Bos *et al.*, 1987; Forrester *et al.*, 1987; Spandidos, 1987; Vogelstein *et al.*, 1988), deletions (Reeve *et al.*, 1984; Fearon *et al.*, 1985 Krontiris *et al.*, 1985) and a restriction fragment length polymorphism (RFLP) (Goldfarb *et al.*, 1982; Capon *et al.*, 1983; Krontiris *et al.*, 1985) of the c-Ha-*ras* locus, were shown to be associated with the development of a variety of human cancers. In this study, we have analyzed the c-Ha-*ras* locus in cervical cancers and shown the loss of one allele in 28% of heterozygous tumors and a mutation at codon 12 in 23% of tumors at advanced stages. A c-*myc* gene activation was found in 86% of tumors containing c-Ha-*ras* mutation and/or deletion. This suggests that the two proto-oncogenes cooperate for the progression of cervical cancers.

MATERIALS AND METHODS

Tissue samples. Carcinoma specimens were obtained by biopsy or by surgical excision from untreated patients with primary invasive carcinoma of the uterine cervix at different clinical stages. Tissue samples were also obtained from carcinoma *in situ* (CIS), lymp node metastasis and local recurrences. (Table 1). Normal cell samples were lymphocytes obtained from 52 of the above patients and normal cervical epithelium from 6 patients who had fibromas of the uterine corpus. All the tissue and cell samples were immediately stored in liquid nitrogen. The human lymphocytes were fractionated in a Ficoll-Hypaque gradient.

Table 1. Carcinoma and tissue specimens from uterine cervix analyzed for c-Ha-*ras* locus.

Origin of tissue	N° of specimens	N° of specimens heterozygous for c-Ha-*ras* (%)
Normal cervix	6	4
Carcinoma *in situ*	3	2
Invasive carcinoma [a] (clinical stage)		
I	28	19
II	29	15
III	31	21
IV	9	6
Metastasis [b]	1	1
Recurrence [c]	3	0
Total	110	68 (62%)

[a] Squamous cell carcinoma
[b] Lymph node metastasis
[c] Recurrence from stage I cancers

DNA analysis   Frozen tissues were ground in liquid
nitrogen and high molecular weight DNA was prepared (Terrier
et al., 1988). DNA preparations were extensively digested
with *Hpa* II and *Msp* I restriction enzymes (Sheng *et al.*,
1988). The c-Ha-*ras* oncogene was characterized by Southern
blot. Hybridization was performed on "GeneScreenPlus"
membrane under stringent conditions with the different
probes labelled by nick translation with ^{32}P dCTP to a
specific activity of 1 to 5×10^8 cpm/μg DNA. Pre-
hybridization, hybridization and washing of the membranes
were done as described previously (Maniatis *et al.*, 1982;
Terrier *et al.*, 1988). The filters were exposed to Kodak
XAR5 film for 1 to 4 days. Fragment sizes were determined by
comparison to those of λ phage DNA-digested with *Hind*III or
*Cla*I enzymes.

Probes   The probes used were the 0.6 kilobase pairs
(kb) (probe 1) and the 2.5 kb (probe 2) DNA fragments
obtained by *Xma*I-*Xma*I and *Bam*HI-*Sac*I cleavage respectively
of the c-Ha-*ras* proto-oncogene (Riou *et al.*, 1988).

RESULTS AND DISCUSSION

A RFLP for *Bam*HI or *Msp*I occurring in a region of
variable tandem repetition (VTR) located at the 3' end of
the c-Ha-*ras* locus was shown in previous studies (Goldfarb
et al., 1982; Capon *et al.*, 1983). When hybridized with c-
Ha-*ras* VTR probe (probe 2), single or two DNA fragments were
revealed in DNA preparations from homozygous and
heterozygous patients (Table 1). Figure 1 displays examples
of hybridization patterns from tumor and lymphocyte DNAs.
The size and the frequency of the different alleles observed
were not found to be different in cancers and lymphocytes of
healthy donors (Data not shown) (Riou *et al.*, 1988).

Identical c-Ha-*ras* VTR alleles were detected in DNAs
from tumors and matched normal lymphocytes. Heterozygosity
for these alleles was found in 62% of cancers analyzed
(Table 1) and the loss of one allele was observed in 28% of
invasive cancers (fig. 1, Table 2). The frequencies of c-Ha-
ras deletion were not found significantly different in
cancers at early stages (29% of stage I and II tumors) from
those at advanced stages (26% of stage III and IV tumors).
An allele loss was observed in the stage I primary cancer
treated by surgery and in the recurrent tumor obtained six
months later from the same patient (Data not shown). The c-
Ha-*ras* deletion was also detected in a stage I primary
cancer and the lymph node metastasis from the same patient.
These results suggest that the c-Ha-*ras* alteration is
present in early stages of cancer development and may
represent a critical step for cancer progression. Therefore,
it would be of great interest to determine if the presence
of c-Ha-*ras* deletion in early stages of cervical cancer
could predict the risk of recurrences. Gene sequences,
located in the vicinity of the c-Ha-*ras* gene on the short

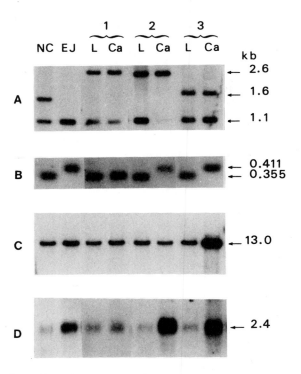

Fig 1. Analysis of c-Ha-*ras* and c-*myc* proto-oncogenes in
invasive squamous cell carcinoma of the uterine
cervix.
DNA and total RNA were both prepared from the same
tissue specimens : normal cervix (NC), EJ cells,
lymphocytes (L) carcinoma (Ca) pairs from patients 1
to 3 DNA preparations (5 μg) were incubated with *Msp*I
(A and B), *Eco*RI (C) and analyzed by Southern blot
hybridization using the c-Ha-*ras* probes (VTR probe 2,
in A; 0.6 kb probe 1, in B) and the c-*myc* probe
(third exon, in C). Total RNA preparations (10 μg)
were analyzed by Northern blot hybridization using
the c-*myc* probe (D). Note mutation in EJ and Ca3 (B),
both c-Ha-*ras* deletion and mutation in Ca2 (A,B),
c-*myc* amplification in Ca3 (C) and overexpression in
Ca2 and Ca3 (D).EJ cells, as most cell lines, showed
elevated levels of c-*myc* transcripts, 3-5 times the
levels usually detected in' non growing cells.

arm of chromosome 11 were also found to be deleted in cervical cancers (Data not shown) as this was previously shown in breast cancer (Ali et al., 1987).

Table 2. C-Ha-*ras* gene alterations in invasive squamous cell carcinomas of the uterine cervix at different clinical stages (UICC)

Clinical stage	No of tumors with gene alteration / Total No of tumors analyzed	
	Allele loss [a] (%)	Mutation at codon 12 (%)
I + II	10/34 (29%)	1/53 (2%)
III + IV	7/27 (NS)[b] (26%)	7/31 (p < 0.01)[b] (23%)
Metastasis [c]	1/1	0/1
Recurrence [c]	0/3	0/3
Total	18/65 (28%)	8/88 (9%)

[a] Allele loss was evaluated from lymphocyte-tumor pairs obtained from patients heterozygous for the c-Ha-*ras* locus.
[b] Statistical analysis for correlations of allele loss and mutation in cervical cancers at stage I and stage II versus cancers at stage III and stage IV was performed by the Chi Square test.
[c] Primary tumors were not obtained.

Recent studies have shown that normal cellular sequences on different chromosomes are lost during the development of human malignancies (reviewed in Klein 1987, Knudson 1987 and Friend et al., 1988). In retinoblastoma, it was suggested that the somatic loss of heterozygosity for these sequences results in homozygosity for a recessive mutant allele on these chromosomes that may contribute to the development of the malignancies.

DNA digestion with *Msp*I restriction enzyme may reveal mutations occurring at codon 12 of the c-Ha-*ras* gene (Goldfarb et al., 1982; Feinberg et al., 1983). Mutated gene generates one DNA fragment of 411 base pairs (bp) as in EJ cells (fig. 1). In non mutated genes, two DNA fragments of 355 and 56 bp are generated. When mutations affect only one c-Ha-*ras* allele, the 411 and 355 bp DNA fragments are observed in electrophoretic gels while the 56 bp DNA fragment is not detectable in our experimental conditions. Of the 88 tumor specimens analyzed, mutations were detected in 8 (9%) cervical cancers affecting either one c-Ha-*ras* allele or both alleles (fig. 1). Mutations affecting c-Ha-*ras* gene in tumors occurred somatically since they are not observed in lymphocytes specimens of these patients. Mutation and allele loss were detected in none of the normal cervices and CIS analyzed. Mutations at codon 12 of the c-Ha-*ras* gene were detected in 1/53 (2%) and in 7/31 (23%) stage I and II tumors and stage III and IV tumors respectively (Table 2), indicating that they are significantly (p < 0.01) associated with cervical cancers of poor prognosis. Moreover, the mutant oncogene was also observed in four carcinomas (one at stage I, two at stage III and one at stage IV) exhibiting a deletion of the c-Ha-*ras* gene on the other allele. C-Ha-*ras* amplification (3-20 fold) was also detected in one and three of these stage I+II and stage III+IV tumors respectively (Data not shown).

Specific point mutations have been shown to occur exclusively in 3 positions of the coding sequence (codons 12, 13 and 61) in the c-Ha-*ras*, c-Ki-*ras* and N-*ras* genes. Using more sensitive methods of analysis, c-Ki-*ras* gene mutations at codon 12 could be observed in a variety of cancers (Reviewed in Bos 1988) and more specifically in colorectal cancers (Bos *et al.*, 1987; Forrester *et al.*, 1987; Vogelstein *et al.*, 1988).

Several other genetic events were also shown to be involved in the progression of cervical cancers. In that respect it is of interest to note that 86% of tumors also contained an amplified and/or overexpressed c-*myc* gene (Table 3 and Data not shown). Activation of the c-*myc* proto-oncogene was previously shown to be involved in cervical cancer progression (Riou *et al.*, 1984; Riou *et al.*, 1985;Riou, 1988a,b) and to be a strong pronostic indicator of recurrence in early cervical cancers (Riou *et al.*, 1987). Consequently our data suggest that the c-*myc* and the c-Ha-*ras* proto-oncogenes cooperate in the progression of cervical cancers. In an experimental model these two proto-oncogenes were already shown to cooperate to transform normal cells into tumor cells (Land *et al.*, 1983). Our study was also extended to the detection of mutations in other members of the *ras* family using the polymerase chain reaction (PCR). Preliminary results seem to indicate that mutation at position 12 of the c-Ki-*ras* was not frequently (2/44) observed contrasting with results obtained in other types of cancer (Data not shown). In several human malignancies (Bos 1988) as well as in experimental tumors (Balmain *et al.*, 1984; Barbacid 1987), the *ras* mutations detected in pre-malignant tumors were considered as being an early event in carcinogenesis. This hypothesis should be tested in pre-cancerous cervical lesions.

Table 3.   C-*myc* gene activation and human papillomavirus
(HPV) in cervical cancers presenting c-Ha-*ras*
alterations.

Clinical stage	No of tumors with deleted and/or mutated c-Ha-*ras* gene and with	
	activated *myc* [a]	HPV
I + II	7/10	9/10 [b]
III + IV	11/11	10/11
Metastasis	1/1	1/1
Total	19/22 [c] (86%)	20 [d]/22 (91%)

[a] C-*myc* amplification and/or overexpression
[b] One stage I cancer presented neither activated c-*myc* nor HPV
[c] C-Ha-*ras* allele loss and mutation were both observed in four cancers.
[d] HPV 16, 18 and 33 were observed in 17, two and one cancers respectively.

It is most likely that several other factors are
involved in the development and progression of cervical
cancers. Among those factors, specific sexually transmitted
types of human papillomaviruses are thought to be implicated
in the genesis of these cancers (Dürst *et al.* 1983; Riou *et
al.*, 1984; reviewed in Riou 1988a). HPV DNA sequences were
found in more than 90% of cervical cancers presenting a c-
Ha-*ras* alteration (Table 3). The role of papillomavirus
sequences in the genesis of cervical cancers is still
unknown; however, in experimental models, recent studies
have shown that type 16 virus DNA cooperates with the
mutated c-Ha-*ras* gene in transforming primary cells
(Matlashewski *et al.*, 1987) and that papillomavirus
sequences integrated near c-*myc* locus would be involved in
the cis-activation of the proto-oncogene in cervical cell
lines (Dürst *et al.*, 1987). This latter situation could
effectively be found in human cervical cancers.

In conclusion, our data consistently show that
alterations at the c-Ha-*ras* locus occur frequently in
cervical cancers and that they contribute, associated with
several other factors, to the genesis and/or progression of
these cancers. The detection of genetic alterations in early

cervical cancers could contribute to accurately predict those cancers which will relapse in order to select the appropriate treatment.

ACKNOWLEDGEMENTS

The investigation was supported by grants from the Association pour la Recherche sur le Cancer (ARC, Villejuif), Ligue Nationale Contre le Cancer (Paris) and INSERM (Paris).

REFERENCES

Ali, I.U., Lidereau, R., Theillet, C., and Callahan, R., 1987, Reduction to homozygosity of genes on chromosome 11 in human breast neoplasia, Science, 238:185.

Balmain, A., Ramsden, M., Bowden, G.T. and Smith, J., 1984, Activation of the mouse cellular Harvey-*ras* gene in chemically induced benign skin papillomas, Nature, 307:658.

Barbacid, M., 1987, *ras* genes, Ann. Rev. Biochem. 56:779.

Bos, J.L., Fearon, E;R., Hamilton, S.R., Verlaan-de Vries, M., Van Boom, J.H., Van der Eb, A.J., and Vogelstein, B., 1987, Prevalence of *ras* gene mutations in human colorectal cancers., Nature, 327:293.

Bos, J.L., 1988, The *ras* family and human carcinogenesis, Mutation Res., 195:255.

Capon, D.J., Chen, E.Y., Levinson, A.D., Seeburg, P.H., and Goeddle, D.V., 1983, Complete nucleotide sequences of the T24 human bladder carcinoma oncogene and its normal homologue., Nature, 303:33.

Dürst, M., Gissman, L., Ikenberg, H., and Zur Hausen, H., 1983, A papillomavirus DNA from a cervical carinoma and its prevalence in cancer biopsy samples from different geographic regions, Proc. Natl. Acad. Sci. USA 80:3812.

Dürst, M., Croce, C.M., Gissman, L., Schwarz, E., and Huebner, K., 1987, Papillomavirus sequences integrate near cellular oncogenes in some cervical carcinomas. Proc. Natl. Acad. Sci. USA, 84:1070.

Fearon, E.R., Feinberg, A.P., Hamilton, S.H., and Vogelstein, B., 1985, Loss of genes on the short arm of chromosome 11 in bladder cancer, Nature, 318:377.

Feinberg, A.P., Vogelstein, B., Droller, M.J., Baylin, S.B., and Nelkin, B.D., 1983, Mutation affecting the 12th amino acid of the c-Ha-*ras* oncogene product occurs infrequently in human cancer, Science, 220:1175.

Forrester, K., Almoguera, C., Han, K., Grizzle, W.E., and Perucho, M., 1987, Detection of high incidence of K-*ras* oncogenes during human colon tumorigenesis, Nature, 327:298.

Friend, S.H., Dryja, T.P., and Weinberg, R.A., 1988, Oncogenes and tumor-suppressing genes, New Engl.J.Med. 318:618.

Goldfarb, M., Shimizu, K., Perucho, M., and Wigler, M., 1982, Isolation and preliminary characterization of a human transforming gene from T24 bladder carcinoma cells, _Nature_, 296:404.

Hall, S.W., and Monaghan, J.M., 1983, Invasive carcinoma of the cervix in younger women, _The Lancet_ ii, 731.

Klein, G., 1987, The approaching era of the tumor suppressor genes, _Science_, 238:1539.

Knudson, A.G., 1987, A two mutation model for human cancer, in _Adv. in Viral Oncol._, G.Klein Ed. Raven Press New York, 7:1.

Krontiris, T.G., DiMartino, N.A., Colb, M., and Parkinson, D.R., 1985, Unique allelic restriction fragments of the human Ha-_ras_ locus in leukocyte and tumour DNAs of cancer patients. _Nature_, 313:369.

Land, H., Parada, L.F., and Weinberg, R.A., 1983, Tumorigenic conversion of primary embryo fibroblasts requires at least two cooperating oncogenes., _Nature_, 304:596.

Maniatis, T., Fritsch, E.F., and Sambrook, J., 1982, "Molecular cloning, A Laboratory Manual", Cold Spring Harbor Laboratory, eds.,New York.

Matlashewski, G., Schneider, J., Banks, L., Jones, N., Murray, A., and Crawford, L., 1987, Human papillomavirus type 16 DNA cooperates with activated _ras_ in tranforming primary cells, _The Embo J._, 6:1741.

Parkin, D.M., Stjernswärd, J., and Muir, C.S., 1984, Estimates of the worldwide frequency of twelve major cancers. _Bulletin of the World Health Organization_, 62:163.

Reeve, A.E., Housiaux, P.J., Gardner, R.J.M., Chewings, W.E., Grindley, R.M., and Millow, L.J., 1984, Loss of a Harvey _ras_ allele in sporadic Wilms' tumour., _Nature_, 309:174.

Riou, G., Barrois, M., Tordjman, I., Dutronquay, I. and Orth, G., 1984, Présence de génomes de papillomavirus et amplification des oncogènes c-_myc_ et c-Ha-_ras_ dans des cancers envahissants du col de l'utérus, _C. Rend. Acad. Sci. Paris_, 299:575.

Riou, G., Barrois, M., Dutronquay, V., and Orth, G., 1985, Presence of papillomavirus DNA sequences, amplification of c-_myc_ and c-Ha-_ras_ oncogenes and enhanced expression of c-_myc_ in carcinomas of the uterine cervix, in Papillomaviruses: Molecular and Clinical Aspects", Howley, P. and Broker, T. eds, A.R. Liss New York, p. 47.

Riou, G., Barrois, M., Lê, M.G., George, M., Le Doussal, V., and Haie, C., 1987, _The Lancet_, ii, 761.

Riou, G., 1988a, Molecular genetics of carcinoma of the uterine cervix., _Medecine Science_, 7:435.

Riou, G., 1988b, Proto-oncogenes and prognosis in early carcinoma of the uterine cervix, in "The prevention of cervical cancer", _Cancer Surveys_, E.G. Knox and C.B.J. Woodman Eds, Oxford University Press, 7 (in press)

Riou, G., Barrois, M., Sheng, Z.M., Duvillard, P., and
    Lhomme, C., 1988, Somatic deletions and mutations of
    c-Ha-*ras* gene in human cervical cancers, <u>Oncogene</u>,
    3:329.
Sheng, Z.M., Guérin, M., Gabillot, M., Spielmann, M.
    and Riou, G., 1988, C-Ha-*ras* -1 polymorphism in human
    breast carcinomas: evidence for a normal distribution
    of alleles, <u>Oncogene Res</u>., 2:245.
Spandidos, D.A., 1987, Oncogene activation in malignant
    transformation: a study of H-*ras* in human breast
    cancer, <u>Anticancer Res</u>., 7:991.
Terrier, P., Sheng, Z.M., Schlumberger, M., Tubiana, M.,
    Caillou, B., Travagli, J.P., Fragu, P., Parmentier,
    C., and Riou, G., 1988, Structure and expression of c-
    myc and c-*fos* proto-oncogenes in thyroid carcinomas,
    <u>Br. J. Cancer</u>,57:43.
Vogelstein, B., Fearon, E.R., Hamilton, S.R., Kern S.E.,
    Preisinger, A.C., Leppert, M., Nakamura, Y., White, R.,
    Smits, A.M.M., and Bos, J.L., 1988, Genetic alterations
    during colorectal-tumor development, <u>New Engl. J. Med</u>.,
    319:525.

HUMAN THYROID NEOPLASMS EXPRESS RAS p21 PROTEIN AT HIGH LEVELS

M. Yiagnisis[1], K. Papadimitriou[2] and D.A. Spandidos[1,3]

[1]Biological Research Center, National Hellenic Research Foundation, 48 Vas. Constantinou Ave., Athens, 11635, Greece.
[2]Ippokration General Hospital, Athens, Greece.
[3]The Beatson Institute for Cancer Research, Garscube Estate, Bearsden, Glasgow, UK.

INTRODUCTION

In vitro and in vivo studies on cell transformation have revealed that recognizable steps in the process of malignant cell transformation can be achieved by quantitative and/or qualitative changes in proto-oncogene expression (for a review see ref. 1).

Several studies employing molecular hybridization (2-5) or immunohistochemical (6-7) analyses have shown elevated expression of ras oncogenes at the RNA and protein levels.

Activated by structural mutations, ras oncogenes have been identified in a wide range of human and animal tumors (for a review see ref. 8). Using a transfection assay with DNA from randomly selected tumors approximately 20% of cases are positive (8). Employing olignucleotide hybridization (9) and RNase A mismatch (10) analyses approximately 40% of colon tumors are positive. However, certain mutations in the ras gene give rise to tumorigenic cells but not to morphologically altered cells in culture, i.e. the formation of foci. Thus, transfection studies may miss a substantial proportion of mutant ras genes.

The presence of activated ras oncogenes in differentiated thyroid follicular and papillary carcinomas has been demonstrated by transfection analysis (11) using the nude mouse tumorigenicity assay (12).

In a previous study we have described our results on the expression of ras p21 in thyroid neoplasms using the ras p21 specific monoclonal antibody Y13 259 (13). Our findings suggest that elevated expression of ras genes may play a role in the development of thyroid neoplasms.

MATERIALS AND METHODS

The rat monoclonal antibody Y13 259 (14) to the ras p21 proteins was prepared as previously described (13). The procedure for immunostaining of paraffin tissue sections has been previously described (7).

RESULTS

The ability of monoclonal antibody to detect enhanced levels of ras p21 was confirmed using appropriate control cells transfected with ras genes as previously described (13). Immunohistochemical analysis of ras p21 expression in thyroid tissues using the monoclonal antibody Y13 259 gave the results shown in Table 1. As shown in the Table both papillary and follicular carcinomas and all types of adenomas contained higher levels of ras p21 as compared to adjacent normal tissue.

DISCUSSION

It has now become clear that qualitative and quantitative changes in the expression of oncogenes could be important in diagnosis and prognosis of specific types of malignant tumors. Our findings suggest that ras p21 is expressed at elevated levels in the majority of thyroid carcinomas and adenomas. It is of interest that a high frequency of ras oncogene activation by point mutation in benign and malignant human thyroid tumors has been found (11,15). Thus, nearly 50% of adenomas, 40% of follicular carcinomas and 60% of undifferentiated carcinomas carry structural mutations on ras genes (11,15). Moreover, activated alleles of all three ras oncogenes (H-, K- and N-ras) were found in thyroid tumors (11,15).

Other oncogenes such as fos and myc (16) and erbB and neu (17) have been found to be transcriptionally activated in thyroid neoplasms. Therefore, multiple oncogene activity may be required for the development of thyroid tumors.

Table 1. Summary of Immunohistochemical Results on Thyroid Tissues with the ras p21 Monoclonal Antibody Y13 259.

Tissue	Total	Staining intensity number of cases		
		-/+	+	++
Papillary carcinomas	13	1	5	7
Follicular carcinomas	7	2	4	1
Normal adjacent to carcinoma	14	6	8	0
Follicular adenomas	15	11	4	0
Fetal adenomas	22	14	7	1
Embryonal adenomas	7	6	1	0
Oxyphil adenoma	4	3	1	0
Atypical adenoma	3	2	1	0
Normal adjacent to adenoma	33	29	4	0

REFERENCES

1. D.A. Spandidos. A unified theory for the development of cancer. Biosci. Rep. 6:691 (1986).
2. D.A. Spandidos and I.B. Kerr. Elevated expression of the human ras oncogene family in premalignant and malignant tumors of the colorectum. Br. J. Cancer 49:681 (1984).
3. J.J. Slamon, J.B. Dekernion, I.M. Verma and M.J. Cline. Expression of cellular oncogenes in human malignancies. Science 224:256 (1984).
4. D.A. Spandidos and N.J. Agnantis. Human malignant tumors of the breast as compared to their respective normal tissue have elevated expression of the Harvey ras oncogene. Anticancer Res. 4:269 (1984).
5. D.A. Spandidos, A. Lamothe and J.N. Field. Multiple transcriptional activation of cellular oncogenes in human head and neck solid tumors. Anticancer Res. 5:221 (1984).
6. A.R.W. Williams, J. Piris, D.A. Spandidos and A.H. Wyllie. Immunohistochemical detection of the ras oncogene p21 product in an experimental tumor and in human colorectal neoplasms. Br. J. Cancer 52:687 (1985).
7. N.J. Agnantis, C. Petraki, P. Markoulatos and D.A. Spandidos. Immunohistochemical study of the ras oncogene expression in human breast lesions. Anticancer Res. 6:1157 (1986).
8. M. Barbacid. Ras genes. Ann. Rev. Biochem. 56:779 (1987).
9. J.L. Bos, E.R. Fearson, S.R. Hamilton, M. Barlaam-de Vries, J.H. Van Boom, A. van der Eb and B. Vogelstein. Prevalence of ras gene mutations in human colorectal cancers. Nature 327:293 (1987).
10. K. Forrester, C. Almoguera, K. Han, W.E. Grizzle and M. Perucho. Detection of high incidence of K-ras oncogenes during human colon tumorigenesis. Nature 327:298 (1987).
11. N.R. Lemoine, E.S. Mayall, F.S. Wyllie, C.J. Farr, D. Hughes, R.A. Padua, V. Thurston, E.D. Williams and D. Wynford-Thomas. Activated ras oncogenes in human thyroid cancers. Cancer Res. 48:4459 (1988).
12. O. Fasano, D. Birnbaum, L. Edlund, J. Fogh and M. Wigler. New human transforming gene detected by a tumorigenicity assay. Mol. Cell. Biol. 4:1695 (1984).
13. K. Papadimitriou, M. Yiagnisis, G. Tolis and D.A. Spandidos. Immunohistochemical analysis of the ras oncogene protein product in human thyroid neoplasms. Anticancer Res. In press.
14. M.E. Furth, L.J. Davis, B. Fleurdelys and E.M. Scolnick. Monoclonal antibodies to the p21 products of the transforming gene of Harvey murine sarcoma virus of the cellular ras gene family. J. Virol. 43:294 (1982).
15. N.R. Lemoine, E.S. Mayall, F.S. Wyllie and D. Wynford-Thomas. High frequency of ras oncogene activation in benign and malignant human thyroid tumors. InP "Ras oncogenes" D.A. Spandidos, ed., Plenum Publ. Co. New York (1989).
16. P. Terrier, Z.-M. Sheng, M. Schlumberger, M. Tubiana, B. Caillou, J.-P. Travagli, P. Fragu, C. Parmentier and G. Riou. Structure and expression of c-myc and c-fos proto-oncogenes in thyroid carcinomas. Br. J. Cancer 57:43 (1988).
17. R. Aasland, J.R. Lillehaug, R. Male, O. Josendal, J.E. Varhaug and K. Kleppe. Expression of oncogenes in thyroid tumours: coexpression of c-erb2/neu and c-erbB. Br. J. Cancer 57:358 (1988).

DETECTION OF MUTATIONS IN THE H-RAS PROTO-ONCOGENE IN LIVER TUMOURS OF THE CF1 MOUSE

A. Buchmann[1], R. Bauer-Hofmann[1], M. Schwarz[1] and A. Balmain[2]

[1]German Cancer Research Centre, 6900 Heidelberg, FRG
[2]Beatson Institute for Cancer Research, Glasgow, UK

The B6C3F1 mouse is characterised by a high spontaneous liver tumour frequency and a high susceptibility to chemical carcinogens. Both spontaneous and carcinogen-induced liver tumours of this mouse strain have been shown to contain specific mutations at either the first or second base of codon 61 of the H-ras proto-oncogene (Reynolds et al. 1986; 1987; Wiseman et al. 1986; Stowers et al. 1988). No such mutations, however, have been reported sofar in liver tumours of rats or other mouse strains. The CF1 mouse is very similar to the B6C3F1 mouse with respect to spontaneous liver tumour formation and sensitivity to chemical carcinogens. We were now interested in whether liver tumours of the CF1 mouse also possess mutations in the H-ras proto-oncogene. For this purpose we analysed samples of liver tumours from the CF1 mouse which occurred spontaneously or after carcinogen or promoter treatment for mutations at codon 61 of the H-ras gene. Since the CF1 liver tumours were available to us only as formalin-fixed, paraffin-embedded specimens, we first tested whether this material could be used as a source for our analyses. For this purpose we used fixed and embedded material from tumours with known mutations in the H-ras proto-oncogene as a model system.

DNA was isolated from the formalin-fixed, paraffin-embedded tissue by prolonged treatment with proteinase K as described by Dubeau et al. (1986). Thereafter, 1 µg of DNA was amplified in vitro by use of the polymerase chain reaction (Saiki et al., 1988) in 100 µl of a reaction mixture containing 67 mM Tris-HCl (pH 8.8), 16.6 mM ammonium sulfate, 6.7 mM magnesium chloride, 10 mM 2-mercaptoethanol,

6.7 mM EDTA, 1.5 mM of each dATP, dCTP, dGTP and dTTP, 170 μg/ml BSA, 1 μM of each primer and 2 units of TAQ–polymerase. Usually, we performed 35 cycles of amplification, each cycle consisting of denaturation (94°C, 1.5 min), primer annealing (37°C, 1.5 min) and chain elongation (72°C, 2 min). The primers used for amplification yielded DNA–fragments of 138 bp for H–ras exon 1 and 130 bp for H–ras exon 2, respectively. After amplification, mutations in the H–ras proto–oncogene were analysed by selective oligonucleotide hybridisation using ^{32}P–labeled oligonucleotide probes (Verlaan–de Vries et al., 1986).

To test whether or not formalin–fixed, paraffin–embedded material can be used as a source for retrospective analyses of mutations in the H–ras proto–oncogene we used fixed and embedded material from two types of tumours which were known from previous experiments with frozen tissue to contain specific mutations in the H–ras proto–oncogene. These tumours were (i.) skin tumours with an A –> T transversion at the second base of codon 61 of the mouse H–ras gene and (ii.) nude mouse tumours which were obtained after injection of epithelial cells with an activated human H–ras gene (G –> T transversion at the second base of codon 12). After isolation of DNA, in vitro amplification and selective oligonucleotide hybridisation we were able to identify the expected mutations in all samples analysed. This finding indicates that archival tumour material can be successfully used as a source for retrospective analyses of gene mutations.

For our analyses of mutations in the H–ras proto–oncogene in liver tumours of the CF1 mouse we used material from spontaneous liver tumours and from tumours induced by administration of aflatoxin B_1 (6 μg/g body weight, single i.p. injection after partial hepatectomy). In addition, liver tumours which occurred after treatment with either phenobarbital (1000 ppm in the diet) or dieldrin (10 ppm in the diet) without prior initiation were analysed. In 6 out of 28 of these tumours we were able to detect mutations at either the first or second base of codon 61 of the H–ras proto–oncogene (Table 1). Two tumours which were induced by administration of aflatoxin B_1 had a C –> A transversion at the first base of codon 61, two tumours which occured after prolonged administration of phenobarbital had the same type of mutation, whereas an A –> G transition at the second base of codon 61 was observed in two spontaneous liver tumours. No such mutations could be detected sofar in dieldrin induced liver tumours. The results obtained by selective oligonucleotide hybridisation were confirmed for two individual tumours by direct sequencing of the amplified DNA–fragments with ^{32}P–labeled sequencing primers as described by Saiki et al. (1988).

The mutations found at codon 61 of the H–ras proto–oncogene in liver tumour samples from the CF1 mouse are very similar to those described in the B6C3F1 mouse. Since both, the B6C3F1 and the CF1 mouse have a high spontaneous liver tumour

Table 1. Analysis of mutations at codon 61 of the H–ras
proto–oncogene in liver tumours of male CF1 mice

Experimental group	No. of tumours analysed	No. of tumours with mutations at codon 61			
		AAA	CGA	CTA	Total
Control	7	0	2	0	2
Aflatoxin B$_1$	8	2	0	0	2
Phenobarbital	7	2	0	0	2
Dieldrin	6	0	0	0	0

frequency and a high susceptibility to chemical carcinogens, it appears conceivable that the increased sensitivity of these mouse strains is correlated with a high mutability in the H–ras proto–oncogene and/or a preferential selection and clonal expansion of H–ras mutated liver cells.

References

Dubeau, L., Chandler, L.A., Gralow, J.R., Nichols, P.W. and Jones, P.A., 1986, Southern blot analysis of DNA extracted from formalin–fixed pathology specimens, Cancer Res., 46: 2964–2969

Reynolds, S.H., Stowers, S.J., Maronpot, R.R. and Anderson, M.W., 1986, Detection and identification of activated oncogenes in spontaneously occurring benign and malignant hepatocellular tumors of the B6C3F1 mouse, PNAS, 83: 33–37

Reynolds, S.H., Stowers, S.J., Patterson, R.M., Maronpot, R.R., Aaronson, S.A. and Anderson, M.W., 1987, Activated oncogenes in B6C3F1 mouse liver tumors: implications for risk assessment, Science, 237: 1309–1316

Saiki, R.K., Gelfand, D.H., Stoffel, S., Scharf, S.J., Higuchi, R., Horn, G.T., Mullis, K.B. and Ehrlich, H.A., 1988, Primer–directed enzymatic amplification of DNA with a thermostable DNA polymerase, Science, 239: 487–491

Stowers, S.J., Wiseman, R.W., Ward, J.M., Miller, E.C., Miller, J.A., Anderson, M.W. and Eva, A., 1988, Detection of activated proto–oncogenes in N–nitrosodiethylamine–induced liver tumors: a comparison between B6C3F1 mice and Fischer 344 rats, Carcinogenesis, 9: 271–276

Verlaan–de Vries, M., Bogaard, M.E., van den Elst, H., van Boom, J.H., van der Eb, A.J. and Bos, J.L., 1986, A dot–blot screening procedure for mutated ras oncogenes using synthetic oligodeoxynucleotides, Gene, 50: 313–320

Wiseman, R.W., Stowers, S.J., Miller, E.C., Anderson, M.W. and Miller, J.A., 1986, Activating mutations of the c–Ha–ras protooncogene in chemically induced hepatomas of the male B6C3F1 mouse, PNAS, 83: 5825–5829

ANALYSIS OF HRAS1-ASSOCIATED POLYMORPHISMS AND SEGREGATION OF TAQ1-DEFINED ALLELES IN DIFFERENT HUMAN TUMORS

Marco A. Pierotti, Paolo Radice, Patrizia Mondini, Virna De Benedetti and Giuseppe Della Porta

Division of Experimental Oncology A
Istituto Nazionale per lo Studio e la Cura dei Tumori
Via G. Venezian 1, 20133 Milan – Italy

INTRODUCTION

Recently, specific restriction fragment length polymorphisms (RFLPs) of a number of different genomic regions have been found to be associated with increased risk for various human diseases (Caskey, 1987). RFLPs are DNA markers defined by a different pattern of restriction fragments obtained in Southern blots with genomic DNA, using probes free of repetitive sequences. Unlike classical expressed markers, RPLPs can be found in genomic sequences irrespective of whether they encode a protein or not. The possibility to approach the problem of a genetic susceptibility to cancer has been made possible by the identification and characterization of oncogenes whose somatic activation has been associated to the development of tumors (Weinberg, 1985). In fact, in addition to somatic changes in oncogene structure and/or expression which occur only in the tumor cell, the possibility must also be considered that constitutional differences in oncogene structure and regulation could influence the individual likelihood of developing tumors. This possibility is now testable since RFLPs have been detected for several of human oncogenes (Pearson et al., 1987; Biunno et al., 1988; Radice et al., 1988).

Here we present an analysis of the different RFLPs that can be detected at the human HRAS1 locus and a study on the frequency of recently identified TaqI-defined HRAS1 alleles (Pierotti et al., 1986) in patients with different types of malignancies.

DIFFERENT POLYMORPHISMS OF THE HRAS1 GENE VTR REGION

In 1983 Capon et al. (Capon et al., 1983) showed that the germ line RFLP found by Goldfarb et al. (Goldfarb et al., 1982) in Southern blot analysis of human HRAS1 gene was due to variation in the copy number of a tandemly-repeated 28 bp sequence located at about 1kb 3' of the polyadenylation signal. The Bam HI restriction enzyme, originally used to analyse the RFLP, yields large polymorhpic fragments and small variation in the number of the repeat units can be missed. However, the sensitivity in the recognition of different HRAS1 allelic variants can be increased digesting the DNA with the endonucleases MspI/HpaII, that have recognition sites closely flanking the HRAS1 variable region, otherwise designated VTR (variable tandem repeats). Using the latter enzymes, Krontiris et al. (Krontiris et al., 1985) were able to categorize the HRAS1 alleles into "common" and "rare" according to their relative ferquencies. In subsequent studies, whereas the same 4 common alleles, named a.1, a.2, a.3 and a.4 in order of increasing size, have been constantely detected, differences in the total number of rare alleles, ranging from 5 to 16, have been reported (Radice et al., 1987). The discrepancy could be due to technical reasons because less resolutive electrophoresis may lead to an underestimate of the rare alleles closely migrating with the common ones. In addition, both MspI and HpaII are methylation sensitive endonucleases and even when used in combination and in large excess they can yield partial digests (Pierotti et al., 1986). In order to overcome these problems, we decided to analyse the HRAS1 polymorphism with the methylation-insensitive TaqI restriction enzyme that, according to the HRAS1 nucleotide sequence (Reddy, 1983) should also yield a fragment containing the entire VTR sequence. The results of this study showed that with TaqI we could define a new RFLP at the HRAS1 locus. In addition to the RFLP, due to the occurrence of variable number of repeats, TaqI identifies a site polymorphism in the VTR region itself (Radice et al., 1987). The latter polymorphism is most likely due to point mutations. In fact, in the VTR region, two tetranucleotide sequences could potentially generate the variant TaqI sites-containing VTRs or, as we designated them, Tp (Tp=TaqI plus) alleles. In all the repetition units, but the last one, the tetranucleotide TCCA can create a TaqI restriction site (TCGA) by a C/G substitution. In addition, in 15 out of 29 repetition units, the sequence TCGC, found in position 5-8, could provide a TaqI cleavage site by a C/A mutation (Reddy, 1983). So far, we have identified 7 different Tp alleles (Table 1). Most of their restriction patterns were compatible with the simultaneous creation of two new TaqI cleavage sites; however, in other instances, we have found Tp alleles with a VTR displaying three TaqI restriction sites or only one (Tp.4 and Tp.6, respectively) (data not shown). In addition, we have also observed that Tp.1, which is the most frequently found among Tp alleles and corresponds to the a.4 MspI common allele, displays a heterogeneity in the electrophoretic mobility of the low molecular weight fragments (fig. 1).

Table 1. TaI restriction fragments of Tp alleles

	Fragment length (bp)				Total size
	I	II	III	IV	
Tp1	2300	850 (±50)	700	–	3850
Tp2	2150	1100	700	–	3950
Tp3	2700	850	700	–	4250
Tp4	2300	850	700	650	4500
Tp5	2700	1700	700	–	4100
Tp6	2300	850 (x2)	–	–	3150/4000
Tp7	2300	1600	–	–	3900

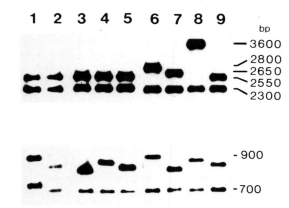

Fig. 1. Microheterogeneity of the Tp1 allele. Eight ug DNA from nine different heterozygous individuals carrying a Tp1 allele were digested with TaqI and run on a 2.0% agarose gel. Heterogeneity of the lower fragments of the Tp1 allele was detected in a size range between 700 and 900 bp.

POLYMORPHISMS OF HRAS1 GENE 5' REGION

Recently, Chandler et al. (Chandler et al., 1987), while studying the allele specific methylation of HRAS1 gene, observed a XhoI site polymorphism located in the 5'promoter region, 195 bp downstream from the BamHI site. They found HRAS1 alleles that had lost this site. Since XhoI will not digest the DNA when the internal cytosine of its recognition sequence (CTCGAG) is methylated, they also used, to digest the same samples, PaeR71, a methylation insensitive isoschizomer of XhoI. They concluded that the failure of XhoI to cut certain alleles was caused by base alteration in the recognition sequence rather than by methylation, defining a new HRAS1 site polymorphism due to the presence versus the absence of a XhoI restriction site. These authors correlated the XhoI polymorphism detected at the 5'of HRAS1 gene with that of the VTR, discussed in the previous section. In particular, they associated the presence of XhoI site with the HRAS1 common alleles a.1 and a.3. This observation prompted them to propose a simple model of evolution for the different VTRs since the complete concordance between the presence or absence of the XhoI site and the length of the VTR suggests that the corresponding mutation and duplication events were somehow related. The technique used to reveal the XhoI site was based on the detection of a shift in the electrophoretic mobility when a DNA sample after a first digestion with BamHI is further treated with XhoI. Since the map distance of BamHI and XhoI is relatively small (about 200 bp),

Table 2. Relationship between VTR and XhoI RFLPs at HRAS1 locus

| VTR alleles | | XhoI site | Incidence |
MspI	TaqI		
A1 (1.03)[1]	B1	+	33/33
A2 (1.45)	B2	-	17/17
A3 (2.10)	B3	+	5/5
A4 (2.60)	Tp1	+	20/20

[1]Size in kb

when this approach is applied to DNA containing large VTR as in example a.4, the shift in the electrophoretic mobility could be missed.

Replacing BamHI with BglII, that maps about 2000 bp 5' to the polymorphic XhoI site, we obtained a more sensitive assay for the analysis of the XhoI polymorphism. By this method we have observed that the XhoI site is not only associated to a.1 and a.3, as found by Chandler et al., but also to a.4. Table 2 summarises the relationship between VTR and XhoI polymorphisms at the HRAS1 locus. It can be noted that a simple correspondance is only found between a.1 and a.3 where a single duplication event of the repeat units could derive an allele from the other. On the contrary, duplication and multiple mutational events have diverged the remaining allelic forms of the human proto-oncogene.

SEGREGATION OF HRAS1 Tp ALLELES IN CANCER PATIENTS

While analysing the HRAS1 MspI/HpaII RFLP, Krontiris et al. observed a difference in the distribution of "rare" alleles between a group of patients with a variety of cancers and one of unaffected controls (Krontiris et al., 1985). In fact, they found that whereas in both groups the majority of alleles fell into the 4 "common" ones, in the cancer patients there was a significant higher prevalence (11.6% vs 3.9%) of the "rare" alleles, some of which were never observed in the unaffected individuals, suggesting the potential usefulness of the analysis of the RFLP of the HRAS1 gene for the detection of individuals at risk of cancer. Other groups have then investigated the association between HRAS1 alleles with VTR of different length and the risk for particular types of cancers with contrasting results. In general a significant segregation of "rare" alleles in a given form of cancer such as lung cancer (Heighway et al., 1986), melanoma (Gerhard et al., 1987), colon adenocarcinoma (Ceccherini-Nelli et al., 1983) and myelodisplasia (Thein et al., 1986) could not be reproduced. In some cases, a difference was noted in the frequency of "common" alleles between cancer

Table 3. Frequency of HRAS1 Tp alleles in malignant melanoma patients and controls

| | Number of subjects | Alleles | | P value |
		Tp	non-Tp	
Patients				
Advanced melanoma	55	19 (17.3%)[1]	91 (82.7%)	
				NS
First stage melanoma	55	14 (12.7%)	96 (83.3%)	
Totals	110	33 (15.0%)	187 (85.0%)	
				<0.02%
Controls	84	12 (7.1%)	156 (92.9%)	

[1] Allelic frequency

Table 4. Frequency of HRAS1 Tp alleles in nasopharyngeal carcinoma patients

	Number of subjects	Alleles		P value
		Tp	non-Tp	
Patients	40	9 (11.2%)[1]	71 (88.8%)	
				NS
Controls	84	12 (7.1%)	156 (92.9%)	

[1]Allelic frequency

patients and normal individuals (Heighway et al., 1986; Wyllie, 1988). We have compared the RFLPs detected at the HARAS1 locus by MspI/HpaII and by TaqI restriction enzymes in 55 melanoma patients with an disease advanced stage and in 53 unaffected individuals without a known cancer familiarity. Our results showed that, whereas no significant difference was observed in the frequency of both "rare" and "common" MspI/HpaII alleles, between melanomas and controls, the frequency of TaqI-defined Tp alleles was found to be significantly higher in melanoma patients than in unaffected individuals (Radice et al., 1987). In a subsequent study, we wanted to verify whether the association with Tp could also be found in patients with melanoma at an earlier stage. Table 3 shows the results: although at a not significant level, the Tp alleles appear less represented in first stage melanoma than in the more advanced forms (12.7% vs 17.3%). However, when the total frequency of Tp alleles found in the two combined groups of melanoma patients (15%) was compared to that of the controls (7.1%), the difference resulted still significant (p< 0.02).

As shown in Table 4, we have also studied a group of patients with a nasopharingeal carcinoma: no difference was found in the distribution of Tp alleles compared to the unaffected controls. Finally, we wanted to analyse a group of patients with different tumors that had in common a suggested pathogenetic mechanism based on loss or inactivation of genes mapping on the short arm of chromosome 11 where the HRAS1 gene is located. They comprised patients with Wilms'tumor, bladder carcinoma and tumor of the testis (Pierotti et al., 1988). Although the number of patients in each group was limited, the total frequency of Tp alleles,

Table 5. Frequency of HRAS1 Tp alleles in urogenital tumor patients

	Number of subjects	Alleles		P value
		Tp	non-Tp	
Patients				
Bladder carcinomas	13	3 (11.5%)[1]	23 (88.5%)	
Wilms' tumors	9	3 (16.7%)	15 (83.3%)	NS
Germinal tumors	20	8 (20.0%)	32 (80.0%)	
Totals	42	14 (16.7%)	70 (83.3%)	
				<0.02%
Controls	84	12 (7.1%)	156 (92.9%)	

[1]Allelic frequency

considering them as single group, was significantly higher in the cancer patients than in the controls (p<0.02) (Table 5).

CONCLUSIONS

Different polymorphisms have been detected at the human HRAS1 locus. The VTR region at the 3' end of the gene shows both length and site polymorphisms suggesting an intrinsic genetic instability most likely due to the repetitive motifs that constitute the region itself. In addition, a XhoI site polymorphism has been detected within the promoter region at the 5'end of the gene.

Taking into account both the polymorphisms detected at the HRAS1 locus, a simple evolutionary relationship between the different allelic forms of the gene cannot be envisaged. It appears that the different alleles could have been generated through multiple mutational and rearrangement events.

Whereas the influence of the XhoI site on the regulation of HRAS1 gene expression has not been suggested, an enhancer-like function of the VTR has been proposed (Seeburg et al., 1984; Spandidos and Holes, 1987). Although the activated allele of the HRAS1 gene is efficiently transforming even without a VTR region, an enhancer function of the VTR would justify the attempt to associate particular VTRs with different forms of cancers. Our results, obtained mainly with melanoma patients, indicate that the polymorphism based on the number of repeat units in the VTR does not yield particular alleles correlated to this form of cancer. On the contrary, the site polymorphism detected with the enzyme TaqI defines in the same population a class of alleles designated Tp that are found with a significative higher frequency in melanoma patients. We are currently considering the possibility that the Tp allelic forms of HRAS1 could provide an indicator for a poor prognosis. The future analysis of the follow-up of our melanoma patients will answer this point.

A further indication of an association between Tp alleles and a general genetic instability comes from the analysis of the urogenital tumors. The frequency of Tp alleles in these patients was similar to that found in melanoma. In both cases a high frequency of allelic losses has been reported. Whereas in melanoma patients the loss does not involve a particular chromosome (Dracopoli et al., 1985), in the tumor of the urogenital tract a significant loss of DNA markers on the short arm of chromosome 11 has been detected (Pierotti et al., 1988). Since both a dominant transforming oncogene, i.e. HRAS1, as well as putative tumor suppressor genes that can be inactivated by loss of genetic material, have been located on chromosome 11p, it is attempting to speculate that the structure of the VTR region could be associated not (only) to the HRAS1 gene expression, but (also) to a particular genetic instability of the short arm of chromosome 11 resulting in gene rearrangement and/or inactivation that could be associated with the tumor progression.

ACKNOWLEDGEMENTS

This work was supported by grants from the Italian National Research Council, Special Project "Oncology", Contract No. 84.00735.44, and the Associazione Italiana Ricerca Cancro.
We thank Mr. Mario Azzini for technical assistance and Miss Anna Grassi for typing the manuscript.

REFERENCES

Biunno, I., Pozzi, M.R., Radice, P., Mondini, P., Pierotti, M.A., Haley, J., Waterfield, M.D. and Della Porta, G., 1988, BglII polymorphisms of the epidermal growth factor receptor (EGF-R) gene, Nucleic Acids Res., 16: 7753.

Capon, D.J., Chen, E.Y., Levinson, A.D., Seeburg, P.H. and Goeddel, D., 1983, Complete nucleotide sequences of the T24 human bladder carcinoma oncogene and its normal homologue, Nature, 302: 33.

Caskey, C.T., 1987, Disease Diagnosis by Recombinant DNA Methods, Science, 230: 1223.

Ceccherini-Nelli, L., De Re, V., Molaro, G., Zilli, L., Clemente, C. and Boiocchi, M., 1983, Ha-ras-1 restriction fragment length polymorphism and susceptibility to colon adenocarcinoma, Br. J. Cancer, 56: 1.

Chandler, L.A., Ghazi, H., Jones, P.A., Boukamp, P. and Fusenig, N.E., 1987, Allele-Specific Methylation of the Human c-Ha-ras-1 Gene, Cell, 50: 711.

Dracopoli, N., Houghton, A. and Lloyd, J.O., 1985, Loss of polymorphic restriction fragments in malignant melanoma: Implications for tumor heterogeneity, Proc. Natl. Acad. Sci. USA, 82: 1470.

Gerhard, D.S., Dracopoli, N.C., Bale, S.J., Houghton, A.N., Watkins, P., Payne, C.E., Greene, M.H. and Housman, D.E., 1987, Evidence against Ha-ras-1 involvement in sporadic and familial melanoma, Nature, 325: 73.

Goldfarb, M., Shimizu, K., Perucho, M. and Wigler, M., 1982, Isolation and preliminary characterization of a human transforming gene from T24 bladder carcinoma cells, Nature, 296:404.

Heighway, J., Thatcher, N., Cerny, T. and Hasleton, P.S., 1986, Genetic predisposition to human lung cancer, Br. J. Cancer, 53: 453.

Krontiris, T.G., DiMartino, N.A., Colb, M. and Parkinson, D.R., 1985, Unique allelic restriction fragments of the human Ha-ras locus in leukocyte and tumor DNAs of cancer patients, Nature, 302: 369.

Pearson, P.L., Kidd, K.K. and Willard, H.F., 1987, Report of the Committee on Human Gene Mapping by Recombinant DNA Techniques, in: "Human Gene Mapping 9", Cytogenet. Cell Genet., 46: 390.

Pierotti, M.A., Radice, P., Biunno, I., Borrello, M.G., Cattadori, M.R. and Della Porta, G., 1986, Detection of two TaqI polymorphisms in the VTR region of the human HRAS1 oncogene, Cytogenet. Cell Genet., 43: 174.

Pierotti, M.A., Radice, P., Lacerenza, S., Mondini, P., Radice, M.T., Sozzi, G., Miozzo, M., Pilotti, S., Gasparini, M., Fossati-Bellani, F. and Della Porta, G., in press, in: "Cancer Cells 7: Molecular Diagnostics of Human Cancer", Cold Spring Harbor Laboratory, New York.

Radice, P., Pierotti, M.A., Borrello, M.G., Illeni, M.T., Rovini, D. and Della Porta, G., 1987, HRAS1 proto-oncogene polymorphisms in human malignant melanoma: TaqI defined alleles significantly associated with the disease, Oncogene, 2: 91.

Radice, P., Donghi, R., Pierotti, M.A., Longoni, A., Fusco, A., Grieco, M., Santoro, M., Vecchio, G. and Della Porta, G., 1988, RFLP for TaqI of the human thyroid papillary carcinoma (PTC) oncogene, Nucleic Acids Res., 16: 9062.

Reddy, P., 1983, Nucleotide Sequence Analysis of the T24 Human Bladder Carcinoma Oncogene, Science, 220: 1061.

Seeburg, P.H., Colby, W.W., Capon, D.J., Goeddel, D.V. and Levinson, A.D., 1984, Biological properties of human c-Ha-ras-1 genes mutated at codon 12, Nature, 312: 71.

Spandidos, D.A. and Holmes, L., 1987, Transcriptional enhancer activity in the variable tandem repeat DNA sequence downstream of the human Ha-ras1 gene, FEB, 218: 41.

Thein, S.L., Oscier, D.G., Flint, J. and Wainscoat, J.S., 1986, Ha-ras hypervariable alleles in myelodysplasia, Nature, 321: 84.

Weinberg, R.A., 1985, The Action of Oncogenes in the Cytoplasm and Nucleus, Science, 230: 770.

Wyllie, F.S., Wynford-Thomas, V., Lemoine, N.R., Williams, G.T., Williams, E.D. and Wynford-Thomas, D., 1988, Ha-ras restriction fragment length polymorphisms in colorectal cancer, Br. J. Cancer, 57: 135.

DNA METHYLATION OF THE HUMAN HRAS1 GENE

Maria Grazia Borrello, Marco A. Pierotti, Daniela Biassoni, Italia Bongarzone, Maria Grazia Rizzetti and Giuseppe Della Porta

Division of Experimental Oncology A
Istituto Nazionale per lo Studio e la Cura dei Tumori
Via G. Venezian 1, Milan – Italy

FUNCTIONAL ANALYSIS: MODULATION OF HUMAN HRAS1 ONCOGENE EXPRESSION BY DNA METHYLATION

Several lines of evidence suggest that DNA methylation may play a role in gene regulation in eukaryotes. Studies of many tissue specific genes using methyl-specific restriction endonucleases have shown a correlation between the methylation levels and the active or inactive status of the genes: most genes are undermethylated in the tissues in wich they are expressed, while they are heavily methylated in non expressing tissues and in the germline (Yisraeli and Szif, 1984). Furthermore "in vitro" DNA methylation of a few specific gene sequences inhibited the activity of these genes when inserted into animal cells (Stein et al., 1982; Busslinger et al., 1983; Kruczek et al., 1983).

On the other hands, constitutively expressed housekeeping genes generally contain CpG islands that are completely unmeyhylated both in the germline and in all somatic cells (Bird et al., 1985).

In this contest, we have previously shown that the transforming activity on NIH/3T3 cells of a mutated allele of the human HRAS1 gene can be significantly lowered by "in vitro" methylation of specific CpG dinucleotides (Borrello et al., 1987). The sequence involved in the inhibitory effect could have been the CpG rich promoter region. To verify this possibility, the plasmid containing a transforming clone of the HRAS1 oncogene (pT24-C3), methylated "in vitro" at both HpaII and HhaI restriction sites, was cotransfected in NIH/3T3 cells with the plasmid pTMDHFR, containing a gene coding for resistance to G-418 antibiotic (Borrello et al., 1988). G-418R colonies, randomly selected normal or transformed morphology.

Table 1. Analysis of a G-418 resistant flat colony (NF291)
before or after 30 uM 5'-Aza-Cyd treatment
(NF291/Aza)

	NF291	NF291/Aza
Morphology	Flat	Spindle shape
Agar growth[1]	0%	100%
In vivo tumorigenicity[2]	–	+
Restriction pattern of human Ha-ras sequences	Bam HI (21.0, 8.8, 6.6 Kb) EcoRI (11.0, 9.5, 3.7 Kb) Hind III (21.0, 18.0 Kb)	
Copies number	Unchanged (about 5)	
Promoter methylation[3]	+	–
Ha-ras m-RNA level	–	+
p21 expression	–	+

[1] Cloning efficiency after 28 days
[2] Assayed after 30 days from the injection of 5×10^5 and 1×10^5 cells in nude mice
[3] Assayed at Hpa II and Hha I restriction sites in the 0.8 Kb Sac I – Sac I fragment

The methylation status of the human oncogene promoter region (SacI-SacI 0.8 kb probe) and its expression were evaluated in eight cell lines, four with normal and four with transformed morphology. The p21 presence was found only in the four transformed lines carrying a demethylated promoter, whereas there was no correlation with the methylation status of the body of the gene (data not shown).

To avoid different behavior of HRAS1 gene in the various colonies due to its copies number or due to different integration sites, a cell line, designated NF291, derived from a flat colony and displaying a normal morphology and a non expressed human HRAS1 oncogene, was treated for 24 hours with 30 um of the demethylating agent 5'-azacytidine (Borrello et al., 1988). Foci of morphologically transformed cells appeared after 10-15 days and were briefly expanded and analysed. Different parameters were evaluated in the colony before and after the treatment with 5'-azacytidine (Table 1). The colony, after the treatment, acquired a transformed phenotype growing in agar and in nude mice, showed a completely demethylated promoter and expressed HRAS1 human gene, whereas the structure and the copy number of the oncogene remained unchanged.

In conclusion, we were able to establish an inverse correlation between the methylation of the CpG rich region of the promoter and the expression of the HRAS1 gene both at mRNA and at p21 protein level.

STRUCTURAL ANALYSIS: DIFFERENT METHYLATION IN TOPOLOGICAL DOMAINS OF HARS1 GENE

To study "in vivo" the methylation status of HRAS1 gene, we analysed, using HpaII and HhaI restriction enzymes, the methylation level of the 5' promoter region, of the coding sequences and of the 3' region in different human tissues.

The methylation level of the 5' region was estimated by using as a probe the same SacI-SacI fragment of 0.8 kb (Sac 0.8) that was completely demethylated in the NIH/3T3 colonies expressing human HRAS1 oncogene. This fragment encompasses the functional promoter, the first non-coding exon and part of the first intron, it is CpG rich and contains 18 HpaII and 18 HhaI sites. The methylation status of the CpG at HpaII and HhaI sites can be easily assessed by using the respective methylation sensitive restriction endonucleases. We digested DNAs from different normal human tissues with SacI only or followed by a further digestion with MspI as a control, or with either HpaII or HhaI to and expanded as short term cultures, showed either a morphologically normal phenotype indistinguishable from the parental untreated NIH/3T3 cells (about 50% colonies) or a transformed morphology with spindle shape and loose attachment to the substrate. The presence of the human oncogene was assessed by a Southern blot analysis in cell lines with a evaluate the methylation level, using as a probe the Sac 0.8 fragment.

Table 2. Methylation status of HRAS1 gene in different human tissue

TISSUE	Promoter region (Sac 0.8) HpaII HhaI	Coding region (Sac 2.9) HpaII/ HhaI		3' region (SacI-ClaI) HpaII HhaI	
HEMATOPOIETIC					
-Spleen	demethylated	Ao	Ao	B2	A1
-Thymus	"	Ao	Ao	A1	Ao
-Tonsil	"	A1	Ao	B1	A1
-Peripheral blood 1 lymphocites	"	A1/A1	Ao	B1	Ao
RESPIRATORY TRACT					
-Lung	"	B1	Ao	B2	A1
GASTROINTESTINAL TRACT					
-Infantile liver	"	A1	Ao	B2	A1
-Adult liver	"	A2	Ao	B2	Ao
GENITOURINARY TRACT					
-Cervix	"	A2	A2	C1	A2
-Placenta	"	B2	A1/A2	B2/C1	A2
-Sperm	"	C2	C2	Ao	Ao
ENDOCRINE SYSTEM					
-Thyroid	"	A2	Ao	C2	A2
SKIN					
-Epidermis	"	A1	Ao	B2	A1
NERVOUS SYSTEM					
-Peripheral nerves	"	A2	Ao	C2	B1

In all the adult human tissues analysed and in two different samples of placenta and sperm cells, the Sac 0.8 region was always completely demethylated (Table 2).

Therefore we wanted to analyse the tissue-specific methylation level of the CpG dinucleotides located in other topological regions of the gene.

The same panel of tissues used for Sac 0.8 methylation analysis was studied for the methylation level of the coding region, using as a probe the SacI-SacI fragment of 2.9 kb that comprises all the coding exons (Sac 2.9). We found different patterns of methylation in the various tissues (Table 2): none showed complete methylation or complete demethylation at the 18 HpaII sites and at the 16 HhaI sites. Most likely, the patterns were complex because of the cellular heterogeneity of the samples analysed. To classify the observed complex patterns of methylation, we defined different classes (Ao, A1, A2, B1, B2, C1, C2), being Ao the most and C2 the least methylated pattern. All the adult tissues belonged to the classes from Ao to A2, therefore showing slight variability, whereas placenta cells and even more sperm cells displayed demethylated patterns.

Since the normal tissues analysed derived from different individuals, we could not exclude that some of the differences in the methylation patterns arise from a genetically determined methylation polymorphism. However, when in the available cases the same tissue from different individuals was analysed, very similar or identical patterns were found.

Finally the methylation level of sites 3' to the coding region of the protooncogene HRAS1, around the VTR (variable tandem repeats) region, was analysed using as a probe the SacI-ClaI fragment. The methylation patterns of this region showed large heterogeneity in the different human tissues at HpaII sites and they were classified in the same way as the Sac 2.9 patterns (Table 2). Among the tissues analysed, the lowest degree of methylation is in thyroid and peripheral nerves, whereas a complete methylation is present only in sperm cells. The methylation of HhaI sites does not deserve further comments because it is similar in the different tissues and all the sites but one are far from the VTR region.

Taking together the functional and structural data on the methylation of HRAS1 gene indicate some conclusions and suggest some hypotheses.

The promoter region of HRAS1 gene displays the properties of an HTF island (Bird, 1986), in fact it is CG rich and it is completely demethylated in different human tissues, including sperm. Consequently, the methylation dependent inhibition of the HRAS1 transcription, shown in transfection experiment, does not appear to play a physiological role, at least in adult normal tissues. Moreover, when we analysed the methylation level of the promoter in tumor tissues of different histological origin, we never found a methylated promoter (data not shown). These observations strongly suggest that the differences in methylation level of HRAS1 gene reported by different authors (Feinberg and Vogelstein, 1983; Chandler et al., 1986; Bustros et al., 1988) by comparing tumor tissues with the normal counterparts, were due most likely to other regions of the gene, i.e. the coding region or the 3' region. In fact, we have shown that, in those regions, tumor tissues displayed the same methylation pattern or a more demethylated pattern than the normal counterparts (data not shown).

The methylation level of the coding region is similar in most adult tissues. Interestingly, the lowest level is found in sperm cells suggesting that the coding region is demethylated to lower the mutation rate, from methylcytosine to thymidine, in the germline.

The methylation of the 3' region shows a clear tissue-specific pattern. It remains to be investigated if the tissue specific methylation of this region, that has been suggested to exert an enhancer-like function (Colby et al., 1986; Spandidos et al., 1987), could have a role in modulating the expression of HRAS1 gene. In fact, the demethylation in all the tissues of the promoter, that allows the expression of the gene, is in agreement with the wide range of tissue-specific expression of the RAS genes reported in the literature (for review see Barbacid, 1987). However, since a systematic study of HRAS1 gene expression in different human tissues is still lacking, it remains to be investigated whether the latter could correlate with the methylation of the 3' region of the gene.

ACKNOWLEDGEMENTS

This work was supported by grants from the Italian National Research Council, Special Project "Oncology", Contract No. 84.00735.44, and the Associazione Italiana Ricerca Cancro.
We thank Mr. Mario Azzini for technical assistance, and Miss Anna Grassi for typing the manuscript.

REFERENCES

Barbacid, M., 1987, ras Genes, Ann. Rev. Biochem., 56: 779.
Bird, A., Taggart, M., Frommer, M., Miller, O.J. and Macleod, D., 1985, A Fraction of the Mouse Genome That Is Derived from Islands of Nonmethylated, CpG-Rich DNA, Cell, 40: 91.
Bird, A.P., 1986, CpG-rich islands and the function of DNA methylation, Nature, 321: 209.
Borrello, M.G., Pierotti, M.A., Bongarzone, I., Donghi, R., Mondellini, P. and Della Porta, G., 1987, DNA Methylation Affecting the Transforming Activity of the Human Ha-ras Oncogene, Cancer Res., 47: 75.

Borrello, M.G., Pierotti, M.A., Donghi, R., Bongarzone, I., Cattadori, M.R., Traversari, C., Mondellini, P. and Della Porta, G., 1988, Modulation of the Human Harvey-RAS Oncogene Expression by DNA Methylation, Oncogene Res., 2: 197.

Busslinger, M., Hurst, J. and Flavell, R., 1983, DNA Methylation and the Regulation of Globin Gene Expression, Cell, 34: 197.

Chandler, L.A., DeClerck, Y.A., Bogenmann, E. and Jones, P.A., 1986, Patterns of DNA Methylation and Gene Expression in Human Tumor Cell Lines, Cancer Res., 46: 2944.

Colby, W.W., Cohen, J.B., Yu, D. and Levinson, A.D., , 1986, Sequences 3' of the Human c-Ha-ras1 Gene Positively Regulate its Expression and Transformation Potential, in: "Gene Amplification and Analysis, Vol. 4, Oncogenes", Papas, T.S. and Vande Woude, G.F., eds., Elsevier Science Publishing Company, Inc., New York.

De Bustros, A., Nelkin, B.D., Silverman, A., Ehrlich, G., Poiesz, B. and Baylin, S.B., 1988, The short arm of chromosome 11 is a "hot spot" for hypermethylation in human neoplasia, Proc. Natl. Acad. Sci. USA, 85: 5693.

Feinberg, A.P. and Vogelstein, B., 1983, Hypomethylation of ras oncogenes in primary human cancers, Biochem. Biophys. Res. Communic., 111: 47.

Kruczek, I. and Doerfler, W., 1983, Expression of the chloramphenicol acetyltransferase gene in mammalian cells under the control of adenovirus type 12 promoters: Effect of promoter methylation on gene expression, Proc. Natl. Acad. Sci. USA, 80: 7586.

Spandidos, D.A. and Holmes, L., 1987, Transcriptional enhancer activity in the variable tandem repeat DNA sequence downstream of the human Ha-ras1 gene, FEB, 218: 41.

Stein, R., Razin, A. and Cedar, H., 1982, In vitro methylation of the hamster adenine phosphoribosyltransferase gene inhibits its expression in mouse L cells, Proc. Natl. Acad. Sci. USA, 79: 3418.

Yisraeli, J. and Szyf, M., 1984, Gene Methylation Patterns and Expression, in: "DNA Methylation: Biochemistry and Biological Significance", Razin, A., Cedar, H. and Riggs, A.D., eds., Springer, New York.

STRUCTURE OF THE DROSOPHILA RAS2 BIDIRECTIONAL PROMOTER

Z. Lev, O. Segev, N. Cohen, A. Salzberg, and R. Shemer

Department of Biology, Technion – Israel Institute of Technology
Haifa 32000, Israel

Introduction

In D.melanogaster three ras genes were isolated and mapped to regions
85D, 64B and 62B, respectively (Neuman-Silberberg et al, 1984). Nucleotide
sequence analysis of genomic and cDNA clones suggests that the putative
Drosophila ras proteins have a molecular weight of 21 to 23 kd. They share
on average 80 percent similarity with vertebrate ras p21 proteins at their
N-terminus and 40 percent at their C-terminus. The specific pattern of
sequence conservation is similar to that found when vertebrate ras genes
are compared among themselves or with yeast ras genes, probabely reflecting
different domains within the ras protein (Neuman-Silberberg et al, 1984;
Mozer et al, 1985; Brock, 1987). Since the Drosophila rasl gene is more
similar to vertebrate ras genes than to the Drosophila ras2 gene, it has
been proposed that the two Drosophila genes have diverged anciently and
that they have different functions (Brock, 1987). The structural and
functional homologies between the Drosophila and vertebrate ras proteins
were demonstrated by precipitating 21 and 27 to 28 kd proteins from
Drosophila cell extract using monoclonal antibodies raised against the v-
Ha-ras p21 protein (Papageorge et al, 1984), and by transforming rat cells
with a chimeric ras comprising the N-terminus of human activated Ha-ras
gene and the C-terminus of Drosophila ras3 gene (Schejter et al, 1985).
The transcription patterns of the three Drosophila ras genes during
development are similar (Lev et al, 1985). Each gene codes for at least two
distinct transcripts. The larger transcript is expressed in similar
abundance during the entire life cycle of the fruit fly. The shorter
transcripts of rasl and ras3, but not of ras2, are maternal/embryonic
specific. All three genes express their normal larval pattern in
transformed neuroblasts derived from lethal(2)giant larvae tumorous larval
brains (Kimchie et al, 1989) The spatial distribution of the RNA
transcripts expressed by the three Drosophila ras genes was examined in
situ (Segal and Shilo, 1986). In the embryo the transcripts are uniformly
distributed, but in the larva they are confined to growing cells in the
imaginal discs, brain and gonads. In the adult most transcripts are found
in the oocytes, but also in differentiated nervous tissues – the brain
cortex and the ventral ganglion. Thus, similarly to vertebrate ras genes,
the Drosophila ras genes possibly participate in several processes at
different developmental levels such as proliferating cells and
differentiated tissue.
In this article we describe the characterization of the Drosophila ras2
gene promoter. Our results reveal that this promoter is eventually a

bidirectional promoter, regulating the ras2 gene and another transcription unit found only 94 bases upstream from the ras2 transcription start site. The promoter may be divided into two separated "mini" promoters. Each one of them controls the transcription of one gene and in addition, enhances the transcription rate of the other gene.

Results

Transcription map of the ras2 promoter region

Preliminary RNA blot hybridizations with a genomic clone containing the Drosophila ras2 gene (Dras2, Newman-Silberberg et al, 1984; Dmras64B, Mozer et al, 1985), detected three major transcripts, 2.9, 2.5 and 1.8 kb long (our unpublished results; Mozer et al, 1985). However, when sequences from the coding region of the gene were cloned and used as a probe only the 1.8 kb transcript was detected (Lev et al, 1985; Segal and Shilo, 1986). A minor 1.4 transcript, found mainly in unfertilized eggs and early embryos, was also observed (Lev et al, 1985). Therefore the larger transcripts are probably coded by another transcription unit, tentatively termed csl, also located in the genomic clone. (These transcripts were erroneously assigned to the ras2 gene by two groups [Mozer et al, 1985; Brock, 1987]). Further experiments showed that sequences downstream from the ras2 gene did not detect these larger transcripts and they are exclusively coded by sequences upstream from the ras2 gene (N. Cohen, unpublished). To establish a fine transcription map of this region, a fragment containing the first two ras2 exons and upstream sequences was cloned. Additional subclones were constructed (Fig. 1) and hybridized with embryonic poly(A$^+$) RNA blots. Clones 1.7 and 1.2 detected all three transcripts. However, clone CH detected only the csl larger transcripts and clone BC only the shorter ras2 transcript. We concluded that the ClaI site separates the ras2 and csl transcription units. Hence, the termini of the csl transcripts are within the 200 bp between the ClaI site and the end of clone 1.2. The 5'-end of the ras2 transcript should be found within the 160 bp separating the ClaI site and the ras2 ATG translation initiation codon (near the DraI site, Fig. 1). Apparently, the two transcription units are not more than 360 bp apart.

To identify the orientation of the csl transcripts relative to the orientation of the ras2 transcript, ras2 sense and antisense RNA probes

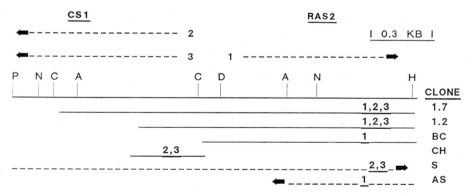

Fig. 1. Localization and orientation of RNA transcripts in the ras2/csl promoter region. DNA subclones and synthetic RNA derived from ras2 genomic clone were labeled and hybridized with 0-16 hr embryonic poly(A$^+$) RNA blots. 1 is the 1.8 kb ras2 transcript and 2&3 are the 2.9 and 2.5 kb csl transcripts. The overlined numbers designate which transcripts hybridized to each probe.( ), DNA; (_ _ _), RNA. S, sense; AS, antisense RNA. A, AvaI; C, ClaI; D, DraI; H, HindIII; N, NcoI; P, PstI.

were synthesized with T7 RNA polymerase (Melton et al, 1984). The first probe reacted with the larger transcripts. The other probe, which had the reverse orientation, detected the ras2 1.8 kb transcript as expected from its known orientation (Newman-Silberberg et al, 1984; Mozer et al., 1985; Brock et al., 1987). Hence, these transcripts and the ras2 transcripts are synthesized from different DNA strands. In other words, the csl transcription unit is expressed in the opposite polarity relative to the ras2 gene (Fig. 1; Cohen et al, 1988).

Determination of ras2 and csl transcription start sites

RNA blot hybridization experiments delimitated the transcription start sites of each transcription unit to within a sequence of less than 200 bp long (Fig. 1). The exact locations of these sites were first determined by external primer extension with T4 DNA polymerase (Hu and Davidson, 1986) and later confirmed by RNase protection analysis (Melton et al, 1984). The external primer extension method is very sensitive and reproducible and we have used it recently to locate the transcription start sites of the Drosophila abl transcripts (Segev et al, 1988). Two csl transcription start sites were obtained, the major one in position 215 and the minor one in position 185.

To confirm these results we carried out an RNase protection analysis using anti-csl synthetic RNA as a probe. Two protected transcripts were obtained corresponding to the same initiation sites (Fig. 2; Cohen et al, 1988).

We repeated this set of experiments to identify the transcription initiation sites of the ras2 gene. The external primer extension assay detected one major start site at 210. The RNase protection analysis revealed a doublet corresponding to sites 211 and 215. As discussed above concerning the csl start site, since only one band was detected in the first assay (and in internal primer extension with reverse transcriptase), we assume that the several RNase-protected transcripts obtained in this experiments represent one major start site (Fig. 2; Cohen et al, 1988).

The results obtained in this section revealed that the distance between the divergently transcribed ras2 and csl genes is extremely short, only 94 bp apart. Apparently, the Drosophila ras2 promoter is bidirectional.

The nucleotide sequence of the Drosophila ras2 promoter region

The nucleotide sequence of the promoter region is shown in Fig. 2. The first sequenced nucleotide was numbered as 1. An AT-rich sequence which may serve as a TATA-like box (Breathnach and Chambon, 1981) to either side is found (at 259) between the ras2 and csl genes, 46 and 43 bp, respectively, upstream from their transcription start sites. The 5'-end of the ras2

```
        ◄ *  O                        ◄ *  O
TGGCCAATTTTTCTGTATTTTTTTCGTTTGATGACTTAGCCTTAGTGATGGGC   222

GAAAGTCGATTGCTGGCTAGCGATAGTTGATATCG ATATTA CTGTCTAGCAGA   276
                                    O*    *  ►
GACGCGCACCCGTCTCAGTGCGAGTGTGGATTT CTCAGTT AACCGAG AACGGT   329
```

Fig. 2. The nucleotide sequence of the ras2/csl promoter region. +1 is the first sequenced nucleotide. Transcription start sites were determined by: O, T4 external primer extension; *, RNase protection. Putative TATA and direct and inverted insect cap boxes are shown. The arrows indicate the direction of transcription;

transcript is within an inverted repeat of the insect cap box (Hultmark et al, 1986). We have recently shown that inverted cap boxes are also found near the Drosophila abl cap sites (Segev et al, 1988) as well as in other genes (Z.Lev, in preparation). The role of this sequence is not known although some evidence suggests that it is involved in transcription (Hultmark et al, 1986). Analysis of direct and inverted repeats in the bidirectional promoter region did not reveal any significant symmetrical features, such as those found in the chorion promoter (Iatrou and Tsitilou, 1983). The ras2 ATG start codon is at 399, in agreement with published sequences (Mozer et al, 1985; Brock et al, 1987). About 160 bp downstream from the major csl transcription start site there is a cluster of three ATG codons (at 5, 25 and 36). The sequence preceding the most distal one, GAAAATG, is very similar to the Drosophila translational start site concensus sequence ([C/A]AA[A/C]ATG; Cavener, 1987).

Sequences required for bidirectional activity of the ras2/csl promoter

To demonstrate a bidirectional RNA transcription activity, the putative ras2 promoter was fused at nucleotide 391 to the bacterial chloramphenicol acetyltransferase (CAT) gene in the promoterless vector pl06 (a derivative of pSVOCAT; Gorman et al, 1982) and the putative csl promoter was fused with CAT at nucleotide 169. Both constructs were tested by transfecting Drosophila Schneider 2 culture cells (DiNocera and Dawid, 1983) and by injecting growing Drosophila embryos (Steller and Pirrotta, 1984). While no CAT activity was detected in cells transfected with the pl06 vector, significant activity was obtained with the ras2-promoter/CAT fusion and with the csl-promoter/CAT fusion.

Further analysis of the promoter region was carried out by introducing progressive deletions into sequences upstream to the ras2 transcription start site. The location of the deletions and the relative CAT activity

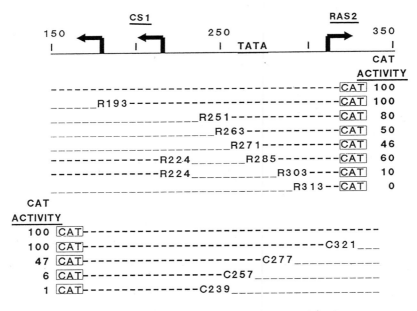

Fig. 3. Deletion analysis of the bidirectional ras2/csl promoter. Schneider 2 cells were transfected with deleted ras2CAT and cslCAT fusions and CAT activity was monitored according to Gorman et al (1982). Clones are termed according to the deletion point. (For nucleotide numbering see Fig. 2) (_ _ _), deleted sequences.

obtained with each deletion are shown in Fig. 3. Apparently, only 40 bases 5'to the cap site are required for 40 percent transcription efficiency in the ras2 direction (deletion R271; see also deletion R224-285). However, all upstream sequences up to the csl cap site are required for maximal transcription efficiency (deletion R193). Similarly 63 bases 5'to the major csl transcription start site are required for 30 percent transcription efficiency in the csl direction (deletion C277), and all the sequences up to the ras2 cap are necessary for achieving maximal efficiency (deletion C321). Thus, the ras2/csl promoter may be divided around base 275 into two separated "mini" promoters. Each one of them controls the transcription of one gene and in addition, enhances the transcription rate of the other gene.

Discussion

In this article we have presented structural and functional evidence suggesting that the promoter of the Drosophila ras2 gene is actually a bidirectional promoter regulating an additional gene, termed csl, in the opposite orientation. Sequences essential for transcription of ras2 are also enhancing transcription of csl in the opposite orientation. Similarly, those sequences which support transcrition of csl are required for full expression of ras2. The cap sites of the two divergent transcription units are separated by only 94 bp. We know of only one eukaryotic bidirectional promoter in which this distance is smaller (73 bp; Williams and Fried, 1986). Interestingly, the promoters of the human c-Ha-ras1 and the mouse c-Ki-ras also exhibit bidirectional activity (Spandidos and Riggio, 1986; Hoffman et al, 1987) but in these cases no specific divergent transcription unit have yet been identified. Vertebrate ras promoters are generally GC rich, do not contain the consensus sequences directing PolII transcription such as the TATA and CAAT boxes but do contain a number of GC boxes (McGrath et al, 1983; Ishii et al, 1985a; Hall and Brown, 1985; Spandidos and Riggio, 1986; Hoffman et al, 1987; Honkawa et al, 1987). These features are very similar to those of the human EGF receptor promoter (Ishii et al, 1985b) and to those of a number of cellular house keeping gene promoters (e.g. HMG coenzyme A reductase, Osborne et al, 1985; mouse and human DHFR, Masters and Atardi, 1985; McGrogan et al, 1985; human superoxide dismutase, Levanon et al, 1985). In this respect, our results show that the Drosophila ras2 promoter is different. It has the usual GC content, may have a TATA box and contains only two imperfect GC boxes.

It is not known how csl, the gene on the other side of the promoter, is related to the ras2 gene. However, in many bidirectional promoters studied so far the two divergently transcribed genes have similar structure and tissue specificity (the Drosophila glue proteins sgs-7 and sgs-8, Garfinkel et al, 1983; the Drosophila yolk proteins yp-1 and yp-2, Garabedian et al, 1985). In other cases there is a regulative interaction between the two genes regulated by the promoter (the SV40 early and late gene products, Gideoni et al, 1985), or there is a functional interaction between them (the yeast gal1 and gal10 proteins, Giniger et al, 1985; the B.mori chorion genes 2132 and 2574 and similar chorion gene pairs, Iatrou and Tsitilou, 1983; the D.melanogaster histones H3 and H4 and H2a and H2b, Parker and Topol, 1984). The Drosophila ras2 gene is not a stage- or tissue-specific gene (Lev et al, 1985; Segal and Shilo, 1986) and the spatial and temporal specificities of the csl gene have not yet been studied. We do know that these two genes are coregulated during development (Mozer et al, 1985). In this context it should be noted that several gene pairs regulated by bidirectional promoters are also ubiquitously expressed (the mouse DHFR gene and small nuclear RNAs, Farnham et al, 1985; the Drosophila t1 and t2 genes, Swaroop et al, 1986; the mouse SURF-1 and SURF-2 genes, Williams and Fried, 1986). In summary, it is possible that some interaction occurs between the ras2 and the csl genes, and further studies are in progress to elucidate this question.

Acknowledgement

This work was supported by the Fund for Basic Research administered by the Israel Academy of Sciences and Humanities. Z.L. is a Research Fellow in Cancer Research supported by the Claire and E.G. Rosenblatt Fellowship of the Israel Cancer Research Fund.

References

Breathnach, R. and Chambon, P. (1981). Ann. Rev. Biochem. 50, 349-383.

Brock, H.W. (1987). Gene 51, 129-137.

Cavener, D.R. (1987). Nucl. Acids Res. 15, 1353-1361.

Cohen, N., Salzberg, A. and Lev, Z. (1988) Oncogene 3, 137- 142.

DiNocera, P.P. and Dawid, I.B. (1983). Proc. Natl. Acad. Sci. USA. 80, 7095-7098.

Farnham, P.J., Abrams, J.M. and Schimke, R.T. (1987). Proc. Natl. Acad. Sci. USA. 82, 3978-3982.

Garabedian, M.J., Hung, M.-C., and Wensink, P.C. (1985). Proc. Nati. Acad. Sci. USA 82, 1396-1400.

Garfinkel, M.D., Pruitt, R.E., and Meyerowitz, E.M. (1983). J. Mol. Biol. 168, 765-789.

Gidoni, D., Kadonaga, J.T., Barrera-Saldana, H., Takahashi, K., Chambon, P. and Tjian, R. (1985). Science 230, 511-517.

Giniger, E., Varnum, S.M., and Ptashne, M. (1985). Cell 40, 767-774.

Gorman, C.M., Merlino, G.T., Willingham, M.C., Pastan, I. and Howard, B.H. (1982). Mol. Cell. Biol. 2, 1044-1051.

Hall, A. and Brown, R. (1985). Nucl. Acids Res. 13, 5255-5268.

Hoffman,E.K., Trusko, S.P., Freeman, N. and George, D.L. (1987). Mol. Cell. Biol. 7, 2592-2596.

Honkawa, H., Masahashi, W., Hashimoto, S. and Hashimoto-Gotoh, T. (1987). Mol. Cell. Biol. 7, 2933-2940.

Hu, M., C.-T. and Davidson, N. (1986). Gene 42, 21-29.

Hultmark, D., Klemenz, R. and Gehring, W.J. (1986). Cell 44, 429-438.

Iatrou, K., and Tsitilou, S.G. (1983). EMBO J. 2, 1431-1440.

Ishii, S., Merlino, G.T. and Pastan, I. (1985a) Science 230, 1378-1381.

Ishii, S., Xu, Y.-H., Stratton, R.H., Roe, G., Merlino, T. and Pastan, I. (1985b) Proc. natl. Acad. Sci. USA 82, 4920-4924.

Kimchie, Z., Segev, O. and Lev, Z. (1989) Cell Diff. Dev., in press.

Lev, Z., Kimchie, Z., Hessel, R. and Segev, O. (1985). Mol. Cell Biol. 5, 1540-1542.

Levanon, D., Lieman-Hurwitz, J., Dafni, N., Wigderson, M., Sherman, L., Bernstein, Y., Laver-Rudich, Z., Danciger, E., Stein, O. and Groner, Y. (1985). EMBO J. 4, 77-84.

Masters, J.N. and Attardi, G. (1985). Mol. Cell Biol. 5, 493-500.

McGrath, J.P., Capon, D.J., Goeddel, D.V. and Levinson, A.D. (1984). Nature 310, 644-649.

McGrogan, M., Simonsen, C.C., Smouse, D.T., Farnham, P.J. and Schimke, R.T. (1985). J. Biol. Chem. 260, 2307-2314.

Melton, D.A., Kreig, P.A., Rebagliati, M.R., Maniatis, T., Zinn, K. and Green, M.R. (1984). Nucleic Acids. Nucleic Acid Res. 12, 7035-7056.

Mozer, B., Marlor, R., Parkhurst, S. and Corces, V. (1985). Mol. Cell Biol. 5, 885-889.

Neuman-Silberberg, F.S., Schejter, E., Hoffmann, F.M. and Shilo, B.Z. (1984). Cell 37, 1027-1033.

Osborne, T.F., Goldstein, J.L. and Brown, M.S. (1985). Cell 42, 203-212.

Papageorge, A.G., DeFeo-Jones, D., Robinson, P., Tememles, G. and Scolnick, E.M. (1984). Mol. Cell. Biol. 4, 23-29.

Parker, C.S., and Topol, J. (1984). Cell 36, 357-369.

Schejter, E. and Shilo, B.-.Z. (1985). EMBO J. 4, 407–412.

Segal, D. and Shilo, B.Z. (1986). Mol. Cell Biol. 6, 2241–2248.

Segev, O., Kimchie, Z. and Lev, Z. (1988). Oncogene, in press.

Spandidos, D.A., and Riggio, M. (1986). FEBS Let. 203, 169–174.

Steller, H. and Pirrotta, V. (1984). EMBO J. 3, 165–173.

Swaroop, A., Sun, J.-W., Paco-Larson, M.L., and Garen, A. (1986). Mol. Cell. Biol. 6, 833–841.

Williams, T.J., and Fried, M. (1986). Mol. Cell. Biol. 6, 4558–4569.

THE *RAP* PROTEINS : GTP BINDING PROTEINS RELATED TO p21 *RAS* WITH A POSSIBLE REVERSION EFFECT ON *RAS* TRANSFORMED CELLS

V. Pizon, P. Chardin, I. Lerosey and A. Tavitian

INSERM U-248, 10, avenue de Verdun 75010 Paris

INTRODUCTION

In a wide variety of organisms, new *ras*-related genes have been identified on the basis of their sequence homologies with the mammalian H-*ras*, K-*ras* and N-*ras* proto-oncogenes. In mammals the proteins encoded by these *ras*-related genes have molecular weights of 21 000-24 000 daltons and share 30% to 50% homologies with the *ras* proteins (Chardin & Tavitian, 1986 ; Lowe et al., 1987 ; Madaule & Axel, 1985 ; Touchot et al., 1987). By mutational analysis different regions of the *ras* proteins have been assigned to three functional domains : i) a carboxy-terminal region necessary for the anchorage of the protein to membranes, ii) an effector region (Willumsen et al., 1986 ; Sigal et al., 1986) that seems necessary for the interaction with the GTPase activating protein GAP (Trahey et al., 1987) and iii) a GTP binding domain (Barbacid, 1987). As *ras* transforming proteins detected in human tumors most frequently contain an amino acid substitution in the GTP binding domain, at either position 12, 13 or 61 (Barbacid, 1987), this domain seems to play an essential role in the control of the biological properties of the protein. Among the four non-contiguous regions that compose this domain, six amino acids, DTAGQE, in positions 57 to 62 of the K-*ras* protein, seemed to be conserved in all the *ras* and *ras*-related proteins. The unique identified protein escaping this consensus sequence was a *ras*-related protein encoded by the Drosophila D*ras* 3 gene (Schejter & Shilo, 1985). This protein 46,7% identical to the human K-*ras* protein possesses a threonine at residue 61 (NTAGTE) instead of the usual glutamine (DTAGQE). As in the H-*ras* protein, it has been shown (Der et al., 1986) that almost every change of the 61st residue of the *ras* protein is associated with the acquisition of the transforming potential, we were interested to understand what could be the biological function of the substitution of the 61st amino acid in a non oncogenic protein. For this purpose we first searched for the existence of a human protein homologous to D*ras*3.

RESULTS

At first a λ gt10 cDNA library constructed from RNA of the Raji human Burkitt lymphoma cell line was screened under low stringency conditions with the complete insert of the Drosophila D*ras*3 cDNA. Out of about 100 000 recombinant phages, fifteen were isolated and analyzed. By cross hybridization of their purified inserts and partial sequencing it appeared that we had identified two categories of clones : fourteen were obviously derived from a same gene while the fifteenth originated from another related one. We named *rap*1A and *rap*2, for *ras*-proximate, the largest clone of the first family and the unique clone of the second family, respectively. In a second step a human λ gt10 pheochromocytoma cDNA library was screened under moderate stringency successively with the complete inserts of the *rap*1A and *rap*2 cDNA. The *rap*1A probe enabled us to isolate another *rap*1A clone and a new clone that we named *rap*1B

The nucleotide sequence of the 1.45 Kbp, 2.1 Kbp and 980 bp inserts of the *rap*1A, *rap*1B and *rap*2 cDNAs were determined. The *rap*1A and the *rap*1B clones code for proteins of 184 amino acids with molecular weights of 20 900 and 20 800 daltons, respectively. The *rap*2 clone encodes a protein of 183 amino acids with molecular weight of 20 700 daltons.

The percentages of amino acid identities between different *ras* related proteins are given in figure 1. The overall identities of the *rap*1A protein with the Drosophila D*ras*3, human K-*ras*, *ral*A and R-*ras* proteins are respectively 66.3%, 53.4%, 43.7% and 40.8%. The percentages of identities of the *rap*1B protein are 66.8% with D*ras*3, 52.3% with K-*ras*, 43.7% with *ral*A and 40.8% with R-*ras*. A similar overall

	K-ras	R-ras	ra11	rap1A	rap1B	rap2	Dras3
K-ras	100%	46,3%	50,5%	53,4%	52,3%	46%	46,7%
R-ras		100%	40,8%	40,8%	40,8%	36,7%	35,3%
ra11			100%	43,7%	43,7%	37,8%	35,9%
rap1A				100%	95,1%	61,9%	66,3%
rap1B					100%	61,4%	66,8%
rap2						100%	48,4%

Figure 1. Percentage of Amino Acid Identity between the
Different ras Related Proteins.

```
                                 26                        51
K-ras  MTEYKLVVVGAGGVGKSALTIQLIQ NHFVDEYDPTIEDSYRKQVVIDGET CLLDILDTAGQEEYSAMRDQYMRTG
rap 1A MREYKLVVLGSGGVGKSALTVQFVQ GIFVEKYDPTIEDSYRKQVEVDCQQ CMLEILDTAGTEQFTAMRDLYMKNG
rap 1B -------------------------- -------------------A-- ----------------------
rap 2  MREYKVVVLGSGGVGKSALTVQFVT GTFIEKYDPTIEDFYRKEIEVDSSP SVLEILDTAGTEQFASMRDLYIKNG
Dras3  MREYKIVVLGSGGVGKSALTVQFVQ CIFVEKYDPTIEDSYRKQVKVNERQ CMLEIVNTAGTEQFTAMRNLYMKNG

        76                       101                       126
K-ras  EGFLCVFAINNTKSFEDIHHYREQI KRVKDSEDVPMVLVGNKCDLPS.RT VDTKQAQDLARSY.GIPFIETSAKT
rap 1A QGFALVYSITAQSTFNDLQDLREQI LRVKDTEDVPMILVGNKCDLEDERV VGKEQGQNLARQWCNCAFLESSAKS
rap 1B -------------------------- ------D---- ---------------N-----
rap 2  QGFILVYSLVNQQSFQDIKPMRDQI IRVKRYEKVPVILVGNKVDLESERE VSSSEGRALAEEW.GCPFMETSAKS
Dras3  SDSC.WSTRSRRNRRLTICRTREQI LRVKDTDDVPMVLVGNKCDLEEERV VGKELGKNLATQF.NCAFMETSAKA

        151                      176
K-ras  RQGVDDAFYTLVREIRKHKEKMSKD GKKKKKKSKTKCVIM
rap 1A KINVNEIFYDLVRQINRKTPVEKKK PKKKS......CLLL
rap 1B -------------------PG-A R--S-......-Q--
rap 2  KTMVDELFAEIVRQMNYAAQPDKDD PCCSA......CNIQ
Dras3  KVNVNDIFYDWSGRSTRSRPRRNRR SRKVP......CVLL
```

THE REFERENCE NUMBERING IS THAT OF K-RAS.
THE GTP BINDING REGIONS ARE BOXED, THE EFFECTOR REGION IS INDICATED
BY HEAVY LINE AND THE MEMBRANE ATTACHEMENT REGION BY DOTTED LINE.

Figure 2

comparison of the rap2 protein gives 48.4% identities with Dras3, 46% with K-ras, 37.8% with ralA and 36.7% with K-ras. When they are compared to each other the rap1A and rap1B proteins show 95% identities, rap1A and rap2 61.9%, rap1B and rap2 61.4%.

Alignments of the rap proteins with the human K-ras and Drosophila Dras3 proteins are shown in figure 2. The four domains implicated in the guanine nucleotide binding site of the ras proteins and corresponding to amino acids 10-17, 57-63, 113-120 and 143-147 are highly conserved among the rap, K-ras and Dras3 proteins. Among the amino acid that had been directly implicated in the GTP binding, Glycine 12, Asparagine 116, Aspartic acid 119, Serine 145 and Lysine 147 are conserved in all the rap and ras proteins. It is noteworthy that the rap proteins, as the Dras3 protein, have a threonine at position 61 instead of the glutamine found in all the ras and ras-related non-oncogenic proteins.

Besides the GTP binding regions, it has been suggested that residues 32 to 42 could define the effector domain of the ras proteins. This region is strictly identical in the rap1A and rap1B proteins whereas there is only one amino acid difference in rap2 at position 39.

The C-terminal region of the ras proteins, characterized by the sequence Cys-A-A-X, where A is an aliphatic amino-acid and X the C terminal amino acid (Powers et al., 1984) is needed for the post translational lipid binding and subsequent anchoring to the plasma membrane (Fujuyama & Tamanci, 1986, Willumsen et al., 1986). The four amino acids terminating the rap1A protein match perfectly such a consensus sequence. In the case of rap1B and rap2, the amino acids following the cystein are a glutamine and an asparagine, respectively.

We further investigated the biochemical properties of the rap1A protein. For this purpose the rap1A cDNA has been cloned in the E. Coli expression vector ptac-c-H-ras, (Tucker et al., 1986) deleted of the c-H-ras, and the rap1A protein produced in a soluble form. Figure 3 shows the GTP binding of the ras, ralA and rap1A proteins. Crude lysates of bacterial cultures expressing the three proteins have been electrophoresed in a SDS polyacrylamide gel, blotted onto nitro-cellulose and incubated with $\alpha^{32}P$ GTP. The profiles of

GTP BINDING

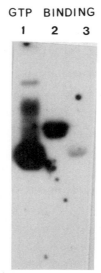

1: RAS
2: RAL
3: RAP1A

Figure 3

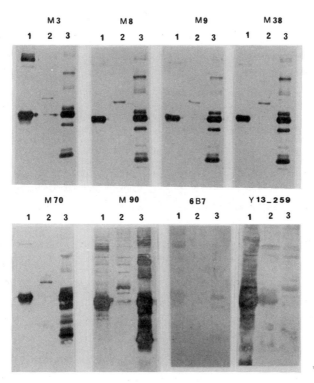

1: RAS   2: RAL   3: RAP1A PROTEINS PRODUCED IN E.COLI

Figure 4

radioactivity remaining after washing show that like *ras* and *ral*A proteins the *rap*1A protein binds GTP. The analysis of the purified *rap*1A protein shows that 1 mole of *rap*1A binds 1 mole of GTP as this is the case for the H-*ras* protein. At 37°C with EDTA, the *rap*1A binding kinetic reached saturation in less than two minutes. At the same temperature, without EDTA and in the presence of Mg^{2+}, about half of the GTP is bound in thirty minutes.

Preliminary results show that the *rap*1A protein has a low GTPase activity. When incubated with a cytoplasmic extract the GTPase activity of the *rap*1A protein was stimulated, suggesting that *rap*1A also interacts with a GAP protein.

Various *ras* monoclonal antibodies have been tested on the *rap*1A protein ; figure 4 shows immunoblotting experiments realised with *ras* monoclonal antibodies. Six monoclonal antibodies M3, M8, M9, M38, M70 and M90 (Lacal & Aaronson, 1986) have been tested on *ras*, *ral*A and *rap*1A proteins. It appears that all of them recognized efficiently the *rap*1A protein on Western blot but work poorly in immunoprecipitation. Monoclonal antibody 6B7 raised against amino acid 29 to 44 of the *ras* proteins (Wong et al., 1986) reacts weakly with *rap*1A and does not react with *ral*A. Monoclonal antibody Y13-259 (Furth et al., 1982) does not recognize *ral*A and *rap*1A.

DISCUSSION

We have isolated human cDNA clones coding for three proteins *rap*1A, *rap*1B and *rap*2 presenting a high degree of homology with the Drosophila D*ras*3 protein. Despite the wide evolutionary distance between fruit-fly and man, human *rap*1 (A and B) and *rap*2 proteins respectively share 78.5% and 63% identities within their first 84 amino acids with Drosophila D*ras*3 protein. These results suggest that they are members of an evolutionary conserved branch of the *ras*-related family. The *rap* and *ras* proteins share numerous structural properties. They have molecular weights of about 21 000 daltons and have a cystein residue four amino acids before their C-terminal end. As for the *ras* protein the *rap* proteins could undergo the palmitoylation of their terminal cystein and therefore could have, in the cell, a sub membrane location. The *ras* proteins have been shown to bind guanosine nucleotides (GDP or GTP) and to possess an intrinsic GTPase activity. An X-ray

88

crystallography structure of the *ras* proteins has recently been obtained (De Vos et al., 1988). According to this model the nucleotide binding pocket of the *ras* protein is constituted by residues 10-17, 57-63, 113-120 and 143-147. The conservation of these amino acid sequences as well as their identical locations in the *rap* proteins suggested that the *rap* proteins were able to bind GTP or GDP. The deduced amino acid sequences of the *rap* proteins show that Glycines 12 and 13 as well as alanine 59 are conserved. The most striking difference between the *ras* and *rap* proteins resides in position 61 where glutamine is replaced by a threonine. *Rap* proteins are the first example of normal human *ras*-related proteins having such a change. Mutational analysis has shown that such a substitution in the *ras* protein does not prevent the binding of the GTP but lowers its GTPase activity and is related with the oncogenic potential of the protein (Der et al., 1986). Analysis of the biochemical properties of the *rap*1A protein produced in E. Coli shows that *rap*1A binds GTP as efficiently as the H-*ras* protein, and that his GTPase activity is low. These results are in good agreement with those observed with the mutated *ras* protein.

In the *ras* proteins the region encompassing amino acids 32 to 42 has been shown to interact with a putative effector molecule that might be the GTPase activating protein known as GAP (Trahey et al. 1987). In the *rap*1A and *rap*1B proteins the region encompassing amino acid 32 to 42 is strictly identical to those of the K-*ras*, N-*ras* and H-*ras* proteins. The *rap*2 protein has only one amino acid difference in position 39. However in the *ras* protein, mutation of the 39 amino acid did not impair the activation of the protein by GAP (Adari et al., 1988). In conclusion it is possible that the *rap*1A, *rap*1B and also the *rap*2 proteins could, like the classical *ras* proteins, have their GTPase activity stimulated by GAP.

Cross reactivities with *rap*1A of different monoclonal antibodies raised against epitopes located all along the *ras* protein indicate an important similarity in the tertiary structure of these proteins. The structural characteristics of the *rap* proteins (same effector domain as *ras* proteins and threonine as the 61st residue) strongly suggest that *rap* and *ras* proteins interact with a common effector and that the *rap* proteins could be in a constitutively active state. By analogy with G proteins where two kinds of α-subunits are activated through the binding of a GTP molecule in response to activating or inhbiting signals and have antagonist effects on the same effector (adenylate cyclase) (Gilman, A., 1984) the *rap* and *ras* proteins could have antagonist effects on a common effector. In this hypothesis, the *rap* genes could be considered as anti-oncogenes. Recent observations by Dr Noda support this idea. From a revertant cell, obtained after transfection of a human cDNA library in K-*ras* transformed cells, Dr Noda group isolated a cDNA clone named K-rev1 which seems responsible of the reversion (personal communication). The amino acid sequence deduced from the K-rev1 clone perfectly matches with the amino acid sequence of the *rap*1A protein. This observation and the structural features of the *rap* proteins let us expect that not only *rap*1A but also *rap*1B and *rap*2 could be considered as anti-oncogenes.

BIBLIOGRAPHIE

Barbacid, M. (1987) Annual Review of Biochemistry **56**, 779-827.

Chardin, P. and Tavitian A. (1986) EMBO J. **5**, 2203-2208.

Der, J., Finkel, T. and Cooper, M. (1986) Cell **44**, 167-176.

De Vos, A.M., Tong, L., Milburn, M.V., Matias, P.M., Jancarit, J., Noguchi, S., Nishimura, S., Miura, K., Ohtsuka, E. and Kim, S-H. (1988) Science **239**, 888-893.

Fujiyama, A. and Tamanci, F. (1986). Proc. Natl. Acad. Sci. USA, **83**; 1266-1270.

Furth, M.E., Davis, L.J., Fleurdelys, B. and Scolnick, E.M. (1982) J. Virol. **43**, 294-304.

Gilman, A. (1984) Cell **36**, 577-579.

Lacal, J.C. and Aaronson, S.A (1986) Proc. Natl. Acad. Sci USA **83**, 5400-5404.

Lowe, D., Capon, D., Delwart, E., Sakaguchi, A., Naylor, S. and Goeddel, D. (1987) Cell **48**, 137-146.

Madaule, P. and Axel, R. (1985) Cell **41**, 31-40.

Powers, S., Kataoka, T., Fasano, O., Goldfarb, M., Strathern, J., Borach, J. and Wigler, M. (1984) Cell **36**, 607-612.

Schejter, E. and Shilo, B.Z. (1985) EMBO J. **4**, 497-412.

Sigal, S., Gibbs, B. Alonzo, S. and Scolnick, M. (1986) Proc. Natl. Acad. Sci. USA **83**, 4725-4729.

Touchot, N., Chardin, P. et Tavitian, A. (1987) Proc. Natl. Acad. Sci. USA **84**, 8210-8214.

Trahey, M. and Mc Cormick, F. (1987) Science, **238**, 542-545.

Tucker, J., Sczakiel, G., Feverstein, J., John, J., Goody, R.S. and Wittinghofer, A. (1986) EMBO J. **5**, 1351-1358.

Willumsen, M., Papageorge, G., Kung, F., Bekesi, E., Robins, T., Johnsen, M., Vass, C. and Lowy, P. (1986) Molec. Cell. Biol. **6**, 2646-2654.

Wong, G., Arnheim, N., Clark, R., Mc Cabe, P., Innis, M., Aldwin, L., Nitecki, D. and Mc Cormik, F. (1986) Cancer Research **46**, 6029-6033.

SUPPRESSION OF TEMPERATURE-SENSITIVE *RAS* IN *S.cerevisiae* BY A PUTATIVE PROTEIN KINASE

Elena Carra, Pietro Masturzo, Alessandra Vitelli, Emanuele Burderi, Irene Lambrinoudaki, Emmanuele De Vendittis, Regina Zahn and Ottavio Fasano

Differentiation Programme, European Molecular Biology Laboratory, Postfach 10.2209, D-6900 Heidelberg, FRG

INTRODUCTION

In the yeast *Saccharomyces cerevisiae*, the RAS1 and RAS2 proteins (collectively denominated RAS) are involved in the regulation of growth, metabolism, and adaptative responses (for a review, see Tamanoi, 1988). These effects are mediated, at least in part, by the adenylate cyclase/protein kinase pathway (Toda et al., 1985; Broek et al., 1985). In fact, the growth arrest caused by the inactivation of the *RAS1* and *RAS2* genes can be bypassed by mutations which constitutively activate the function of the adenylate cyclase (*CYR1*) and of three distinct but homologous cyclic AMP-dependent protein kinases (*TPK1*, *TPK2*, and *TPK3*) (Matsumoto et al., 1982; Toda et al., 1985; Kataoka et al., 1985; De Vendittis et al., 1986; Toda et al., 1987a-b).

Our work is aimed at determining the number and the identity of the cellular components that mediate *RAS*-dependent cellular responses. By using an approach that is not biased by *ad hoc* assumptions about the identity of the cellular elements regulated by *RAS*, we also expect to determine whether or not all the effects of *RAS* are mediated by cyclic AMP. For this purpose, we have previously constructed yeast strains in which the *RAS1* gene was deleted and the *RAS2* gene was replaced by mutant alleles encoding a defective gene product (De Vendittis et al., 1986; Fasano et al., 1988). In one of these strains (*ras1⁻ ras2-ts31*), the impaired *RAS* function did not allow growth at 37 ºC. Subsequently, we have isolated and characterized yeast genomic sequences on the basis of their ability to suppress the temperature-sensitive phenotype of *ras1⁻ ras2-ts31* strains, upon expression on multicopy plasmids. This strategy is based on the assumption that cellular components that constitute a physiological target for *RAS* could be functionally activated by overexpression, even in the absence of functional RAS proteins. Therefore, it is likely that some of the genes involved in suppression could encode positive regulatory elements of the growth control pathway. Indeed, using this strategy we previously isolated a truncated adenylate cyclase gene (De Vendittis et al., 1986). We report now the identification of another gene, that we have called *KOM1*. This gene potentially encodes a protein that is structurally related to the catalytic subunits of the mammalian and yeast cyclic AMP-dependent protein kinases as well as to mammalian protein kinase(s) C.

RESULTS

Yeast strains with a disrupted *RAS1* gene and with a mutated chromosomal *RAS2* gene encoding a protein with amino acid substitutions at position 40, 82 and 84 cannot grow at 37 °C on glucose-containing medium (Fasano et al., 1988). To identify cellular genes which, by overexpression, could bypass the requirement for functional *RAS1* and *RAS2* genes, we transformed temperature-sensitive yeast cells with a library of yeast DNA made into a high copy number yeast vector. Since this vector (YEP13, Broach et al., 1979) carries a selectable marker, we could isolate about 2×10^3 transformant yeast colonies, using an appropriate selective medium. The transformants were subsequently tested for growth at 37 °C, using a replica plating technique. Most of them, like the original temperature-sensitive strain, could not grow at 37 °C. In nine out of ten colonies that showed a variable degree of growth at this temperature, loss of plasmid sequences upon growth on nonselective medium was associated with reversion to temperature-sensitivity. The plasmids that we isolated from the transformants, introduced into *ras1⁻ ras2-ts31* cells, conferred to this strain the ability to grow at high temperature. Therefore, since the plasmid vector alone was ineffective, we concluded

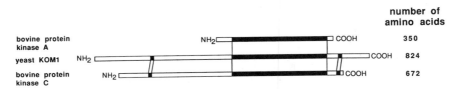

Figure 1. Schematic sequence comparison between the *KOM1* gene product, the catalytic subunit of the bovine cyclic AMP-dependent protein kinase (Shoji et al., 1983), and the bovine protein kinase C (Parker et al., 1986). Dark boxes indicate regions with conserved structural features. Details are shown in Figure 2.

that suppression of temperature-sensitivity was caused by the DNA inserts carried by the plasmids. One of the DNA inserts was completely sequenced. We found that the largest open reading frame initiated by ATG could encode a protein of 824 amino acids. This amino acid sequence was compared to the protein sequence databases, using the FASTP computer program (Lipman and Pearson, 1985). The highest homologies detected were for the catalytic subunit of cyclic AMP-dependent protein kinases from yeast and mammals, as well as for mammalian protein kinase(s) C. Figure 1 schematically illustrates the sequence similarity existing between the *KOM1* gene product, a catalytic subunit of the bovine cyclic AMP-dependent protein kinase (Shoji et al., 1983) and a bovine protein kinase C (Parker et al., 1986). The amino acid sequence similarity between *KOM1* and the catalytic subunit of the cyclic AMP-dependent protein kinase was distributed over about 310 amino acid residues, corresponding to the carboxy terminal segment of the two proteins (shown in detail in Figure 2). The same degree of structural similarity was found by comparing the *KOM1* gene product with the yeast cyclic AMP-dependent protein kinases *TPK1*, *TPK2* and *TPK3* (Lisziewicz et al., 1987; Toda et al., 1987a, results not shown). The region of homology between *KOM1* and the bovine protein kinase C was slightly longer than that observed with the cyclic AMP-dependent protein kinases, and was distributed over about 340 carboxy terminal amino acid residues. Interestingly, additional amino acid sequence similarity between *KOM1* and a bovine protein kinase C was detected over short amino and carboxy-terminal segments of the two proteins (filled boxes in Figure 1, underlined sequences in Figure 2). The amino-terminal sequence, underlined in the top panel of Figure 2, is highly conserved in the protein kinase C family, and connects two duplicated cysteine-rich domains (Parker et al., 1986) that were not present in *KOM1*.

```
KOM1 167 REAAAAAYGPDTDIPRGKLEVTIIEARDLVTRSKDSQPYVVCTFESSEFI
              ..|  |||||  ||.|
bKC   84 FSCPGADKGPDTDDPRSKHKFKIHTYGSPTFCDHCGSLLYGLIHQGMKCF

---------------------------------------------------------------

bKA    1 .......GNAAAAKKGSEQESVKEFLAKAKEDFLKKWENPAQNTAHL..D
              | |                  .      .|.| ||           |
KOM1 361 RLYPMIHNLAHASQHQWHSLKPRVIDEVVSGDILIKWTYKQTKKRHYGPQ
              |   ...| |  ||
bKC  288 VPIPEGDEEGNVELRQKFEKAKLGPAGNKVISPSEDRRQPSNNLDRVKLT

bKA   42 QFERIKTLGTGSFGRVMLVKHMETGNHYAMKILDKQKVVKLKQIEHTLNE
              .|| ..  ||  |  ||   ||  .|  |||||.|  |  .|| .| ||. |
KOM1 411 DFEVLRLLGKGTFGQVYQVKKKDTQRIYAMKVLSKKVIVKKNEIAHTIGE
              ||. .|||||.| ||.   ...  | |
bKC  338 DFNFLMVLGKGSFGKVMLADRKGTEELYAIKILKKDVVIQDDDVECTMVE

bKA   92 KRIL...QAVNFPFLVKLEFSFKDNSNLYMVMEYVPGGEMFSHLRRIGRF
              .  ||       ||.|  |  |||   .||.|   .|.  |||.|  ||  .  |||
KOM1 461 RNILVTTASKSSPFIVGLKFSFQTPTDLYLVTDYMSGGELFWHLQKEGRF
              .. |      ||.    |||   ||.|. |.|..| |
bKC  388 KRVLALL..DKPPFLTQLHSCFQTVDRLYFVMEYVNGGDLMYHIQQVGKF

bKA  139 SEPHARFYAAQIVLTFEYLHSLDLIYRDLKPENLLIDQQGYIQVTDFGFA
              ||  .|.|| ||..|| . ||   .|| |      ||||||..|.  |  .|. | |||.
KOM1 511 SEDRAKFYIAELVLALEHLHDNDIVYRDLKPENILLDANGNIALCDFGLS
              |  |   ||  ||. ..|   ||        |.|||||  .|..||  .|.|.  |||..
bKC  436 KEPQAVFYAAEISIGLFFLHKRGIIYRDLKLDNVMLDSEGHIKIADFGMC

bKA  189 K.RVKG..RTWTLCGTPEYLAPEIILSK.GYNKAVDWWALGVLIYEMAAG
              |  .|.   ||  |.|||  |||||.|   ||    ||.|  |||||.||   |
KOM1 561 KADLKD..RTNTFCGTTEYLAPELLLDETGYTKMVDFWSLGVLIFEMCCG
              |  .|   ||||||  .|.||.   |  ||.|  |  |||.|  |||..||   |
bKC  486 KEHMMDGVTTRTFCGTPDYIAPEIIAYQ.PYGKSVDWWAYGVLLYEMLAG

bKA  235 YPPFFADQPIQIYEKIVSGKVRFPSH.FSSDLKDLLRNLLQVDLTKRFGN
              . ||||.    .|.||  |||.|    .|    .|..   ||         .|.|
KOM1 609 WSPFFAENNQKMYQKIAFGKVKFPRDVLSQEGRSFVKGLLNRNPKHRLG.
              ||  .|...  |     |.|  |.  |.  ||        ...|     |||
bKC  535 QPPFDGEDEDELFQSIMEHNVSYPKS.LSKEAVSICKGLMTKHPGKRLGC

bKA  284 LK.DGVNDIKNHKWFATTDWIAIYQRKVEAPFIPKFKGPGDTSNFDDYEE
              ||  ...  |  .||  ||  ||.|.|.  .|| |  .   .   |||||| |
KOM1 658 AIDDG.RELRAHPFFADIDWEALKQKKIPPPFKPHLVSETDTSNFDP.EF
              . .... |..| |  ||||  |...  |.|  |||||||.  |... |||  ||
bKC  584 GPEGE.RDVREHAFFRRIDWEKLENREIQPPFKPK.VCGKGAENFDK.FF

bKA  333 EEIRVSINEKCGKEFSEF*
              |  .|
KOM1 706 TTASTSYMNKHQPMMTATPLSPAMQAKFAGFTFVDESAIDEHVNNNRKFL
              |.        |     . |   | ||.|.
bKC  631 TRGQPVLTPPDQLVIANID.....QSDFEGFSYVNPQFVHPILQSAV*
```

Figure 2. Amino acid sequence similarity between conserved regions of *KOM1*, of the catalytic subunit of the bovine cyclic AMP-dependent protein kinase (bKA, Shoji et al., 1983), and of the bovine protein kinase C (bKC, Parker et al., 1986). Gaps were introduced to maximize homology. Solid lines between sequences indicate identical amino acids, while points indicate conservative amino acid substitutions.

DISCUSSION

We have isolated a suppressor of temperature-sensitive *RAS* gene function that potentially encodes a protein kinase. This possibility is supported by the very high degree of amino acid sequence similarity between *KOM1* and the catalytic domain of known protein kinases. Among these, the cyclic AMP-dependent protein kinase(s) from mammalian and yeast, and protein kinase C, show the highest degree of amino acid sequence similarity to *KOM1*. Amino acid identities appear to be distributed along most of the catalytic domain of these proteins (see Figures 1 and 2). In general, residues that are conserved in cyclic AMP-dependent protein kinase(s) and in protein kinase(s) C are also present in *KOM1*. A potentially discriminating feature is represented by the correspondence between residues 730-740 of *KOM1* and residues 650-670 of protein kinase(s) C (underlined in Figure 2, lower panel). This region is not present in the shorter cyclic AMP-dependent protein kinase. In addition, the alignment of residues 392-397 of *KOM1* with residues 25-30 of the cyclic AMP-dependent protein kinase appears also to be a distinctive feature (Figure 2). Since the structural characteristics of the catalytic domain, that determine the substrate specificity for cyclic AMP-dependent protein kinases and for protein kinase(s) C are not known, we cannot assign *KOM1* to any of the two classes of protein kinases. In the future, the biochemical characterization of *KOM1* might be helpful to elucidate the relationship between structural properties of protein kinases, and their substrate specificity.

Another point that is raised by our findings is the relationship between *KOM1* and *RAS*. The amount of RAS1 and RAS2 proteins present in yeast cells was about the same in *KOM1*+ and *kom1*- strains (Fasano et al., unpublished results). This argues against the possibility that *KOM1* is a positive regulator of the intracellular level of the RAS proteins. Therefore, the suppression of temperature sensitive *RAS* by an overexpressed *KOM1* gene might reflect a role for the latter as a target of *RAS*. Alternatively, *KOM1* might be an element of a parallel pathway. It is interesting that, while *ras1*- *ras2*- yeast mutants are not viable, strains with a partially defective *RAS* function grew more poorly on nonfermentable carbon sources that on glucose (O. Fasano et al., 1988). This phenotype resembles that of *kom1*- cells, (Carra et al., manuscript in preparation) and further suggests that *KOM1* might be a downstream element of the *RAS* effector pathway.

ACKNOWLEDGEMENTS

We wish to thank Petra Riedinger for art work. This research was carried out in the framework of Contract ST2J-0388-C from the Commision of the European Communities to O.F. P.M. was partially supported by a fellowship from the Commision of the European Communities.E.B. was supported by a fellowship from Ministero della Pubblica Istruzione, Roma Italy. E.D.V. was on leave from the Istituto di Scienze Biochimico-Fisiche, II Facolta' di Medicina e Chirurgia , Universita' di Napoli, Italy, and was supported by a Fellowship from the Commission of European Communities. A.V. was supported by a fellowship from the Associazione Italiana per la Ricerca sul Cancro.

REFERENCES

Broach, J.R., Stratern, J.N. and Hicks, J.B.,1979, Transformation in Yeast: Development of a Hybrid Cloning Vector and Isolation of the *CAN1* gene. Gene 8:121.

Broek, D., Samiy, N., Fasano, O., Fujiyama, A., Tamanoi, F., Northup, J.and Wigler, M.,1985, Differential Activation of Yeast Adenylate Cyclase by Wild-Type and Mutant *RAS* Proteins. Cell 41:763.

De Vendittis, E., Vitelli, A., Zahn, R. and Fasano, O.,1986, Suppression of Defective *RAS1* and *RAS2* Functions in Yeast by an Adenylate Cyclase Activated by a Single Amino Acid Change. EMBO J. 5:3657.

Fasano, O., Crechet, J.B., De Vendittis, E., Zahn, R., Feger, G., Vitelli, A, and Parmeggiani, A., 1988, Yeast Mutants Temperature-sensitive for Growth after Random Mutagenesis of the Chromosomal *RAS2* gene and Deletion of the *RAS1* gene. EMBO J 7:3375.

Kataoka, T., Broek, D. and Wigler, M. ,1985, DNA Sequence and Characterization of the *S. cerevisiae* Gene Encoding Adenylate Cyclase. Cell, 43:493.

Lipman, D.J. and Pearson, W.R., 1985, Rapid and Sensitive Protein Similarity Searches. Science, 227, 1435.

Lisziewicz, J., Godany, A., Forster, H. and Kuntzel, H., 1987, Isolation and Nucleotide Sequence of a *Saccharomyces cerevisiae* Protein Kinase Gene Suppressing the Cell Cycle Start Mutant *cdc25*. J. Biol. Chem., 262:2549

Matsumoto, K., Uno, I., Oshima, Y. and Ishikawa, T., 1982, Isolation and Characterization of Yeast Mutants Defective in Adenylate Cyclase and cAMP-dependent protein kinase. Proc. Natl. Acad. Sci. USA, 79:2355.

Parker, P.J., Coussens, L., Totty, N., Rhee, L., Young, S., Chen, E., Stabel, S., Waterfield, M.D. and Ullrich, A. 1986, The Complete Primary Structure of Protein Kinase C - the Major Phorbol Ester Receptor. Science, 233:853.

Shoji, S., Ericson, L.H., Walsh, K.A., Fischer, E.H. and Titani, K.,1983, Amino Acid sequence of the Catalytic Subunit of Bovine Type II Adenosine Cyclic 3' 5'-Phosphate Dependent Protein Kinase. Biochemistry, 22:3702.

Tamanoi, F., 1988, Yeast *RAS* Genes. Biochim. Biophys. Acta, 948:1.

Toda, T., Uno, I., Ishikawa, T., Powers, S., Kataoka, T., Broek, D., Cameron, S., Broach, J., Matsumoto, K. and Wigler, M., 1985, In Yeat, *RAS* Proteins Are Controlling Elements of Adenylate Cyclase. Cell, 40:27.

Toda, T., Cameron, S., Sass, P., Zoller, M. and Wigler, M., 1987, Three Different Genes in *S. cerevisiae* Encode the Catalytic Subunits of the cAMP-Dependent Protein Kinase. Cell, 50:277.

Toda, T., Cameron, S., Sass, P., Zoller, M., Scott, J.D., McMullen, B., Hurwitz, M., Krebs, E.G. and Wigler, M., 1987, Cloning and Characterization of *BCY1*, a Locus Encoding a Regulatory Subunit of the cAMP Dependent Protein Kinase in Saccharomyces cerevisiae. Mol. Cell. Biol., 7:1371.

THE FUNCTION OF THE MAMMALIAN RAS PROTEINS

Alan Hall, Jonathan D.H. Morris, Brendan Price,
Alison Lloyd, John F. Hancock, Sandra Gardener[+],
Miles D. Houslay[+], Michael J.O. Wakelam[+], and
Christopher J. Marshall

Institute of Cancer Research, Chester Beatty
Laboratories, Fulham Road, London SW3 6JB, U.K.

[+]Molecular Pharmacology Group, Department of Biochemistry
University of Glasgow, Glasgow G12 8QQ, U.K.

INTRODUCTION

Single amino acid alterations in one of the three ras proteins have been detected in 25-50% of human cancers and it is believed that the somatic mutational event which generated these amino acid substitutions was an important step in the development of these malignancies (Barbacid, 1986; Bos et al., 1987; Paterson et al., 1987).

The ras proteins, p21^{ras}, are expressed in most if not all cells. They are located on the cytoplasmic side of the plasma membrane, they bind GTP and GDP and have an intrinsic GTPase activity. This, together with some sequence similarities to other proteins, has suggested that p21^{ras} functions as a regulatory guanine nucleotide binding protein (see Barbacid, 1986 for review). It is still not clear, however, which processes are being regulated and with which molecules p21 interacts.

It has been observed that some of the oncogenic mutations of p21^{ras} detected in human malignancies (located at codons 12,13 and 61) lead to a decrease in the intrinsic GTPase activity of the protein. This has led to the proposal that increased levels of the active GTP form account for its transforming properties (McGrath et al., 1984). However, there is not always a correlation between the observed reduction in GTPase activity of a particular mutant and the strength of its transforming activity (Trahey et al., 1987). Recently a cytoplasmic protein, GAP, has been identified that increases the GTPase activity of normal but not val12 or asp12 p21^{ras} (Trahey and McCormick, 1987) and it is now clear that all oncogenic mutations have a dramatic effect on the in vivo GTPase activity of the protein.

The biochemical function of p21^{ras} has been very difficult to study. Microinjection of E. coli produced oncogenic p21 into quiescent cells induces DNA synthesis and cell division (Stacy and Kung, 1984; Feramisco et al., 1984; Trahey et al., 1987), whilst microinjection of neutralising anti-p21^{ras} antibodies into quiescent fibroblasts blocks proliferation after addition of growth factors (Mulcahy et al., 1985).

This strongly suggests that ras is involved in regulating growth factor induced cell proliferation and that the oncogenic mutations lead to a breakdown in this regulatory process. In S. cerevisae, the yeast ras proteins directly regulate adenylate cyclase activity (Toda et al., 1985), but this seems not to be the case in mammalian cells and xenopus oocytes (Birchmeier et al., 1985; Beckner et al., 1985). We report here some of our efforts to look for a possible target for p21^{ras} regulation, and in particular for changes in another known second messenger system, namely the breakdown of phosphatidyl inositol lipids by a phospholipase C.

RESULTS

Analysis of stably transfected cell lines

In order to see if p21^{ras} can affect the inositol phospholipid (PI) breakdown pathway, cells expressing a variety of ras gene constructs were labelled overnight in ^{3}[H]inositol, washed extensively and the rate of generation of inositol phosphates measured in the presence of Li$^+$. Since these measurements are made in the absence of added growth factors, we refer to these conditions as basal rates. Clones constitutively expressing mutant N-ras genes or clones containing N-ras linked to the heterologous inducible mouse mammary tumour virus promoter were used. Induction of the T15 cell line leads to the expression of around 10^6 molecules of normal p21^{ras} (McKay et al., 1986), but no change in the basal rate of production of inositol phosphates is observed compared to uninduced cells or parental NIH-3T3 cells (Wakelam et al., 1986). In contrast, induction of the cell line pLHT leads to the expression of around 10^5 molecules of oncogenic p21^{N-ras}, and we observe a 3-4 fold increase in the rate of inositol phosphate production (Hancock et al., 1988). All clones constitutively expressing activated ras genes have a significantly higher basal rate of PI metabolism than parental NIH-3T3 cells. We have observed the same results with NIH-3T3 cells constitutively expressing oncogenically activated Ha- and Ki- ras genes (Hancock et al., 1988).

Although we observed no change in PI metabolism in fully induced T15 cells, we did find a striking increase in bombesin stimulated PI breakdown (Wakelam et al., 1986). Addition of bombesin led to a small (20%) increase in the rate of production of inositol phosphates in uninduced T15 cells and in NIH-3T3 cells. However, addition of bombesin to induced T15 cells gave a 2-3 fold stimulation. Measurement of the binding of radiolabelled bombesin to induced and uninduced T15 cells showed no change in bombesin receptor number. We have recently confirmed that IP$_3$ is the primary breakdown product of this synergistic reaction and that in induced T15 cells bombesin gives a 2-3 fold stimulation of intracellular Ca^{2+} release compared to NIH-3T3 cells (AL, CJM, M Whittaker, SG, MJOW, unpublished results).

An analysis of two independent clones constitutively expressing high levels of normal p21^{N-ras}, however, showed no increase in bombesin stimulated PI breakdown. It is unclear to us whether some other change has occurred in T15 cells or whether the use of an inducible clone allows subtle changes in PI metabolism to be analysed, which are not detectable in cell lines stably transformed by overexpression of normal N-ras protein.

We have also isolated transformed cell lines constitutively expressing normal Ha and normal Ki p21^{ras}. Clones expressing normal Ha are not significantly stimulated by any of the ligands we have tried. Although preliminary work suggested an increased PDGF response in normal Ha-ras overexpressing clones, subsequent isolation of other clones, including an inducible p21^{Ha-ras} containing clone, has not confirmed these original findings. In agreement with others (Parries et al., 1987) we have observed an increased bradykinin stimulated response in inositol phosphate formation in clones overexpressing Ki-ras. However, these effects probably arise through an increase in bradykinin receptor numbers on the cells (Parries et al., 1987). The mechanism by which normal (and mutant) Ki-ras increases bradykinin receptor numbers is unknown.

Analysis of transient ras expression in COS-1 cells

COS-1 cells were transfected with plasmids containing mutant or normal Ha-ras cDNAs under the control of the SV40 early promoter. Since the plasmid used (pEXV) also contains the SV40 origin of replication, a large number of plasmid copies are produced. This coupled with the high level expression from the strong promoter generates large amounts of p21^{Ha-ras} and its effect on PI turnover can be examined (Hancock et al., 1988). We found that oncogenic but not normal ras leads to a significant increase in the basal rate of PI turnover. Although the effect does not appear to be as large as that obtained with the stably transfected cell lines, only around 20% of the COS-1 cells actually take up DNA after transfection. When this is taken into account and assuming that both transfected and untransfected cells contribute to the background basal rate, then the increases observed in the two systems are comparable.

Analysis of scrape-loaded cells

The experiments described so far have made use of ras DNA introduced into cells to generate ras protein. It is not possible, therefore, to look for very early ras-induced events. Although microinjection of E. coli purified ras protein would overcome this problem, it is difficult if not impossible to inject sufficient cells to look at biochemical changes. We have made use of the scrape-loading technique (McNeil et al., 1984) to introduce E. coli produced proteins into large numbers (10^5) of Swiss 3T3 cells. Cells, scrape-loaded with oncogenic p21^{ras} become morphologically transformed after 15h and, if quiescent cells are scrape-loaded, oncogenic ras in the presence of insulin stimulates DNA synthesis (Morris et al., 1988).

We have used this approach to look for changes in second messengers induced by oncogenic p21^{ras}. 1,2-diacylglycerol (DAG) formation was monitored by looking for phosphorylation of an 80kD substrate of protein kinase C. We find that phosphorylation of the 80kD protein occurs 5-10 min after scrape-loading val12 p21^{ras}. Scrape-loading the biologically inactive val12ser186 ras protein does not produce this effect. When cells, prelabelled with ^{3}H inositol, were scrape-loaded with val12 p21^{ras}, we could find no increase in inositol phosphate production, whereas PDGF stimulated a 4-fold increase in inositol phosphates over 20 minutes (Morris et al., 1988).

DISCUSSION

It is clear that normal ras proteins play a crucial role in regulating cell proliferation and that oncogenic mutations lead to a breakdown in this control. However, the biochemical processes regulated by mammalian p21ras are unknown and much effort has been expended looking for the effects of ras on second messenger signalling systems.

We have found, in agreement with others (Fleischman et al., 1986), that oncogenically activated ras increases the basal rate of inositol phosphate production in NIH-3T3 cells (Hancock et al., 1988). Normal ras, even when expressed at 10 fold higher levels, does not increase this basal rate (Wakelam et al., 1986; Hancock et al., 1988). Since it is likely that oncogenic mutations in ras lead to constitutive activation of the process(es) normally regulated by p21ras, we have also looked for effects of normal p21 on growth factor stimulated PI metabolism. We have isolated one cell line, T15, which contains an inducible normal N-ras gene in which the ras protein is 50-fold over-expressed. Induction of p21N-ras leads to an increase in bombesin stimulated inositol phosphate production (Wakelam et al., 1986). This effect has not, however, been observed in clones overexpressing normal N-ras constitutively, making it difficult to draw any firm conclusions concerning the role of the normal ras protein, if any, in PI metabolism. We do not know at what level the co-operation between the activated bombesin receptor and normal p21N-ras occurs in the T15 cells or why this effect is only observed in this inducible cell line. Indeed, using this kind of approach with transfected cell lines, it is very difficult to determine whether changes in PI metabolism, or any other signalling system, are an immediate effect of ras function or whether they are a consequence of the chronic production of abnormal ras proteins.

In an attempt to overcome these problems, we have used a technique, scrape-loading, capable of introducing purified ras protein into a large number of cells quickly. The effects of oncogenic ras protein, introduced into quiescent Swiss 3T3 cells in this way, on morphology and DNA synthesis are identical to those observed after microinjection of protein (Stacey and King, 1984). Since more than 10^5 cells can be scrape-loaded, however, it should be possible to determine early ras-induced biochemical changes using this technique.

We have found activation of an 80kD protein 5-10 min after scrape-loading val12 p21ras indicating the rapid formation of 1,2-diacyl-glycerol and the activation of protein kinase C. Surprisingly, however, we could find no indication for a concomitant increase in inositol phosphates. We are led to conclude that the DAG is coming from a source other than phosphatidyl inositoside breakdown and we are currently looking for changes in other phospholipids. Furthermore, since changes in inositol phosphate production has clearly been observed by many groups in cell lines expressing ras genes, we suppose that this increase in phospholipase C activity is a consequence of the chronic expression of oncogenic p21ras in cells and not of a direct activation of this secondary messenger system.

ACKNOWLEDGEMENTS

The work described in this paper was supported by the Cancer Research Campaign (UK) and the Medical Research Council (UK).

REFERENCES

Barbacid, M. (1986) ras Genes. Ann. Rev. Biochem. 56: 779.

Beckner, S.K., Hattori, S., and Shih, T.Y., 1985, The ras oncogene product is not a regulatory component of adenylate cyclase, Nature, 317:71.

Birchmeier, C., Broek, D., and Wigler, M., 1985, Ras proteins can induce meiosis in xenopus oocytes, Cell, 43:615.

Bos, J.L., Fearon, E.R., Hamilton, S.R., deVries, M.W., van Bloom, J.M., Van der Eb, A.J., and Vogelstein, B., 1987, Prevalence of ras mutations in human colorectal cancer, Nature, 327:293.

Feramisco, J.R., Gross, M., Kamata, T., Rosenberg, M., and Sweet, R.W., 1984, Microinjection of the oncogene form of the human H-ras (T24) protein results in rapid proliferation of quiescent cells. Cell, 38:109.

Fleischman, L.F., Chawala, S.B., and Cantley, L., 1986, Ras transformed cells: Altered levels of phosphatidylinositol 4,5 bisphosphate and catobolites, Science, 231:407.

Hancock, J.F., Marshall, C.J., McKay, I.A., Gardener, S., Houslay, M.D., Hall, A. and Wakelam, M.J.O., 1988, Mutant but not normal p21^{ras} elevates inositol phospholipid breakdown in two different cell systems, Oncogene, 3:187.

McGrath, J.P., Capon, D.J., Goeddel, D.V., and Levinson, A.D., 1984, Comparative biochemical properties of normal and activated human ras p21 protein, Nature, 310:644.

McKay, I.A., Marshall, C.J., Cales, C., and Hall, A., 1986, Transformation and stimulation of DNA synthesis in NIH-3T3 cells are a titratable function of p21^{N-ras} expression, The EMBO J., 5: 2617.

McNeil, P.L., Murphy, R.F., Lanni, F., and Taylor, D.L., 1984, A method for incorporating macromolecules into adherent cells. J. Cell. Biol., 98:1556.

Morris, J.D., Price, B., Lloyd, A., Self, A.J., Marshall, C.J. and Hall, A., 1988, Scrape-loading of Swiss 3T3 cells with ras protein induces rapid activation of protein kinase C followed by DNA synthesis, in press.

Mulcahy, L.S., Smith, M.P., and Stacey, D.W., 1985, Requirement for ras proto-oncogene function during serum stimulated growth of NIH-3T3 cells, Nature, 313:241.

Parries, G., Hoebel, R., and Racker, I., 1987, Opposing effects of a ras oncogene on growth factor stimulated phosphoinositide hydrolysis: Desensitization to PDGF and enhanced sensitivity to bradykinin, Proc. Natl. Acad. Sci. USA, 84:2648.

Paterson, H., Reeves, B., Brown, R., Hall, A., Furth, M., Bos, J., Jones, P., and Marshall, C.J., 1987, Activated N-ras controls the transformed phenotype of HT1080 human fibrosarcoma cells, Cell, 51:803.

Stacy, D.N. and Kung, H.F., 1984, Transformation of NIH-3T3 cells by microinjection of Ha-ras p21 protein, Nature, 310:508.

Toda, T., Uno, I., Ishikawa, T., Powers, S., Kataoka, T., Broek, D., Broach, J., Matsumoto, K., and Wigler, M., 1985, Yeast RAS proteins are controlling elements in the cyclic AMP pathway, Cell, 40:27.

Trahey, M. and McCormick, F., 1987, A cytoplasmic protein stimulates normal N-ras p21 GTPase, but does not affect oncogenic mutants, Science, 238:542.

Trahey, M., Milley, R.J., Cole, G.E., Innis, M., Paterson, H., Marshall, C.J., Hall, A., and McCormick, F., 1987, Biochemical and biological properties of the human N-<u>ras</u> p21 protein, <u>Molec. Cell. Biol.</u>, 7:556.

Wakelam, M.J.O., Davies, S.A., Houslay, M.D., McKay, I., Marshall, C.J., and Hall, A., 1986, Normal p21^{N-ras} couples bombesin and growth factor receptors to inositol phosphate production, <u>Nature</u>, 323:173.

RAS PROTEINS AS POTENTIAL ACTIVATORS OF PROTEIN KINASE C FUNCTION

Janet E. Jones and Juan Carlos Lacal*

Laboratory of Cellular and Molecular Biology
National Cancer Institute
Bethesda, Maryland 20892

SUMMARY

The structural and functional relationships of the products of the *ras* genes with known G proteins have suggested that *ras*-p21s may be important mediators of signal transduction pathways involved in cellular proliferation and differentiation. One of the best characterized pathways constitutes the activation of the Ca^{++}-sensitive, phospholipid dependent protein kinase C (PKC), following receptor mediated hydrolysis of inositol phospholipids by phospholipase C (PLC). We have investigated the putative involvement of the *ras*-p21 proteins in the modulation of PLC, responsible for the generation of 1,2-diacylglycerol (DAG) and inositol phosphates (IPs), as well as the consequent activation of PKC by the generated DAG. Utilizing *Xenopus laevis* oocytes we have detected a rapid increase in the generation of DAG after microinjection of the transforming H-*ras* p21 protein. Elevated basal levels of DAG were also observed in NIH-3T3 cells transformed by a variety of mutated *ras* proteins without increased levels of the corresponding IPs. These results suggest a novel source for the generated DAG other than the hydrolysis of phosphatidyl inositol phosphates. We have also observed that microinjection of mutated *ras* proteins into cells previously devoided of endogenous PKC by chronic treatment with phorbol esters, lacked its otherwise potent mitogenic activity. The mitogenic function of *ras*-p21 was regenerated under identical conditions by co-microinjection with PKC . These results suggest the requirement of functional PKC for the mitogenic activity of *ras*. Finally, our previous findings that microinjection of *ras*-p21 into 3T3 cells induces a rapid increase in the intracellular pH of 0.15-0.2 pH units (Hagag et al. 1987)

(*)Present address: Instituto de Investigaciones Biomédicas, Facultad de Medicina, Universidad Autónoma, Madrid 28029, Spain.

demonstrate that *ras*-p21 can induce the activation of PKC under normal conditions. Taken together, all these results imply that *ras*-p21 proteins may function as regulators of PKC activity.

INTRODUCTION

The *ras* genes were first isolated and characterized as the oncogenes of two strains of acute rat retroviruses designated as Harvey- and Kirsten-MSV (1,2). Both Harvey- and Kirsten-*ras* genes encode 21 Kda proteins, designated as p21, responsible for their transforming activities (3). The retroviral *ras*-p21 molecules were found to be guanine-nucleotides binding proteins (4,5), located in the internal leaflet of the cytoplasmic membrane (6).

Analysis of a number of human tumors from a diversity of tissues revealed the presence of activated *ras* genes with high incidence (reviewed in 7, 8). Most of these were found to be homologues of either Harvey- or Kirsten-*ras* genes. A new member of this family designated N-*ras* was detected in a human neuroblastoma cell line. Comparison of the sequence of the normal *ras* genes to those of their activated counterparts showed that single point mutations within the coding sequence were sufficient to confer the transforming phenotype. Most of the activating lesions from naturally occuring tumors have been localized at either residue 12 or 61 of the *ras* p21 molecule. Other mutated positions have been also found in tumors from human tissues and experimental animal systems (9,10). *In vitro* mutagenesis also demonstrated that ras genes can be efficiently activated by mutations at codon 59, 63, 116, and 119 (reviewed in 7,8).

One of the most important clues to the physiologic function of ras-p21 proteins was the finding of their ability to bind guanine-nucleotides with high affinity (4,5). The availability of highly purified ras p21 proteins led to the discovery of another important property, its ability to hydrolyze GTP (11-13). Initial characterization indicated that *in vitro* GTP hydrolysis rates of mutated proteins were significantly slower (5 to 10-fold) than that of the normal protein (11-13). Recently, Trahey and McCormick (14), based on inconsistencies between *in vitro* GTPase activity and biological activities of a series of *ras* p21 mutants, discovered a factor which enhances at least 100-fold the GTPase activity of the normal p21 protein with no effect on the activity of mutated proteins. This new factor, designated as GAP for GTPase Activating Protein, is present in a large variety of tissues and established cell lines (15), works both *in vivo* and *in vitro,* and constitutes a good candidate for the effector molecule of *ras* function (15-17).

The GDP/GTP binding activity of *ras* proteins implicated them as putative signal-transducing proteins involved in the transmission of information from the extracellular milieu into the cell (18). Moreover, such proteins like the G_S and G_i of the adenylate cyclase system (19), transducin (20), or the elongation factor of protein synthesis, EF-Tu (21), show a certain degree of homology to the *ras* p21 in regions known to be associated with the GTP-binding domain. Since mutated *ras* genes have a dominant effect on cell proliferation when transfected into NIH-3T3 cells, the *ras* p21 product likely plays an important role in the regulation of cell proliferation in this system.

A number of studies have implicated the metabolism of phosphatidyl-inositols (PIs) in the function of growth factors, hormones, and neurotransmitters (reviewed in 22). Phospholipase C (PLC) is the enzyme responsible for the hydrolysis of phosphatidyl-inositol bisphosphate (PIP_2), to generate 1,4,5 inositol trisphosphate (IP_3) and 1,2 diacylglycerol (DAG). The former is the second messenger responsible for the Ca^{++} mobilization from endoplasmic reticulum stores. The latter is the intracellular activator of protein kinase C (PKC), an important cellular component thought to be involved in a large number of regulatory events, including cell proliferation (23). It has been postulated that

PLC is regulated by a G protein in a number of systems (23-26) . These results lead to the hypothesis that *ras*-p21 could be such a G protein.

We have investigated the putative role of *ras* proteins in the regulation of the activity of PLC and PKC in a number of systems. The results obtained demonstrate that in 3T3 cells, *ras* is not a direct regulator of the PLC responsible for the generation of IPs and DAG. However, *ras* p21 can induce the generation of DAG from a novel source, to efficiently activate PKC. Moreover, functional PKC is absolutely required for the mitogenic activity of *ras* p21. These results implicate *ras* proteins as putative signal-transducers involved in the regulation of PKC function.

RESULTS

PRODUCTION OF DIACYLGLYCEROL INDUCED BY *RAS*-P21 IN *XENOPUS LAEVIS* OOCYTES

Bacterially expressed and purified ras-p21 proteins have proven to be good tools for the characterization of their biochemical and biological properties. Purified ras proteins are potent mitogens when microinjected into a number of mammalian cells (27), can induce differentiation of the pheochromocytoma cell line PC-12 (28) and maturation of *Xenopus laevis* oocytes (29). Due to the feasibility of the analysis of phospholipid metabolites using a small number of oocytes, we utilized this system to investigate the involvement of *ras* proteins in the regulation of phospholipase C activity. Fig. 1 represents the results obtained when oocytes previously labelled with (^{3}H)glycerol were microinjected with either normal or mutated Harvey-*ras* p21 proteins. While no effects were observed after microinjection of the normal protein, a rapid elevation in DAG levels was observed with the transforming protein, which increased fivefold within 20 minutes of microinjection.

It has been reported that microinjection of the transforming *ras*-p21 protein into *Xenopus laevis* oocytes induces maturation, while the normal protein is much less efficient for this activity (29). We investigated the ability of our purified proteins to induce maturation under the same conditions in which we observed induction of DAG. Figure 2A shows the results obtained when different amounts

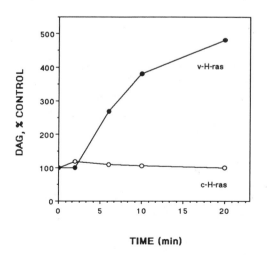

FIGURE 1. Induction of DAG production in *Xenopus laevis* oocytes by microinjection of normal (o) or transforming (●) *ras*-p21 proteins.

of normal or transforming ras proteins were microinjected into oocytes. Induction of maturation was dose-dependent. While 10 ng of the transforming protein were sufficient to fully induce maturation, the normal protein was ineffective even at doses of 150 ng per oocyte. Figure 2B shows the effect of microinjection of different amounts of normal and transforming *ras*-p21 proteins on the levels of DAG, as a time-course. Even at the minimal levels required to induce maturation, microinjection of the transforming protein induced a rapid increase in the DAG levels within 20 min. Therefore the ability of the ras protein to induce maturation correlated with its ability to induce elevation of the DAG levels.

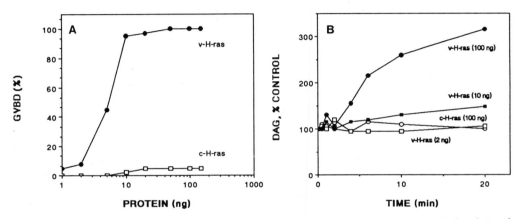

FIGURE 2. (A) Induction of maturation of *X. laevis* oocytes by microinjection of different amounts of normal (o) or transforming (●) *ras*-p21 proteins. (B) Relationship of oocyte maturation with DAG production after microinjection of different amounts of purified *ras* proteins.

Cleavage of phosphatidylinositols 4,5-bisphosphate (PIP$_2$) by phospholipase C (PLC) produces DAG and inositol (1,4,5)-trisphosphate (InsP$_3$). DAG is an important cofactor of PKC activation, while IP$_3$ is responsible for the release of Ca^{++} from nonmitochondrial intracellular stores (reviewed in 18,22). Therefore we investigated whether the elevation of DAG levels in *Xenopus* oocytes was due to the activation of PLC activity. When oocytes were labelled with (^{3}H)myo-inositol prior to microinjection, the levels of inositol phosphates (InsPs) could be determined as a function of time.

Fig 3 represents the levels of inositol phosphate (InsP), inositol bisphosphate (InsP$_2$), and inositol tris- and tetraphosphate (InsP$_3$ and InsP$_4$) after microinjection of either normal or mutated Harvey-*ras* p21 proteins. A rapid increase in the levels of both InsP and InsP$_2$ was observed 4-6 min after microinjection of the transforming H-*ras* protein. However, no alteration was observed with the normal protein even 20 min after microinjection.

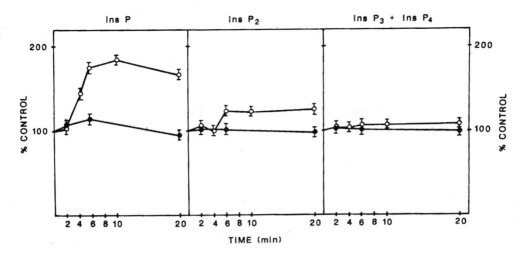

FIGURE 3. Inositol phosphates levels in *X. laevis* oocytes after microinjection of normal (●) or transforming (o) ras-p21 proteins.

INTRACELLULAR LEVELS OF DIACYLGLYCEROL IN *RAS*-TRANSFORMED CELLS

The above results suggested that *ras* p21 could be involved in the activation of a PLC specific for the generation of DAG and InsPs. Therefore we investigated the basal levels of both metabolites under conditions of steady-state labelling in a series of NIH-3T3 cell lines expressing high levels of *ras* proteins. We first generated a cell line of NIH-3T3 cells by transfection with an expression vector, pLTR-41, carrying the highly transforming viral H-ras p21 protein driven by the Abelson-MuSV LTR promoter (30). We also generated cell lines carrying an

CONSTRUCTIONS	EXPECTED PRODUCTS	Ffu/ ug DNA
pLTR-41	1 — 189	$10^{4.1}$
pSR-41	1 — 189	$10^{3.2}$
pLTR-Ser[186]	1 — 189	NO
pSR-Ser[186]	1 — 189	$10^{3.0}$
pLTR-p18d	1 — 164	NO
pSR-p18d	1 — 164	$10^{2.5}$

FIGURE 4. Schematic representation and biological activity of generated *ras* mutants carrying the *src*-related sequences.

TABLE 1. Steady-state levels of (3H)glycerol-labelled DAG and (3H)myo-inositol-labelled Ins Ps in normal and *ras*- or *sis*-transformed NIH-3T3 cells.

Cell line	Transforming phenotype	Ins Ps (%control)	DAG (%control)
NIH-3T3	-	100	100
+serum	-	140±5	125±8
Transfectants			
pLTR-41	+	100±1	153±13
pSR-41	+	61±3	145±14
pSV2-neo	-	90±6	108±6
pLTR-p18d+pSV2-neo	-	81±3	113±3
pSR-18d	+	38±3	138±3
v-sis	+	420±9	147±7

expression vector for a mutated ras protein, pLTR-Ser[186], in which the critical Cys[186] residue was substituted by a Ser. As previously demonstrated (31), this mutated ras protein was not able to translocate to the plasma membrane and was transformation deficient (Fig 4). Transforming activity of the Ser[186] mutant was reverted by construction of a chimeric vector, pSR-Ser[186], coding for a protein in which the first 15 amino acids of the p60[v-*src*] protein were fused in frame to the full length *ras* product. This amino terminal peptide is responsible for the myristoylation and membrane localization of the p60 protein (32).

In addition, we also generated a cell line expressing high levels of a deletion/substitution mutant of *ras* p21, in which the 25 amino acids at the carboxy terminal of the molecule were removed. The truncated protein, designated as p18d, was transformation deficient (Fig 4) due to the lack of residue Cys[186]. Fusion of this protein to the peptide from p60[v-*src*] regenerated its transforming activity (plasmid pSR-p18d). Finally, as a further control, we generated a cell line expressing a chimeric protein of the viral H-ras product fused to the amino terminal peptide from p60[v-*src*].

The results of the analysis of the basal levels of DAG and InsPs in the generated transformed, and non-transformed cell lines are summarized in Table1. While the transforming phenotype of the different *ras* mutants correlated with elevated levels of DAG when compared to the normal NIH-3T3 parental cell lines, no increase in the basal levels of InsPs was observed in any of the cell lines expressing the different *ras* products. By contrast, normal NIH-3T3 cells and NIH-3T3 transformed by v-*sis* showed an increase of both DAG and Ins Ps. Therefore, DAG in the various *ras*-transformed cell lines may have been generated from a different source than the hydrolysis of phosphatidylinositols. Table 2 summarizes the results obtained when the levels of the products of a PLC attack of other major membrane phospholipids were analyzed. All cell lines expressing transforming-deficient *ras* mutants, presented levels of phosphocholine, phosphoethanolamine, and phosphoserine comparable to those of their normal counterparts. By contrast, those cell lines expressing *ras*-p21 transforming products, showed increased levels of phosphocholine and phosphoethanolamine, but normal levels of phosphoserine, and cells transformed by v-*sis* oncogene showed lower than normal levels of phosphocholine and phosphoethanolamine. These results strongly suggested an altered metabolism of phosphatidylcholine and phosphatidylethanolamine in *ras* transformed cells. The fact that v-*sis* increased the production of Ins Ps without increasing the levels of phosphocholine or phosphoethanolamine also

TABLE 2. Steady-state levels of phosphocholine (P-cho), phosphoethanolamine (P-eth), and phosphoserine (P-ser) in control NIH-3T3 and *ras*- or *sis*-transformed cells.

Cell line	Transformed phenotype	P-cho (%)	P-eth (%)	P-ser (%)
NIH-3T3	-	100	100	100
+serum	-	100±1	100±1	100±2
Transfectants:				
pLTR-41	+	247±2	230±1	100±1
pSR-41	+	162±5	190±2	ND
pSV2-neo	-	87±3	83±5	ND
pLTR-p18d+pSV2-neo	-	81±2	91±7	ND
pSR-p18d	+	142±2	160±5	ND
v-sis	+	40±4	40±3	ND

suggests that different oncogenes affect phospholipid metabolism according to different patterns.

FUNCTIONAL RELATIONSHIP BETWEEN *RAS*-p21 AND PROTEIN KINASE C IN NIH 3T3 CELLS

It is very well established that activation of PKC can induce proliferation of a number of cell lines, in particular rodent fibroblasts (33,34). This cell system has the advantage that PKC can be efficiently removed by chronic treatment with phorbol esters (TPA or PDBu) for 48-72 hr. As a result, cells become unresponsive to phorbol ester challenge. Figure 5 illustrates the downregulation effect on the mitogenic activity of PDBu in Swiss 3T3 cells, induced after 72 hr exposure to the same agent. While PDBu (200 nM) was able to induce DNA-synthesis in 60 % of the untreated cells, it was inactive when the cells were previously treated with PDBu for 72 hr. This blockage was specific to the mitogenic activity of PDBu, since addition of serum induced a mitogenic response similar to that of the control, untreated cells.

Since the elevated levels of DAG induced by *ras*-p21 in both *X. laevis* oocytes and mammalian cells could be a mechanism of endogenous activation of protein kinase C (PKC), these results implicated the possibility of a functional relationship between *ras*-p21 and PKC. Therefore, we examined the requirement for PKC in the mitogenic activity of *ras*-p21 after microinjection into 3T3 cells. When bacterially expressed and purified *ras*-p21 was microinjected into quiescent Swiss 3T3 cells, an induction of DNA-synthesis was observed in 55% of microinjected cells (Fig 6A). This *ras*-p21 mediated activity was markedly reduced to about 10% of microinjected cells when cells were pretreated with PDBu for 72 hr (Fig 6B). The reduction in the activity is equivalent to an 80% inhibition over control, untreated cells. Under identical conditions, microinjection of purified PKC did not induce any significant increase in DNA-synthesis over background levels, indicating the efficient removal of PDBu in the assay conditions. However, when PKC was comicroinjected with *ras*-p21, an almost complete recovery of the mitogenic activity of the *ras* protein was observed. These results suggest that functional PKC is required for the mitogenic activity of *ras*-p21.

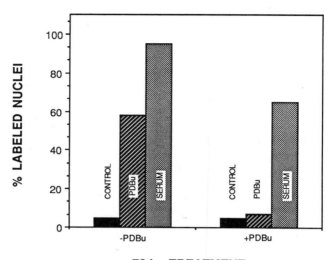

72 hr TREATMENT

FIGURE 5. Effect on induction of DNA-synthesis by phorbol esters and serum after chronic treatment with phorbol esters. Cells were treated for 72 hr with 200 nM PDBu (+PDBu) or DMSO alone (-PDBu), and then challenged with 10% serum or 200 nM PDBu. DNA synthesis was estimated after 24 hr labelling in medium containing (3H)methyl thymidine as indicated in Materials and Methods.

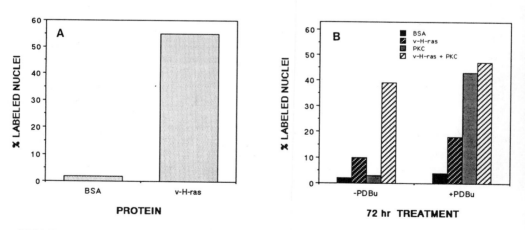

FIGURE 6. (A) DNA synthesis induced in Swiss 3T3 cells after microinjection of BSA or viral H-*ras* p21 protein (B) DNA synthesis induced in Swiss 3T3 cells pretreated with PDBu for 72 hr before microinjection with BSA, viral H-*ras* alone, PKC alone or viral H-*ras* and PKC.

FUNCTIONAL RELATIONSHIP BETWEEN RAS-p21 AND PROTEIN KINASE C IN PC-12 CELLS

In the pheochromocytoma cell line PC-12, oncogenic *ras* proteins induce differentiation to sympathetic neurons (28,35) in a way similar to that induced by NGF (nerve growth factor). Moreover, phorbol-esters are not sufficient but seem to be synergetic with NGF for induction of differentiation.(36). We investigated whether the same functional relationship between *ras* proteins and PKC found in mouse fibroblasts prevails in the neuronal differentiation system. UR61/J is a subclone of PC-12 which carries the activated N-*ras* gene under the control of the glucocorticoid-inducible promoter (a gift from A. Pellicer). After chronic treatment of UR61/J cells with phorbol esters, a drastic downregulation of the immunoreactive PKC was detected (Fig.7). Addition of dexamethasone at concentrations of 0.01 to 1 μM after treatment with 200 nM PDBu for 48 hr, was more potent in inducing neuronal differentiation than similar concentrations added to cells untreated with phorbol esters (Fig 8). These results suggest that removal of PKC in PC-12 cells does not block the ability of *ras*-p21 to induce differentiation. Experiments are in progress to investigate the apparent paradox between the fibroblast and the neuronal systems.

DISCUSSION

Our studies show that *ras*-p21 proteins can activate PKC, an ubiquitous protein kinase whose activity seems to be involved in a number of cellular functions, including regulation of proliferation and differentiation (34). Activation of PKC is apparently mediated by the constitutive generation of DAG induced in the *ras*-transformed cells by activation of the metabolism of two major phospholipids: phosphatidylcholine (PC), and phosphatidylethanolamine (PE). Both phospho-choline and phosphoethanolamine, as well as CDP-choline and CDP-ethanolamine levels are increased in *ras*-transformed cells. However this

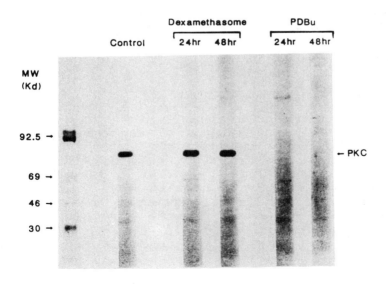

FIGURE 7. Downregulation of PKC in PC-12 cells by treatment with PDBu

is not enough evidence to suggest a direct activation of a PLC specific for these phospholipids mediated by *ras*-p21 proteins. In fact, similar effects have been observed in cells treated with phorbol esters (37-39), in a mechanism that seems to involve translocation to the membrane of the CDP-choline transferase, responsible for the synthesis of CDP-choline. Therefore, further characterization of the pathways involved in the synthesis and degradation of PC and PE will be required to establish the step(s) altered by *ras*-p21 proteins.

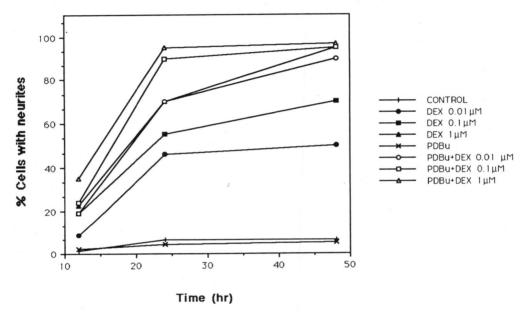

FIGURE 8. Neuronal differentiation induced in UR61/J cells by addition of dexamethasone in control cells and in cells treated with PDBu for 48 hr

It has been previously shown that activation of PKC leads to an activation of the Na^+/H^+ antiporter, which induces a rapid increase in the intracellular pH (40). Cytoplasmic alkalinization has been proven to be an essential step required for serum and growth factors function in fibroblasts (41). We have investigated the effects of microinjection of purified *ras*-p21 proteins on intracellular pH in NIH-3T3 cells, as a measurement of the activation of PKC (42). Transforming *ras* protein was able to induce a rapid increase in the intracellular pH after microinjection. No effect was observed after microinjection of the normal protein under identical conditions. Moreover, this activity was substantially reduced in the presence of amiloride, an inhibitor of the Na^+/H^+ antiporter, and under extracellular Na^+ concentrations at which the antiporter system is not functional. Finally, the mitogenic activity of the ras-p21 protein was reduced to 50% of its maximun effect in the presence of amiloride during the first 4 hr after microinjection, but it did not interfere with this activity when added to the cells at later times. All these results indicate that *ras*-p21 was able to activate the Na^+/H^+ antiporter system and this activity was required for its mitogenic function.

Our results of a lack of mitogenic activity of *ras*-p21 in NIH-3T3 cells devoid of endogenous PKC, suggest the requirement of functional PKC for *ras* activity. Moreover, the ability to regenerate *ras*-induced mitogenic activity in downregulated cells by comicroinjection of *ras* and PKC, suggest that *ras* function is mediated by PKC in mouse fibroblasts. The fact that serum can overcome the blockage imposed by deprivation of PKC, implies the existence of an alternate pathway for the mitogenic signal to that mediated by PKC. Since a neutralizing antibody to *ras*-p21 activity, Y13-259, inhibits serum-induced DNA-synthesis in this system (43), and *ras*-p21 microinjected in Xenopus laevis oocytes induces a rapid increase in the production of DAG, it seems that *ras* is located upstream of PKC in the signal transduction pathway. However recent studies indicated that Y13-259 is also able to block the mitogenic activity of phorbol esters (44), suggesting a more complex mechanism.

Previous reports suggested the cooperative effect of phorbol-esters and *ras*, in the induction of transformation (45). These results suggested a role of PKC in the regulation of cell proliferation in rodent fibroblasts. Recent studies indicate that there is a family of genes codifying for different PKC isozymes (46-49), in agreement with the observation that several PKC isozymes could be resolved from rat and rabbit brain (50,51). The diversity of PKCs, their cell-type specific expression (52-55), and their differences in biochemical properties (56-57), constitute the basis for the hypothesis that different PKC enzymes might serve for specific cellular functions.

Overexpression of PKC proteins in murine fibroblasts alter the growth properties of the cells with very little effect on their tumorigenic properties (58,59). However a complete description of the biological properties of the various PKC isozymes has not been carried out. Understanding of the functional interaction of *ras* proteins with specific PKCs might provide important clues for the regulation of fibroblasts proliferation and neuronal differentiation.

ACKNOWLEGMENTS

We thank P. Blumberg for the purified PKC, P. de la Peña for helping in the oocytes microinjection, A. Pellicer for the UR61/J inducible cell line, and S. Aaronson for his support.

EXPERIMENTAL PROCEDURES.

Determination of lipid metabolites in *Xenopus* oocytes

Stage VI oocytes were manually dissected and selected from ovarian fragments by standard procedures (60). Oocytes were then microinjected with 40 nl of 20 mM Tris-HCl, pH 7.2 containing 10 µCi/ml of either (3H)glycerol (Amersham, 450 mCi/mmol) or (3H)myo-inositol (Amersham, 3.5 Ci/mmol). After incubation at 23°C for 5 hrs, 5-20 oocytes were homogenized and lipid extracted as follows. For (3H)inositol-containing lipids, oocytes were homogenized in 440 µl of methanol:1N HCl (10:1;v/v), and phases were split by addition of 0.6 ml of chloroform and 160 µl of deionized water. Organic phases were saved, and aqueous phases extracted with 400 µl of chloroform. Both organic phases were collected and washed with 220 µl of chloroform:1N HCl (10:1, v/v), and dried under nitrogen. Samples were resuspended in 200 µl of chloroform and radioactivity contained in phosphatidyl-inositols (PIs) estimated by scintillation counting in 10 ml Aquasol. Aqueous phases were loaded into Dowex 1X8 columns (formate form), and eluted with 8 ml of 0.2 M ammonium formate plus 0.1 M formic acid (InsP), 0.4 M ammonium formate plus 0.1 M formic acid (InsP2), and 1.2 M ammonium formate plus 0.1 M formic acid (InsP3+InsP4). For (3H)glycerol-containing lipids, oocytes were homogenized in 160 µl deionized water before adding 0.6 ml of chloroform:methanol (1:2, v/v). Phases were split by addition of 0.2 ml chloroform and 0.2 ml deionized water. Organic phases were saved, and aqueous phases washed with 0.5 ml chloroform. Organic phases were collected and dried under nitrogen. Samples were resuspended in chloroform and resolved by thin-layer chromatography (TLC) in hexanes:diethyl ether:acetic acid (60:40:1, v/v/v). Bands corresponding to diacylglycerol were visualized by iodine vapor exposure along with corresponding standards. After scraping off the plates, radioactivity was estimated by scintillation counting in Aquasol.

Determination of oocyte maturation

Stage VI oocytes were selected as indicated above and microinjected with the indicated amounts of normal or transforming *ras*-p21 proteins in 40 nl of 20 mM MES, pH7.0. Twenty-four hr after microinjection, maturation was quantitated by fixing in 10% TCA, splitting the oocytes (25 per protein concentration) and visualization of the germinal vesicle breakdown.

Determination of phospholipid metabolites in *ras*-transformed cells

Cell lines containing each transforming construct or normal counterparts were grown in Dulbecco's modified Eagle's medium (DMEM) supplemented with 10% calf serum. Cultures were labelled to equilibrium for 70 hr in serum-containing medium in 35 mm dishes with either 10 μCi of (3H)glycerol, (3H)myo-inositol, (3H)methyl choline (Amersham, 70-90 Ci/mmol), (14C)ethanolamine (40-60 mCi/mmol), or (14C)serine (150 mCi/mmol). Cultures were washed and maintained for 4 hr in serum-free medium, and incubated for 1 hr in serum-free or serum containing medium. In experiments designed for determinations of Ins Ps, LiCl (20 mM) was present during the 1 h incubation period. Reactions were terminated by addition of TCA to a final concentration of 16%. Cells were scraped and centrifuged to remove precipitated material. Supernatants were washed with four volumes of water-saturated diethyl-ether, and the extraction repeated three more times. Total levels of Ins Ps were determined as indicated for oocytes eluting from the Dowex columns with 1.2 M ammonium formate, 0.1 M formic acid. Water-soluble metabolites from PC, PE, and PS were fractionated by TLC in silica gel G plates in methanol:0.5% NaCl:NH3 (100:100:2, v/v/v). Samples were run with corresponding standards, and radioactivity determined by scintillation counting in Aquasol. For DAG determinations, lipids were extracted and fractionated basically as described for oocytes.

DNA-synthesis stimulation of Swiss-3T3 cells by microinjection of *ras*-p21

Swiss 3T3 cells were grown in 60 mm dishes in DMEM supplemented with 10% calf serum. When the cells were semiconfluent, plates were washed five times with DMEM , and DMEM containing 1 mg/ml BSA added. Cells were grown in this medium containing or not 200 nM PDBu for 72 additional hrs. Where indicated, cells were microinjected with H-*ras* p21, PKC, or both, as previously described (30). DNA synthesis was estimated by incubation in the presence of (3H)methyl-thymidine for 24 hr. Cells were fixed in methanol at -20ºC for 5 min, and placed under a sensitive emulsion (nuclear tracking emulsion, Kodak). After 48 hr at 4ºC, emulsions were developed by standard procedures and percentage of cells actively synthesizing DNA estimated. Where indicated, cells were treated with serum or PDBu and then incubated for 24 hr in the presence of 2 μCi/ml (3H)methyl thymidine. Incorporation was stopped by addition of TCA (10%, final concentration). Cells were washed twice with ethanol, and resuspended in 0.2 N NaOH, 0.1% SDS. Radioactivity was estimated by scintillation counting in Aquasol.

Western-blot analysis

UR61/J cells were grown in RPMI medium supplemented with 10% horse serum and 5% fetal calf serum. When 80% confluent, cells were treated with dexamethasone (1 µM), or PDBu (200 nM) for indicated times. After incubation, cells were washed in PBS, and lyzed in 50 mM phosphate buffer, pH 7.0, containing 100 mM Na Cl, 0.1% SDS, 1% Triton X-100, and 0.1% deoxycholate. Equivalent amounts (100 µg) from each extract were resolved in a 10% PAGE, transfered to nitrocellulose paper in 50 mM Tris, glycine, methanol buffer. Paper was equilibrated in Blotto buffer (Collaborative Research), and probed against anti-PKC monoclonal antibody (Amersham).

REFERENCES

1. Harvey, J.J. (1964). Nature 204: 1104-1105.
2. Kirsten, W.H. and Mayer, L.A. (1967) J. Natl. Cancer Inst. 39:311-334.
3. Shih, T.Y., Weeks, M.O., Young, H.A., and Scolnick, E.M.(1979) Virology 96: 64-79.
4. Scolnick, E.M., Papageorge, A.G., Stokes, P.E., and Shih, T.Y.(1979) Proc. Natl. Acad. Sci. USA 76: 5355-5359.
5. Shih, T.Y., Papageorge, A.G., Stokes, P.E., Weeks, M.O., and Scolnick, E.M. (1980). Nature 287: 686-691.
6. Willingham, M.C., Pastanm, I., Shih, T.Y., and Scolnick, E.M.(1980) Cell 19: 1005-1014.
7. Barbacid, M. (1987) Ann. Rev. Bioch. 56: 779-827.
8. Lacal, J.C., and Tronick, S.E. (1988) The ras oncogene. In The Oncogene Handbook. Reddy, P., Curran, T, and Skalka,A. edts. Elsvier, Holland.
9. Bos, J.L. et al. (1985) Nature 315:726-730.
10. Reynolds, S. et al (1987) Science 237:1309-1316.
11. McGrath, J.P., Capon, D.J., Goeddel, D.V., and Levinson, A.D. (1984) Nature 310: 644-649.
12. Gibbs, J.B., Sigal, I.S., Poe, M., and Scolnick, E.M. (1984) Proc. Natl. Acad. Sci. USA 81: 5704-5708.
13. Sweet, R.W., Yokoyama, S., Kamata, T., Feramisco, J.R., Rosenberg, M., and Gross, M. (1984) Nature 311: 273-275.
14. Trahey, M., and McCormick, F. (1987) Science 238: 542-545.
15. Adari, H., Lowy, D.R., Willumsen, B.M., Der, C.J., and McCormick, F. (1988) Science 240: 518-521.
16. Calés, C., Hancock, J., Marshall, C.J., and Hall, A. (1988) Nature 332: 548-551.
17. Di Donato, A., Srivastava, S.K., and Lacal, J.C. (1988) In "The guanine-nucleotides binding proteins:common structural and func-tional properties". Bosch, L., Kraal , B., and Parmeggiani, A. Plenum Publishing Co. (in the press).
18. Berridge, M.J., and Irvine, R.F. (1984) Nature 312: 315-319
19. Gilman, A.G. (1984) Cell 36: 577-579.
20. Stryer, L. (1986) Ann. Rev. Neurosc. 9: 787-819.
21. Ochoa, S. (1986) Arch. Biochem. Biophys. 223: 325-349.
22. Berridge, M.J. (1984) Biochem. J. 220: 345-
23. Cockcroft, S, and Gowperts, B.D. (1985) Nature 314:534-536.
24. Melin, P., Sundler, R., and Jergil, B. (1986) FEBS Lett. 198:85-88.
25. Smith, C.D., Cox, C.C., and Snyderman, R. (1986) Science 232:97-100.
26. Uhing, R.J., Prpic, V., Jiang, H, and Exton, J.H. (1986) J. Biol. Chem. 261:2140-2146.
27. Stacey, D.W. and Kung, H. (1984) Nature 310:508-511.
28. Bar-Sagi, D., and Feramisco, J. (1985) Cell 42:841-848.
29. Birchmeier, C., Broek, D., and Wigler, M. (1985) Cell 43:615-621.
30. Lacal, J.C. et al (1986) Cell 44:609-617.
31. Willumsen, B.M. et al (1984) Nature 310:583-586.
32. Garber, E.A. et al (1983) Nature 302:161-163.
33. Rozengurt, E. (1986) Science 234:161-166.
34. Nishizuka, Y. (1986) Science 233:305-312.
35. Noda, M. et al. (1985) Nature 318: 73-75.
36. Hall, F.L. et al. (1988) J. Biol. Chem. 263: 4460-4466.
37. Muir, J.G. and Murray, A.W. (1987) J. Cell Physiol. 130: 382-391.
38. Kolesnick, R.N. (1987) J. Biol. Chem. 262: 14525-14530.
39. Glatz, J.A. et al. (1987) Carcinogenesis 8: 1943-1945.
40. Shuldiner, S., and Rozengurt, E. (1982) Proc. Natl. Acad. Sci.USA 79:7778-7782.
41. Pouyssegur, J (1985) Trends Biochem. Sci. 10:453-455.
42. Hagag, N. et al (1987) Mol. Cell. Biol.7, 1984-1988.
43. Mulcahy, L.S., Smith, M.R., and Stacey, D.W. (1985) Nature 313: 241-243.

44. Yu, C., Tsai, M., and Stacey, D.W. (1988) Cell 52: 63-71.
45. Dotto, G.P., Parada, L.F., and Weinberg, R.A. (1985) Nature 318: 472-475.
46. Knopf, J.L. et al. (1986) Cell 46: 491-502.
47. Coussens, L. et al. (1986) Science 233: 859-866.
48. Parker, P.J. et al. (1986) Science 233: 853-858.
49. Ono, Y. et al. (1987) FEBS Lett. 226: 125-128.
50. Huang, K., Nakabayashi, H., and Huang, F.L. (1986) Proc. Natl. Acad. Sci. USA 83: 8535-8539.
51. Jaken, S., and Kiley, S.C. (1987) Proc. Natl. Acad. Sci. USA 84: 4418- 4422.
52. Brandt, S.J. et al. (1987) 49: 57-63.
53. Yoshida, Y. et al. (1988) J. Biol. Chem. 263: 9868-9873.
54. Ohno, S. et al. (1987) Nature 325: 161-164.
55. Hidaka, H. et al. (1988) J. Biol. Chem. 263: 4523-4526.
56. Huang K. et al (1986) Proc. Natl. Acad. Sci. USA 83: 8535-8539.
57. Ono, Y. et al (1988) J. Biol. Chem. 263: 6927-6932.
58. Housey, G.M. et al. (1988) Cell 52: 343-354.
59. Persons, D.A. et al. (1988) Cell 52: 477-458.
60. M. Zasloff (1983) Proc. Natl. Acad, Sci. USA 80: 6436-6440

NOVEL PHOSPHORYLATION OF ras p21 AND MUTATIONAL STUDIES

Thomas Y. Shih, Pothana Saikumar*, David J. Clanton and
Linda S. Ulsh

Laboratory of Molecular Oncology, National Cancer Institute
Frederick Cancer Research Facility, Frederick, MD. 21701
U. S. A.

INTRODUCTION

Since ras p21 of Harvey and Kirsten murine sarcoma viruses was
identified in 1979 and activated ras oncogenes were found in human tumors
in 1982, the growth of ras literature has been a truly impressive phenomenon
(Fig. 1). Considering a family of three human ras genes, each of which
encodes a protein of only 189 amino acid residues, the number of articles
dealing with this subject has reached more than 200 per year since 1986,
and before the end of 1988, the total number will soon reach the one thou-
sand mark. Beyond doubt this group of ras proteins have afforded us the
opportunity to peer into the most intimate secret of how cells control
their growth and differentiation, and the machinery of how cells communi-
cate among others. But the urgent drive on ras oncogene research has
been fueled by the promise that we eventually come to grip on the molecular
secret of the most dreadful disease, cancer. As it is always the case
in any area of scientific research, the more we open up the Pandora's
box, the more we want to know what else is in the box. Do we still leave
any stone untouched?

It is well known that v-ras p21 proteins of Harvey and Kirsten murine
sarcoma viruses are phosphorylated by their autokinase activities at the
threonine-59 accepter sites (Shih et al., 1982). From the recent three
dimensional structure determined by X-ray crystallography of ras proteins
(de Vos et al., 1988), it is apparent that this phosphorylation may help
maintaining the active GTP-bound conformation of ras proteins from relaxa-
tion into an inactive GDP-bound form by holding in place the γ phosphoryl
group at the threonine site after its cleavage from GTP by the GTPase
activity of p21. The oncogenic potential of v-ras may thus be enhanced
by this threonine mutation of the alanine residue of c-ras p21, in addition
to position 12 mutations of the v-ras p21 that further depress its GTPase
activity. It is interesting to note in this connection, that our earlier
observation indicates that the autophosphorylated p21 has a much longer
half life of metabolic turnover of 56 hr as compared to that of 20 hr
found for c-ras p21, suggesting a tighter association with its putative
targets (Ulsh and Shih, 1984).

---

*Present address: Wistar Institute of Anatomy and Biology, Philadelphia,
  PA 19104, U. S. A.

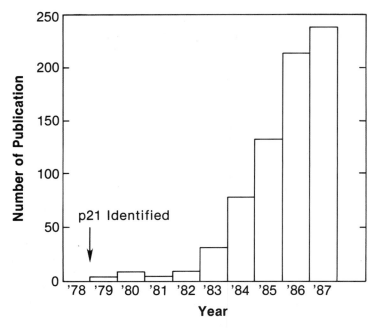

Figure 1. Growth of ras p21 literature.

In contrast to the strong phosphorylation of v-ras p21 of Harvey and Kirsten sarcoma viruses, very little phosphorylation was observed for c-ras p21, which lacks the autokinase site. We wish to report here, however, that in cells expressing high levels of ras proteins, novel phosphorylation on serine residues was found, and this phosphorylation was particularly prominent in p21 of K-ras(4B) gene. Both protein kinase C (PKC) and protein kinase A (PKA) appear to phosphorylate p21 in vitro and in vivo at common serine residues in the hypervariable regions close to p21 C-termini of, respectively, K-ras and H-ras genes (Saikumar et al., 1988).

STIMULATION OF c-ras p21 PHOSPHORYLATION BY PHORBOL ESTER AND cAMP

The murine early myeloid cell line, 416B, expresses a p21 of the c-K-ras gene, with exon 4B, at a level approximately 20 times higher than most ras-transformed cells (Scolnick et al., 1981; George et al., 1986). Fig. 2A shows that when this cell was incubated with [^{32}P]orthophosphate, p21 was labeled, and this phosphorylation was stimulated by phorbol dibuty-rate (PDBu), a PKC activator, by 3 to 5 fold. The increase was detectable within 5 min of treatment, and the protein level was unchanged (Fig. 2C). Definitive phosphophorylation was also detected in p21 of a NIH3T3 cell line, 18A, transformed by a LTR-linked c-H-ras gene (Fig. 2B). The effect of PDBu stimulation was, however, much smaller than that of K-ras p21. Interestingly, phosphorylation of the K-ras p21 in 416B cells was also strongly stimulated by activators of PKA, cAMP-dependent protein kinase. Fig. 3 shows the increased phosphorylation of p21 by treatment with permeable 8-bromo-cAMP or dibutyryl-cAMP. Again, the p21 level remains unchanged during this treatment.

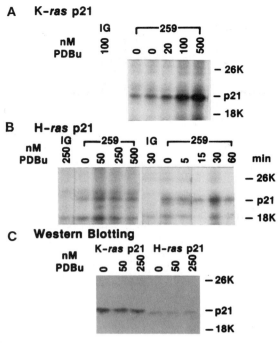

A K-*ras* p21

IG ┌──259──┐
nM 100 0 0 20 100 500
PDBu

— 26K
— p21
— 18K

B H-*ras* p21

IG ┌─259─┐ IG ┌───259───┐
nM 250 0 50 250 500 30 0 5 15 30 60 min
PDBu

— 26K
— p21
— 18K

C Western Blotting

K-*ras* p21 H-*ras* p21
nM 0 50 250 0 50 250
PDBu

— 26K
— p21
— 18K

Figure 2. Phosphorylation of c-ras p21s - stimulation by PDBu.
(A) 416B cells, expressing high levels of c-K-ras p21, were
labeled with [^{32}P]orthophosphate and p21 was immunoprecipitated
with monoclonal antibody, Y13-259, or with normal IgG as a
control. Cells were treated with 20, 100 and 500 nM PDBu for
30 min. Controls were treated with 100 nM 4- β -phorbol.
(B) H-ras p21 from 18A cells treated either with 0 to 500 nM
PDBu for 30 min (left) or with 100 nM PDBu for 0 to 60 min
(right). The p21 was immunoprecipitated with Y13-259 or with
control IgG.
(C) Western blot of p21 in 416B and 18A cells treated with 0 to
250 nM PDBu for 30 min. Equal amounts of proteins in each lysate
(180 ug) were resolved by NaDodSO$_4$-PAGE, and p21 was detected by
monoclonal antibody 259 and peroxidase-conjugated goat anti-rat
IgG after Western transfer.

BOTH PKC AND PKA PHOSPHORYLATE p21 AT THE SAME SITE(S) IN VITRO AND IN VIVO

Stimulation of p21 phosphorylation by phorbol ester and cAMP suggests
that PKC and PKA in some way mediate this phosphorylation. To determine
whether or not these protein kinases can directly phosphorylate p21, we
undertook detailed peptide mapping of the p21 phosphorylated in vitro with
purified protein kinases and compared these peptide maps to p21 phosphorylated in vivo by [^{32}P]orthophosphate labeling. Fig. 4 shows phosphorylation
by PKC of a recombinant H-ras p21 purified from E. coli overproducing this
protein under non-denaturing condition. It is important in our study that
p21 to be isolated without denaturation, otherwise, phosphorylation pattern
is completely different from that seen in p21 phosphorylated in vivo.
Phosphorylation was also obtained by PKA, a catalytic subunit of cAMP-
dependent protein kinase. Similar in vitro phosphorylation was performed
with K-ras p21 immunoprecipited from 416B cell lysates with monoclonal

121

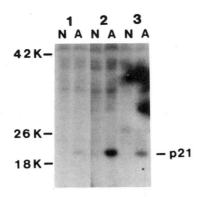

Figure 3. Phosphorylation of K-ras p21 - stimulation by
permeable cAMP derivatives. 416B cells were labeled for 4 hr
with [^{32}P]orthophosphate in the presence of 0.5 mM of 8-bromo-
cAMP or dibutyryl-cAMP. p21 was immunoprecipitated from equal
amounts of cell lysates with monoclonal antibody 259 (lanes A)
or normal IgG (lanes N). Lanes 1, control without added cAMP;
lanes 2, 8-bromo-cAMP; lanes 3, dibutyryl-cAMP.

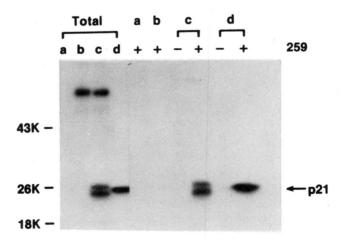

Figure 4. Phosphorylation of the recombinant H-ras p21 by PKC.
Purified p21 overproduced in E. coli was phosphorylated with
PKC using [γ -^{32}P]ATP as phosphoryl donor. Lanes a, reaction
mixture withough PKC; lanes b, reaction mixture without p21;
lanes c, complete reaction mixture; lanes d, p21 autophosphorylated by [γ -^{32}P]GTP. The left panel shows total phosphorylation products, and the right panel shows p21 immunoprecipitated
with monoclonal antibody 259 (+), or with normal IgG (-). The
marked p21 position indicates the p21 band stained with Coomassie
blue.

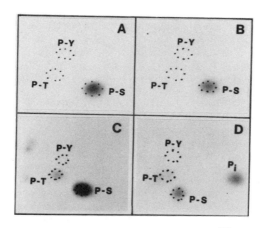

Figure 5. Phosphoamino acid analysis of [^{32}P]labeled p21.
(A) K-ras p21 in 416B cells. (B) H-ras p21 in 18 A cells.
(C and D) Recombinant H-ras p21 phosphorylated in vitro
with PKC; lower band (C) and upper band (D) of the doublet.

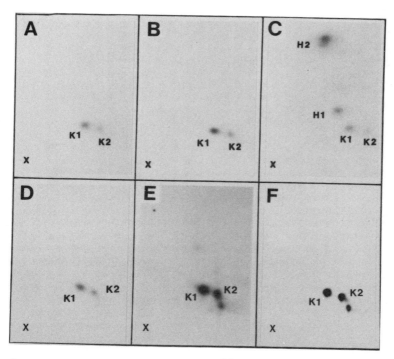

Figure 6. Tryptic peptide mapping of [^{32}P]labeled K-ras p21. (A) p21
phosphorylated in vivo in [^{32}P]labeled 416B cells. (B) PDBu-stimulated
phosphorylation in 416B cells. (C) A mixture of 416B K-ras p21 and
H-ras p21 phosphorylated by PKC. (D) p21 immunoprecipitated from 416B
·lysate and phosphorylated in vitro with [γ -^{32}P]ATP by PKC. (E)
Equal mixture of 416B p21 phosphorylated in vivo and in vitro by PKC.
(F) p21 from 416B lysate phosphorylated in vitro by PKA.

antibody 259 in the protein A-Sepharose complexes. As it is seen in Fig. 5, these phosphorylation reactions by protein kinases occurred at serine residues as were the case for p21 phosphorylated in vivo. Fig. 6 shows that tryptic peptide map of K-ras p21 phosphorylated in vitro by PKC (D) is identical to p21 phosphorylated in vivo at the basal level (A) and in cells treated with PDBu (B) by a mixing experiment (E). Intriguingly, the peptide map of K-ras p21 phosphorylated by PKA (F) is similar to that by PKC (E), suggesting that both protein kinases phosphorylate the same site(s) on p21.

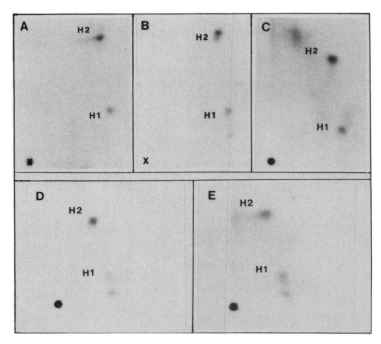

Figure 7. Tryptic peptide mapping of [^{32}P]labeled H-ras p21. (A) p21 phosphorylated in vivo in 18A cells labeled with [^{32}P]orthophosphate. (B) p21 phosphorylated in vitro by PKC, lower band of the doublet. (C) p21 phosphorylated in vitro by PKA. (D and E) From a separate set of experiments comparing p21 phosphorylated by PKC (lower band of the doublet) (D) to an equal mixture of the products of PKA and PKC (E). x indicates the origin of the two dimensional map. Horizontal, electrophoresis. Vertical, chromatography.

Observation on K-ras p21 can also be extended to H-ras p21. Fig. 7 shows that the tryptic peptide map of H-ras p21 phosphorylated in vivo (A) is similar to that phosphorylated in vitro by PKC (B) and PKA (C). The identity of PKC and PKA products are demonstrated by a mixing experiment (D and E). Phosphopeptides of H-ras p21, however, are different from those of K-ras p21 (Fig. 6C).

STRUCTURAL SIGNIFICANCE OF THE NOVEL PHOSPHORYLATION SITE(S)

There are 11 serine residues in H-ras p21 and 10 residues in K-ras (4B) protein. Most of these residues are located in p21 sequences that are identical between these two proteins. The preliminary peptide sequencing of H-ras p21 phosphorylated by PKC indicated that a major phosphorylation site, 7 residues following a trypsin cleavage, suggesting that serine-177 was the major site. We, therefore, constructed mutants altering the serine-177 into an alanine residue (clone 177A) or a cysteine residue (clone 177C). After overproducing these proteins in E. coli, purified mutant proteins were subjected to phosphorylation by protein kinases. Fig. 8 shows that mutation on serine-177 eliminates phosphorylation of H-ras p21 by either PKC (A) or PKA (B), while the wild type protein of c-H-ras was all phosphorylated. This observation confirms the results obtained by chemical sequencing of the phosphopeptides. This phosphorylation site lies in the hypervariable region close to the C-terminus of p21. By studying K-ras deletion mutants, Ballester et al. (1986) also

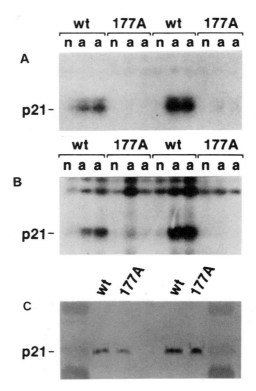

Figure 8. Lack of phosphorylation by PKC and PKA of H-ras p21 mutation at serine-177. Purified p21s of wild type (wt) c-H-ras and mutant 177A, derived from the former, were phosphorylated with [γ -^{32}P]ATP by PKC in (A) and by PKA in (B). The phosphorylated products were immunoprecipitated with monoclonal antibody 259 (lanes a) and with a normal IgG (lanes n). Results at the left side of both (A) and (B) are from reaction mixtures containing 0.1 ug p21, and those at the right contain 1 ug p21. (C) Western blot of p21 (0.2 ug at the left and 1 ug at the right) used in (A) and (B).

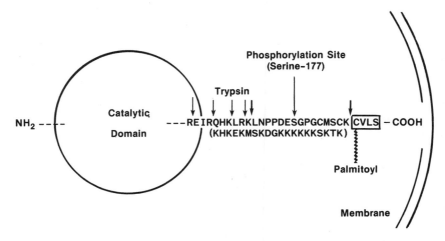

Figure 9. Novel phosphorylation site(s) of ras p21 common for both
protein kinases A and C. The amino acid sequence around the phosphorylation site at serine-177 of H-ras p21 region is shown. Small
arrows indicate trypsin cleavage sites. The homologous sequence of
K-ras(4B) p21 is shown in parenthesis with the phosphorylation
site at serine-182. This hypervariable region is flanked by a
membrane attachment site involving the palmitoylated cysteine-186
at the C-terminal region and the large globular catalytic domain
comprising the 164 amino acid residues of the p21 N-terminal region.

suggested serine-182 in the analogous hypervariable region of K-ras(4B)
p21 to be the major phosphorylation site (Fig. 9). Although p21 is a
highly conserved protein, amino acid sequences in this region differ
among the family of the three ras genes, and the hypervariable sequences
of yeast ras proteins are much longer than those of mammalian species.

The potential significance of the novel phosphorylation can best be
appreciated from the three dimensional structure of ras proteins (de Vos
et al., 1988; Shih et al., 1988). Fig. 9 shows that the hypervariable
region of p21 is the continuation of a long α helix extending outward
from the globular catalytic domain in the opposite direction from the
GTP-binding site. Conceivably, the hypervariable region of p21 functions
as a flexible joint connecting the catalytic domain to the membrane
attachment site, which involves palmitoylation of cysteine-186 at the p21
C-terminal region (Willumsen et al., 1984; Chen et al., 1985). Phosphorylation of serine residues in this region of ras proteins close to the
membrane attachment site can potentially modulate the transmission of
signals derived from interaction of ras proteins with membrane components
to the business region of the catalytic domain which interacts with other
effector molecules. Although the interaction of p21 with GAP, the GTPase
activating protein (Trahey and McCormick, 1987), at the residues 32-40
region of the catalytic domain is clear (Adari et al., 1988; Cales et al.,
1988; Sigal, 1988), the interaction with the putative receptor molecules
of the plasma membrane is not yet identified. The hypervariable regions
of ras proteins are the prime candidates. It is intriguing to notice that
the PKC phosphorylation site of epidermal growth factor receptor is at an
analogous site close to the transmembrane domain (Hunter et al., 1984;

Gentry et al., 1986); and phosphorylation sites by PKC and PKA of pp60^{src} are close to the N-terminal myristoylation site (Gould et al., 1985; Collett et al., 1979; Cross and Hanafusa, 1983). We hope that physiological studies of ras mutants of the novel phosphorylation sites will allow us to evaluate the biological significance of p21 phosphorylation by protein kinases. If the novel phosphorylation indeed has any physiological significance, it is interesting to note that multiple protein kinases can phosphorylate the same site(s). PKC and PKA have been construed to mediate cellular signaling pathways very often of opposite physiological consequences (Nishizuka, 1986). Phosphorylation of identical target sites by these two protein kinases suggests a potential mechanism of cross-talk between different signaling pathways.

ACKNOWLEDGMENTS. We thank Drs. Takis S. Papas and Richard Ascione for their support of this study, Drs. James Lautenberger and Takis Papas for the generous use of the pJL6 expression vector, Dr. Kuo-Ping Huang for his help on C-kinase studies, Dr. Arun Seth for help in constructing mammalian expression vectors, and Dr. Donald Blair for advice on DNA transfection.

REFERENCES

Adari, H., Lowy, D. R., Willumsen, B. M., Der, C. J. and McCormick, F., 1988, Guanosine triphosphatase activating protein (GAP) interacts with the p21 ras effector binding domain, Science, 240: 518-521.

Ballester, R., Furth, M. E. and Rosen, O. M., 1987, Phorbol ester and protein kinase C-mediated phosphorylation of the cellular Kirsten ras gene product, J. Biol. Chem., 262: 2688-2695.

Cales, C., Hancock, J. F., Marshall, C. J. and Hall, A., 1988, The cytoplasmic protein GAP is implicated as the target for regulation by the ras gene product, Nature, 332: 548-551.

Chen, Z. Q., Ulsh, L., DuBois, G. and Shih, T. Y., 1985, Post-translational processing of p21 ras proteins involves palmitylation of the C-terminal tetrapeptide containing cysteine-186, J. Virol., 56: 607-612.

Collett, M. S., Erikson, E. and Erikson, R. L., 1979, Structural analysis of the avian sarcoma virus transforming protein: sites of phosphorylation, J. Virol., 29: 770-781.

Cross, F. R. and Hanafusa, H., 1983, Local mutagenesis of Rous sarcoma virus: the major sites of tyrosine and serine phosphorylation of pp60^{src} are dispensible for transformation, Cell, 34: 597-607.

de Vos, A. M., Tong, L., Milburn, M. V., Matias, P. M., Jancarik, J., Noguchi, S., Nishimura, S., Miura, K., Ohtsuka, E. and Kim, S. H., 1988, Three-dimensional structure of an oncogenic protein: catalytic domain of human c-H-ras p21, Science, 239: 888-893.

Gentry, L. E., Chaffin, K. E., Shoyab, M. and Purchio, A. F., 1986, Novel serine phosphorylation of pp60^{c-src} in intact cells after tumor promoter treatment, Mol. Cell. Biol., 7: 735-738.

George, D. L., Glick, B., Trusko, S. and Freeman, N., 1986, Enhanced c-Ki-ras expression associated with Friend virus integration in a bone marrow-derived mouse cell line, Proc. Natl. Acad. Sci., USA, 83: 1651-1655.

Gould, K. L., Woodgett, J. R., Cooper, J. A., Buss, J. E., Shalloway, D. and Hunter, T., 1985, Protein kinase C phosphorylates pp60^{src} at a novel site, Cell, 42: 849-857.

Hunter, T., Ling, N. and Cooper, J. A., 1984, Protein kinase C phosphorylation of EGF receptor at a threonine residue close to the cytoplasmic face of the plasma membrane, Nature, 311: 480-

Scolnick, E. M., Weeks, M. O., Shih, T. Y., Ruscetti, S. K. and Dexter, T. M., 1981, Markedly elevated levels of an endogenous sarc protein in a hemopoietic precursor cell line, Mol. Cell. Biol., 1: 66-74.

Shih, T. Y., Clanton, D. J., Saikumar P., Ulsh, L. S. and Hattori, S., 1988, Structure and function of ras p21: studies by site-directed mutagenesis, In THE GUANINE-NUCLEOTIDE BINDING PROTEINS, Eds. Bosch, L., Kraal, B. and Parmeggiani, A., NATO ASI SERIES, Plenum Press, in press.

Shih, T. Y., Stokes, P. E., Smythers, G. W., Dhar, R. and Oroszlan, S., 1982, Characterization of the phosphorylation sites and the surrounding amino acid sequences of the p21 transforming proteins coded for by the Harvey and Kirsten strains of murine sarcoma viruses, J. Biol. Chem., 257: 11767-11773.

Sigal, I. S., 1988, A structure and some function, Nature, 332: 485-486.

Trahey, M. and McCormick, F., 1987, A cytoplasmic protein stimulates normal N-ras p21 GTPase, but does not affect oncogenic mutants, Science, 238: 542-545.

Ulsh, L. S. and Shih, T. Y., 1984, Metabolic turnover of human c-ras[H] p21 protein of EJ bladder carcinoma and its normal cellular and viral homologs, Mol. Cell. Biol., 4: 1647-1652.

Willumsen, B. M., Norris, K., Papageorge, A. G., Hubbert, N. L. and Lowy, D. R., 1984, Harvey murine sarcoma virus p21 ras protein: biological and biochemical significance of the cysteine nearest the carboxy terminus. EMBO J., 3: 2581-2585.

THE ROLE OF p21*ras* IN PROLIFERATIVE SIGNAL TRANSDUCTION

Dennis W. Stacey, Chun-Li Yu, Fu-Sheng Wei, and Men-Hwei Tsai

The Department of Molecular Biology
The Cleveland Clinic Foundation
9500 Euclid Ave.
Cleveland, Ohio 44106

While numerous evidences indicate that oncogenes or their cellular proto-oncogene counterparts play critical roles in the control of cellular proliferation, there is little indication how these molecules function or interact with each other. Our goal is to relate the biochemical and molecular characteristics of the *ras* proto-oncogene to its biological function within living cells. This effort relies upon a monoclonal antibody able to neutralize the biological activity of *ras* following microinjection into living cells. With this antibody we have effectively deprived the injected cells of *ras* activity and then studied the biological consequences of this treatment. The data obtained not only confirm the critical importance of *ras* activity during cellular proliferation; but suggests that p21*ras* might have a role in the function of other proto-oncogenes. Finally, recent evidence might even indicate the means by which the activity of p21*ras* is controlled during normal mitogenic stimulation. The implications of this possibility are discussed.

IDENTIFICATION OF A NEUTRALIZING ANTI-*ras* ANTIBODY

Neutralization of a cellular protein by microinjected antibody followed by the characterization of the biological consequences of this treatment is a relatively new approach. While the potential is great, there are several obvious and perhaps other not-so-obvious cautions which must be considered. To begin with, it is critical to ensure that the antibody not only recognizes the protein in question but that it inactivates the biological activity of the intracellular target protein completely, and for the duration of the experimental procedure. Then it must be shown that the biological consequences of this inactivation are due to inactivation of the target protein *alone*. We were fortunate to have identified such an antibody for the study of *ras* activity. This monoclonal antibody, Y13-259 (antibody 259) was originally prepared by Furth and coworkers (1982). It was one of several monoclonals prepared to precipitate p21*ras*, but had the ability to recognize different members of the *ras* gene family. Recent studies indicate that the epitope recognized by antibody 259 is located near the amino-terminus of the protein in a sequence which forms an extended loop on the external surface of the protein near its unalterable effector domain (Adari, et al., 1988; Cales, et al., 1988).

When antibody 259 is injected into cells transformed by a variety of *ras* proteins, the cells revert from the morphology of transformed cells to that of nontransformed cells and stop actively proliferating. This inhibition of the phenotype of transformation persists for 30 to 40 hrs until the antibody is effectively removed from the recipient cell (which then returns to the previous transformed phenotype). The antibody is able to neutralize the transforming capacity of coinjected, bacterially synthesized *ras* protein, but only if the antibody is in molar excess (Kung, et al., 1986). Biochemical experiments by others suggest that the antibody has the ability to block nucleotide exchange by *ras* protein (Lacal and Aaronson, 1986; Hattori, et al., 1987). Other monoclonal antibodies isolated at the same time as 259 but without neutralizing potential lack this ability to inhibit nucleotide exchange. The fact that the antibody is present within injected cells for at least 24 hrs is indicated by staining of injected cultures with a fluorescent anti-immunoglobulin antibody.

As evidence that the antibody promotes these effects by virtue of its interaction with *ras* protein alone, it was injected into cells transformed by a viral *ras* gene engineered to lack the epitope recognized by the antibody. In these cells little inhibition was observed (Papageorge, et al., 1986). Finally, several tumor cells were studied. The ability of the injected antibody to inhibit the proliferation of these cells correlated closely with a mutation in the cellular *ras* gene (Stacey, et al., 1987). These two latter observations would not be expected if the antibody were interacting with another gene product within the cell. Taken together these observations tend to support the requirements outlined above for an antibody able to specifically and efficiently inhibit an intracellular function. Unfortunately, in a new area such as this it might be impossible to anticipate all potential problems. For this reason caution must be exercised in the intrepretation of these data until these results are verified by an independent approach and the general procedure of antibody microinjection is better understood. For the purpose of this analysis, however, there will be no further discussion of the injected antibody. It will be assumed that its injection specifically and efficiently eliminates *ras* activity from the recipient cell for at least 24 hrs.

THE REQUIREMENT FOR *ras* ACTIVITY DURING MITOGENIC STIMULATION

Because it appears that antibody 259 is able to eliminate *ras* activity from within injected cells, experiments were performed to determine the role of p21*ras* during mitosis. For this experiment NIH3T3 cells were deprived of serum and thereby rendered quiescent. Antibody 259 was microinjected into cells within a defined region of the culture and the culture was treated with 10% serum to initiate a cycle of DNA synthesis. After thymidine labelling for the next 18 to 24 hrs it was apparent that the antibody had efficiently inhibited the injected cells from synthesizing DNA. Control antibody had no effect upon thymidine labelling. This experiment was then performed by adding serum to quiescent cells followed by microinjection of anti-*ras* antibody at various times thereafter. Even when the antibody was injected up to 6 hrs after addition of serum it was able to inhibit thymidine labelling with good efficiency. In these cultures DNA synthesis began at 8 hrs after the addition of serum and by 10 hrs most cells had entered S-phase. After 8 to 10 hrs following serum addition the injected antibody lost its ability to inhibit thymidine labelling (Mulcahy, et al, 1985).

It appears therefore that *ras* activity is required for the initiation of a cycle of DNA synthesis but is no longer needed after DNA synthesis has begun. This observation was confirmed when anti-*ras* antibody was injected into cells of a nonsynchronized culture. Cells labelled within 10 hrs of such injections were able to incorporate labelled thymidine efficiently, while those labelled at later times were efficiently inhibited from incorporating thymidine. The cells labelled soon after injection would most likely be those already in S-phase at the time of injection. Since *ras* is apparently not needed for the process of DNA synthesis these cells were not affected. At later times, however, a cycle of DNA synthesis in progress at the time of injection would have terminated, cells would have to initiate a new cycle of DNA synthesis to be able to incorporate labelled thymidine at this time. This reinitiation had apparently been efficiently inhibited. This type of observation has been made in numerous lines of non-transformed epithelial and fibroblast-like cells (Mulcahy, et al., 1985).

On the basis of these observations we concluded that *ras* activity is required for the initiation of a new cycle of DNA synthesis and that its requirement occurs just prior to this initiation (Figure 1). It is also clear that *ras* is not needed during DNA synthesis. It is not clear, however, if there is a requirement for *ras* activity at only a single time during the cell cycle. The data above do not rule out the possible requirement for *ras* during G-2 phase or G-0 phase of the cell cycle.

ras ACTIVITY AND THE FUNCTION OF OTHER ONCOGENES

In order to understand the function of cellular p21*ras*, it will probably be necessary to understand its relationship to other oncogenes and proto-oncogenes. We explored the possibility that an interrelationship exists between the action of *ras* and other oncogenes by obtaining NIH3T3 cells transformed by a variety of retroviral

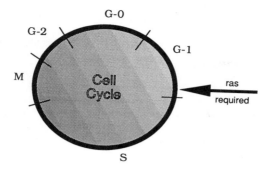

Fig. 1. Requirement for *ras* activity during the cell cycle. The cell cycle is illustrated here with a circle divided arbitrarily into segments representing its different phases. The experiments described with anti-*ras* injections indicate that *ras* activity is required just prior to the initiation of the DNA synthetic or S-phase. It is clear that *ras* activity is not required during S-phase but it is possible that *ras* activity is also required at other times in the cell cycle.

oncogenes. These cells then received anti-*ras* injections and were pulsed with thymidine at 18-24 hrs after injection. This experiment was performed on cells transformed by oncogenes of three different types. First, tyrosine kinase, plasma membrane-localized oncogenes were tested. These are similar to (or are derived from) the genes for growth factor receptors. In each case, the injected anti-*ras* antibody induced a reversion from the transformed morphology to a morphology similar to non-transformed cells. In addition, the thymidine labelling of the injected cells was reduced by approximately 80%. It was apparent that the antibody had blocked the oncogene from expressing its transforming potential within these cells. This result was similar to that observed following injection into *ras*-transformed cells, except that in the tyrosine kinase-containing cells the inhibition of thymidine labelling was perhaps not quite as efficiently inhibited as observed in nontransformed NIH cells or in cells transformed by *ras*. It is therefore possible that the activity of these oncogenes is not completely inhibited by the antibody (Smith, et al., 1986).

Cells transformed by the *sis* oncogene (which is derived from the platelet derived growth factor gene) behaved exactly as the tyrosine kinase genes described above. This might not be surprising since a growth factor would be expected to work through its receptor. If the receptor requires *ras* in order to function, it is not surprising that the growth factor also is *ras*-dependent. The third class of soluble, serine kinase-containing oncogenes, however, yielded entirely different results. When anti-*ras* was injected into these cells there was no apparent alteration in the transformed cell morphology, nor in the ability of the cells to incorporate thymidine. When the actual percentages of incorporation of thymidine were calculated the results were clear. The inhibition of tyrosine kinase oncogenes was near 80% while that of the serine kinase oncogenes was consistently below 10%. The anti-*ras* antibody had clearly distinguished between the activity of these two different classes of oncogenes (Smith, et al., 1986). The basis for this distinction is not known but might suggest the differences in the mechanism of actions of these two types of oncogenes.

The distinction between the groups of oncogenes described above in their requirements for cellular *ras* activity is the basis of a model of the interaction of proto-oncogenes during mitogenic stimulation. We propose that proliferation in normal cells is controlled by the sequential action of different classes of proto-oncogenes. Proliferation is initiated by the interaction between a growth factor and its tyrosine kinase-containing, plasma membrane-localized receptor. The occupied receptor initiates a signal which then passes to cellular *ras* proteins also located on the plasma membrane (Figure 2). This part of the model is based upon direct experimental evidence. It is directly shown by the experiments above that the complex mixture of growth factors present in serum require cellular *ras* activity to induce proliferation. This is also the case with an oncogene related to a growth factor and to the oncogenes related to growth factor receptors (so far tested). It is possible, however, that rather than functioning directly upstream of *ras*, that the tyrosine kinases and *ras* function in parallel but are mutually required for the initiation of a proliferative signal (Figure 3).

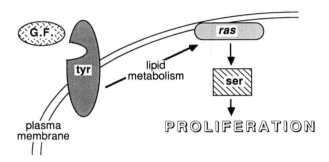

Fig. 2. Model for proliferative signal transduction. Our model is presented to explain the data obtained when anti-*ras* antibody was microinjected into cells transformed by different oncogenes. We propose that proliferation in normal cells is controlled by a cascade of proto-oncogenes. The signal is initiated by the interaction of a growth factor with its tyrosine kinase-containing, transmembrane receptor proto-oncogene. The proliferative signal then proceeds to *ras* proteins which function to transfer the signal to cytoplasmic serine kinase proto-oncogenes (from Stacey, et al, 1978).

The role of the serine kinase proto-oncogenes, however, is a matter of more speculation since there is no direct experimental evidence concerning them at present. We postulate that they function downstream from cellular *ras* to receive its proliferative signal (Figure 2). This would explain why they do not require cellular *ras* activity to transform cells. It is possible, however, that these genes function in an entirely different proliferative signalling system than *ras* or growth factors (Figure 3). More data will be required to distinguish between these possibilities and to determine more fully the interrelationship between the actions of different proto-oncogenes. This study would be greatly facilitated by an antibody able to neutralize the cellular *raf* gene. Unfortunately, considerable effort to obtain such an antibody has so far been unsuccessful.

If our model of proliferative signal transduction (Figure 2) is incorrect, and multiple pathways for proliferative control do exist (Figure 3), it is not apparent from the work described initially where anti-*ras* antibody efficiently blocked proliferation in numerous cell types. If there were a separate *ras*-independent means of inducing proliferation it apparently was not active to a great extent in any of the normal cells tested. On the other hand, to a limited extent there is clear evidence for a *ras*-independent signalling system. In the tyrosine kinase-transformed cells, as mentioned above, there was consistently less inhibition by injected anti-*ras* antibody than in non-transformed cells. The differences were only 10% but were observed with such consistency that the differences might be important (Smith, et al., 1986). In addition, tumor cells often proliferate independently of *ras* activity (Stacey, et al., 1987). This might be due to the activation of oncogenes downstream of *ras* or the activation of

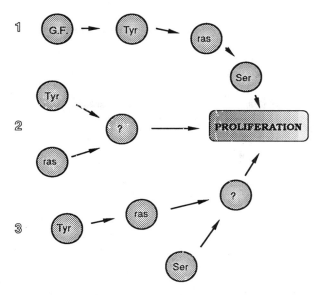

Fig. 3. Multiple possible signalling pathways. While the model presented in Figure 2 (represented here as pathway 1) adequately explains the data obtained, the data do not rule out other explanations. Some of these other possibilities are presented. It is possible, for example, that *ras* and tyrosine kinases (tyr) function in separate but mutually required pathways leading to proliferation (pathway 2). Even if the tyrosine kinases and *ras* do function in the same pathway, it is possible that the serine kinase (ser), cytoplasmic oncogenes function in a pathway separate from that containing *ras* (pathway 3).

oncogenes in a separate pathway. If two pathways do function, however, the problem is then to find the proteins which function upstream of the serine kinase oncogenes and those which function downstream of *ras*.

ROLE OF LIPIDS IN PROLIFERATION

It is clear from the data described above that mitogenic stimulation normally involves the action of cellular *ras* proteins. It is also clear from many other studies that mitogenic stimulation often involves dramatic and rapid changes in the metabolism of lipids, primarily but not exclusively phosphatidylinositol (Berridge, 1984). Because of the alterations in lipid metabolism during mitogenesis, it has been postulated that *ras* proteins might actually function to control the activity of a phospholipase. With the anti-*ras* antibody, we were in a position to test this possibility. To do so, we attempted to treat cells in such a way as to duplicate the action of several different lipases.

For example, phospholipase C generates diacylglycerol and inositol phosphates (which activates protein kinase C and induces calcium mobilization within the cell). The action of phospholipase C was duplicated by the addition to cells of phorbol esters and a calcium ionophore (which duplicate the action of diacylglycerol and inositol phosphates respectively). In addition, the action of phospholipase A_2 was duplicated by the addition of prostaglandin F2α. This phospholipase functions to cleave arachidonic acid from phospholipids. Arachidonic acid production is a rate limiting step in the production of prostaglandins. The experiment was performed by microinjecting anti-*ras* antibody into cells followed by the treatments described above. If *ras* proteins function by controlling the activity of either of these phospholipases, we expected that the injected antibody would no longer be able to block the proliferation induced by treatments designed to duplicate their action. In other words, if we were able to bio-chemically duplicate the action of *ras* proteins, then *ras* activity would not be needed.

The results suggested, however, that *ras* does not function to control the action of either of these phospholipases. Instead of overcoming the requirement for *ras* activity, the treatments described induced proliferation which was efficiently inhibited by injected antibody. In fact, the degree of inhibition observed was even greater than seen with serum-treated cells. It is known that phosphatidic acid is highly mitogenic in certain cell types. We therefore tested the possibility that *ras* might function to control the production of this lipid. Again cells were treated with phosphatidic acid following injection of anti-*ras* antibody. As before, the phosphatidic acid-induced proliferation was efficiently inhibited by the injected antibody. In the case of phosphatidic acid, the highly efficient proliferation observed was almost totally inhibited (99%) by the antibody (Yu, et al., 1988). We were therefore unable to eliminate the requirment for *ras* activity by treatment of cells with lipid related materials. It therefore appeared that *ras* does not function to control the activity of any of the phospholipases tested. In fact, the dramatic inhibition by anti-*ras* antibody of lipid-induced proliferation suggested that rather than controlling the action of lipases, *ras might itself be controlled by phospholipid metabolites.*

One of the attractive aspects of our model in which *ras* functions to transfer a proliferative signal between tyrosine and serine kinase proto-oncogenes is its similarity to other signalling systems (Gilman, 1984). If this model were correct it would fit into the general pattern of signalling systems involving a ligand (growth factor) binding to a transmembrane receptor (tyrosine kinase proto-oncogene) which initiates a signal involving the action of a GTP binding and hydrolyzing protein (*ras*) which transfers the signal to a cytoplasmic kinase (serine kinase proto-oncogene). In other systems of this type, however, the GTP binding (G) protein is associated with other peptides which interact with the receptor and participate in the activation of the G protein. In the case of *ras*, however, there are no known associated proteins. It is unclear, there-fore, how *ras* might be activated. It is therefore likely that a novel means of activa-tion, such as that suggested above, might operate.

The possibility that *ras* proteins might be controlled by the action of phospho-lipases is a novel suggestions which obviously must be substantiated by alternative procedures. If, however, tyrosine kinase oncogenes do function by initiating a prolif-erative signal which must be received by *ras* proteins, and if *ras* proteins are controlled by the action of phospholipases, then it might be possible that the means of signalling between tyrosine kinases and *ras* is at the level of lipid metabolism. There are inde-pendent suggestions that tyrosine kinases might function by altering lipid metabolism. It is known, for example that tyrosine kinase oncogenes and growth factor receptors do at times bind tightly to a phosphatidylinositol kinase (Whitman, et al., 1988). In addi-tion, the nucleotide sequence of one phospholipase C molecule is similar to a sequence known to be involved in the activity and presumably the control of some tyrosine kinase oncogenes (Meyer, et al., 1988; Stahl, et al., 1988). It is known that phospho-lipase C is phosphorylated on tyrosine as a result of growth factor addition in one cell line (Wahl, et al., 1988). Finally, one of the primary substrates of tyrosine kinases are lipocortins or calpactins which might function to control lipid metabolism (Brugge, 1986). While the biological significance of these observations are not known, it does make a direct correlation between tyrosine kinases and lipid metabolism as suggested.

LIPIDS AND GAP ACTIVITY

The data discussed above suggests that tyrosine kinases, growth factor receptors, *ras* proteins, and lipid metabolism might function together during mitogenic stimula-tion. This is only a suggestion, however, which must be verified by independent means. On the basis of these observations we considered the possibility that *ras* proteins might be directly influenced by certain lipids. It is known that mutations of *ras* which induce transformation are those which tend to increase the likelyhood that the protein will be associated with GTP rather than GDP. This can be accomplished by increasing the basal rate of nucleotide exchange or by reducing the normal rate of GTP hydrolysis. In our initial efforts to demonstrate an influence of lipids we tested their ability to alter either of these two biochemical properties. These efforts, how-ever, were unsuccessful. While this work was in progress it was reported that the basal rate of GTPase was strongly induced by a common cellular protein called GTPase acti-vating protein (GAP), whose identification and characteristics are discussed in detail in other chapters of this volume. The identification of GAP added a new dimension to the studies of the influence of lipids upon *ras* proteins. We addressed the possibility that lipids might directly influence the interaction between *ras* and GAP.

To test this possibility crude GAP preparations were incubated with purified, bacterially synthesized *ras* proteins in the presence and absence of various lipids. The results were quite clear. Some of the lipids tested did indeed inhibit the activity of GAP. The experiments involved binding ^{32}P-labelled GTP to *ras* protein followed by incubations with GAP. The extent of GTP hydrolysis was then determined by precipitating the *ras* protein with an antibody and determining if the bound nucleotide had been hydrolyzed. The GAP protein alone induced the rapid and often nearly complete conversion of bound GTP to GDP. In the presence of phosphatidic acid, however, this conversion was greatly diminished (Tsai, et al., in press). The activity of GAP had been inhibited and the *ras* protein was much more likely to be linked to GTP following the incubation. If this reaction were an indication of the situation existing within the cell, it is clear that the effect of the lipid would be to increase the likelyhood that cellular *ras* proteins were bound to GTP and were therefore active biologically. There is no direct evidence, however, that the interaction between GAP and phosphatidic acid has any biological significance.

Further analyses were performed to determine the types of lipids able to influence GAP activity. Only phosphatidic acids with unsaturated fatty acids had activity in the assay described above. While phosphatidic acid with saturated side chains is most active in biologically stimulating mitosis in NIH3T3 cells (Yu, et al., 1988), this molecule had no apparent effect upon the in vitro interaction between *ras* and GAP. Other phospholipids were also tested. The most common cellular lipids were inactive, including phosphatidylcholine and phosphatidylserine. Phosphatidylinositol was inhibitory to GAP but not as inhibitory as were phosphatidylinositol mono- and diphosphates. Phosphatidylinositol normally contains arachidonic acid in the #2 position as did the most inhibitory phosphatidic acid. The apparent correlation between inhibition of GAP and the presence of an arachidonic acid prompted the analysis of free fatty acids. We found that free arachidonic acid effectively inhibited GAP activity. Other fatty acids were also inhibitory but with less than 50% the activity of arachidonic acid. As with the phosphatidic acids tested, the free fatty acids with unsaturated side chains were totally inactive in the GAP assay (Tsai, et al., in press).

THE POSSIBLE SIGNIFICANCE OF LIPID-GAP INTERACTIONS

The possible significance of the above observations are clear. If lipids do normally control the activity of GAP they may be directly involved in controlling the activity of cellular *ras* proteins and thereby of controlling mitosis. This would fit with the predictions made on the basis of the anti-*ras* antibody injections described above. In fact, the apparent interaction between GAP and lipids serves to substantiate the model formulated to describe those injection experiments (Figure 1). It is clear, however, that much further study will be require before any conclusions can be made regarding the biological significance of the apparent lipid-GAP interaction. Even in the absence of further data, it is possible to consider some of the consequences and uncertainties concerning these results.

The first and most serious consideration regarding the biochemical studies with lipids is the possibility that there is a trivial explanation for the results. The possibility of such a problem is emphasized by the early studies of lipid interaction with purified *ras* protein. Before the identification of GAP we had treated purified *ras* proteins with lipids and found as we expected that the lipids did influence the GTP binding properties of the protein. This experiment was repeated numerous times. The rate of binding was increased as was the extent of binding. The ratio of GTP to GDP was altered as predicted, and there was apparently some influence of the lipid on the GTPase activity of the protein. This interaction was totally dependent upon the presence of calcium ions in the reaction. The initial excitement of this first observation was dashed, however, when we found that a low concentration of phosphate in the *ras* protein preparation was responsible for the results seen. We had used a filter binding assay to determine the extent of GTP binding to *ras* protein. The slight amount of phosphate in the enzyme solution had apparently reacted with the calcium in the reaction mixture to yield a precipitate. This precipitate had apparently included not only the phosphatidic acid but the GTP in solution to yield a labelled complex which was retained on the filter. This artifact was further complicated by the fact that the kinetics and even the extent of the inclusion of labelled GTP in the complex fit exactly with those expected for the binding of *ras* protein to GTP. The reaction required the phosphatidic acid, no other lipid would suffice.

There are several observations with GAP that suggest the possibility of such errors even in this present study. It is known for example that salt interferes with the activity of GAP protein. This might be an indication of the physical requirements for the interaction between *ras* and GAP proteins. On the other hand, ammonium ions seem to be particularly effective in inhibiting GAP activity. It is possible that some unidentified component of the reaction mixture is responsible for GAP inhibition. In an attempt to rule out this possibility the lipids used have been purified and carefully characterized. In addition, care was taken to ensure that the ionic strength of the reaction is near physiological levels. Furthermore, there seems to be a definite and reproducible requirement in the structure of the active lipids. If a contaminant were responsible for the inhibition observed, it must have a mobility on silica TLC plates identical to phosphatidic acid, and be present only in lipid preparations with unsaturated fatty acids.

The nature of the lipids found to be inhibitory is consistent with their participation in the control of mitogenesis. All the lipids found to be inhibitory have been found to be rapidly metabolized during mitogenic stimulation. This includes arachidonic acid. Furthermore, many of the active lipids are mitogens themselves. It is important to point out, however, that the most mitogenically active lipid, phosphatidic acid with saturated fatty acids, is totally inactive in inhibiting GAP. Furthermore, the fact that these lipids are mitogenically active does not confirm that they are present in high enough amounts and in the proper membrane locations to affect the activity of *ras* and GAP proteins within cells. Phosphatidylinositol is inhibitory but present in the plasma membrane all the time. This lipid must not, therefore, be directly involved in the control of *ras* activity. It is possible that this lipid, like salt concentrations within the cell, functions to set a basal level of GAP activity from which changes must occur.

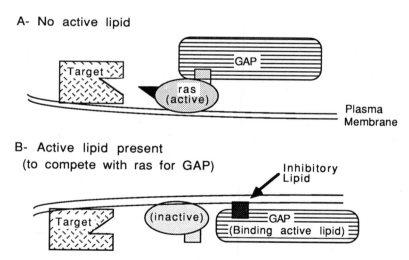

Fig. 4. Model for the action of GAP to regulate *ras* activity. The observation that lipids might alter the natural interaction between GAP and *ras* proteins might be interpreted to indicate that GAP functions to control *ras* activity. In this way, GAP would induce *ras* protein GTPase activity and thereby inactivate *ras* biologically. The production of critical lipids might then block this activity of GAP leading to the activation of *ras* and its positive interaction with a target which then functions to propagate the proliferative signal into the cell. In this scheme, GAP functions only to regulate the activity of *ras*.

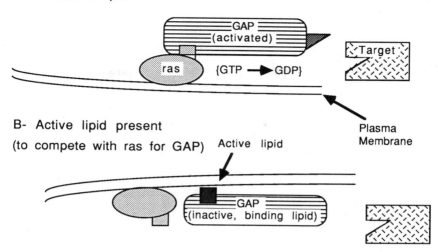

A- No active lipid

B- Active lipid present
(to compete with ras for GAP) Active lipid

Fig. 5. Model for activation of GAP by *ras* proteins. The model presented in Figure 4 is consistent with the data described where lipids interfere with the interaction between GAP and *ras*. The data presented do not, however, rule out the possibility that it is *ras* proteins which function to regulate the activity of GAP. In this way, GAP is able to interact with the downstreatm effector only when associated with a *ras* protein bound to GTP. This interaction renders GAP able to send a positive proliferative signal, and might also result in the hydrolysis of GTP by *ras* proteins. The active lipids then might complete with GAP for binding to *ras* proteins. When bound to the lipid the GAP would be inactive.

We have not observed any effect of lipids upon the exchange rate of GTP or GDP bound to *ras* proteins. This rate is quite slow. It might be argued that if only the rate of hydrolysis is altered during mitogenic stimulation then this stimulation would need to be very slow. For example, assume that 50% of all *ras* proteins must be bound to GTP to stimulate a biological response. Even if GTPase were totally inhibited, it might take 20 min for enough bound GDP to be replaced by GTP to initiate a positive response. While the rate of action of *ras* is not known, there are several suggestions that it occurs more rapidly than this. On the other hand, if a lower percentage of *ras* must be bound to GTP to be active, then the time constraints would be different. If only 1% of cellular *ras* need be associated with GTP to initiate a positive response, then this response might occur within one minute of the inhibition of GTPase. In direct observations reported by F. McCormick and coworkers very little *ras* was found to be bound to GTP even in mitogenically stimulated cells. It might therefore be possible that only a small amount of *ras*-GTP is required to induce proliferation, and that a rapid biological response from cellular *ras* proteins might be induced by simply altering the rate of hydrolysis. This would require that the rate of hydrolysis be much greater than the rate of exchange in the absence of mitogenic stimulation.

A final consideration relates to the possible sequence of events in the action of *ras* and GAP during mitogenic stimulation. It has been proposed that GAP functions downstream of *ras* because mutations in the effector domain of *ras* normally render the protein unresponsive to GAP. GAP might therefore be the effector of *ras* (Adari, et al, 1988; Cales, et al, 1988). On the other hand, numerous other mutations including many of those which activate *ras* to a transforming oncogene also block its interaction with GAP. It is possible that each of these mutations include an alteration in another region of the *ras* protein which is involved in GAP binding, or that GAP and another effector bind to similar areas of the protein. It has therefore not been conclusively demonstrated whether *ras* functions to modulate GAP activity or visa versa. Even if lipids do normally function to control the activity of *ras* by inhibiting GAP, it is possible that GAP might function either upstream or downstream of *ras*. Thus it is possible that GAP functions to simply down-regulate *ras* in a controllable way. When GAP activity is blocked by a lipid, for example, *ras* would become active and would interact with the appropriate effector molecule (Figure 4). If GAP were the downstream effector of *ras*, however, interaction between the two proteins, presumably resulting in GTP hydrolysis, would still be required. If this interaction were disrupted by a lipid, then the lipid could control production of the resulting proliferative signal (Figure 5).

While the distinction between GAP acting prior to or following *ras* during mitogenic stimulation cannot be resolved by the experiments with lipids described above, these two possibilities do have important implications. If it is assumed that GAP is directly involved with *ras* in proliferative signalling, there are two functions which this interaction might have. First, the two proteins might act simply to biochemically amplify the original signal. If GAP acts to modify *ras* activity, a few molecules acting upon GAP might result in the activation of many *ras* molecules and an amplification of the signal. This possibility would be most likely if GAP does not act as the effector of *ras*. On the other hand, the signal might be sufficiently strong upon reaching *ras*, but GAP might function to modify the response. Consider for example that there might be multiple positive and even perhaps negative proliferative signals generated within a cell. GAP might function to process all the signals within a cell and determine on the basis of these whether or not the cell should divide. If the accumulated signals are positive, GAP interacts with *ras* resulting in GTP hydrolysis (in the case of normal *ras* proteins) and passage of the signal. In this case, one *ras* protein might be required to activate a single GAP protein, thereby resulting in no amplification of the signal. The fact that multiple lipids interact with GAP might favor the latter possibility. It is possible that different mitogenic stimuli might generate different lipids which together interact with GAP. GAP then would respond by generating a signal based upon the cumulative input it receives.

SUMMARY

The study of oncogenes will require both a molecular understanding of their gene products and a biological understanding of how these function. We have studied the biological function of the cellular *ras* protein by microinjecting a neutralizing antibody into living cells. The results suggest that *ras* is required for proliferation. It appears that *ras* might function by receiving proliferative signals originating from growth factor receptors. A model to explain how proto-oncogenes function together to control proliferation is based upon this observation and suggests that *ras* proteins might function to transfer a proliferative signal from growth factor receptors to cytoplasmic proto-oncogenes with serine kinase activity. The action of phospholipids appears to be directly related to activity of *ras*. We proposed, therefore, that lipid metabolism might be the biochemical link between the action of growth factor receptors and *ras* proteins. In support of this possibility we have found that the interaction between GAP and *ras* is strongly influenced by a specific set of lipids.

REFERENCES

Adari, H., Lowy, D.R., Willumsen, B.M., Der, C.J. and McCormick, F., 1988, Guanosine triphosphatase activating protein (GAP) interacts with the p21 *ras* effector binding domain, Science, 240:518.

Berridge, M.J., 1984, Inositol triphosphate and diacylglycerol as second messengers, Biochem. J., 220:345.

Brugge, J.S., 1986, The p35/p36 substrates of protein-tyrosine kinases as inhibitors of phospholipase A_2, Cell, 46:149.

Cales, C., Hancock, J.F., Marshall, C.J. and Hall, A., 1988, The cytoplasmic protein GAP is implicated as the target for regulation by the *ras* gene product, Nature, 332:548.

Furth, M.E., Davis, L.J., Fleurdelys, B. and Scolnick, E.M., 1982. Monoclonal antibodies to the p21 products of the transforming gene of the Harvey murine sarcoma virus and of the cellular *ras* gene family, J. Virol.,43:294.

Gilman, A., 1984, G proteins and dual control of adenylate cyclase, Cell, 36:577.

Hattori, S., Clanton, D.J., Satoh, T., Nakamura, S., Kaziro, Y., Kawakita, M. and Shih, T.Y., 1987, Neutralizing monoclonal antibody against *ras* oncogene product p21 which impairs guanine nucleotide exchange, Mol. Cell. Biol., 7:1999.

Kung H.-F., Smith, M.R., Bekesi, E., Manne, V. and Stacey, D.W., 11986, Reversal of transformed phenotype by monoclonal antibodies against Ha-*ras* p21 proteins, Exptl. Cell Res., 162:363.

Lacal, J.C. and Aaronson, S.A., 1986, Activation of *ras* p21 transforming properties associated with an increase in the release rate of bound guanine nucleotide, Mol. Cell. Biol., 6:4214.

Mayer, B.J., Hamaguchi, M., Hanafusa, H., 1988, A novel oncogene with structural similarity to phospholipase C, Nature, 332:272.

Mulcahy, L.S., Smith, M.R., and Stacey, D.W., 1985, Requirement for *ras* proto-oncogene function during serum-stimulated growth of NIH3T3 cells, Nature, 313:241.

Papageorge, A.G., Willumsen, B.M., Johnsen, M., Kung, H.-F., Stacey, D.W., Vass, W.C., and Lowy, D.R., 1986, A transforming *ras* gene can provide an essential function ordinarily supplied by an endogenous *ras* gene, Mol. Cell. Biol., 6:1843.

Smith, M.R., DeGudicibus, S.J,, and Stacey, D.W., 198ᶠ Requirement for c-*ras* proteins during viral oncogene transformation, Nature, 320:540.

Stacey, D.W., DeGudicibus, S.R., and Smith, M.R., 1987, Cellular *ras* activity and tumor cell proliferation, Exptl. Cell. Res., 171:232.

Stahl, M.L., Ferenz, C.R., Kelleher, K.L., Kriz, R.W., and Knopf, J.L., 1988, Sequence similarity of phospholipase C with the non-catalytic region of *src*, Nature, 332:269.

Tsai, M.-H., Yu, C.-L., Wei, F.-S., and Stacey, D.W., 1988, The effect of GTPase activating protein upon *ras* is inhibited by mitogenically responsive lipids, Science (in press).

Wahl, M.I., Daniel, T.O., and Carpenter, G., 1988, Antiphosphotyrosine recovery of phospholipse C activity after EGF treatment of A-431 cells, Science, 241:968.

Whitman, M., Downes, C.P., Keeler, M., Keller, T., and Cantley, L., 1988, Type 1 phosphatidylinositol kinase makes a novel inositol phospholipid, phosphatidyl-inositol-3-phosphate, Nature, 332:644.

Yu, C.-L., Tsai, M.-H., and Stacey, D.W., 1988, Cellular *ras* activity and phospholipid metabolism, Cell, 52:63.

CHARACTERIZATION OF RAS PROTEINS PRODUCED AT HIGH LEVELS IN THE BACULOVIRUS EXPRESSION SYSTEM

Peter N. Lowe, Susan Bradley, Alan Hall*, Vivienne F. Murphy[+],
Susan Rhodes, Richard H. Skinner, and Martin J. Page[+]

Dept. of Molecular Sciences, The Wellcome Research Laboratories
Langley Court, Beckenham, Kent, BR3 3BS, U.K
[+]Dept. of Molecular Biology, Wellcome Biotech, Beckenham, Kent
*Chester Beatty Laboratories, Institute for Cancer Research
London, U.K.

INTRODUCTION

The p21 protein products of Ras oncogenes have been produced in a variety of expression systems. Most workers have used bacterial systems, which although producing good yields have several drawbacks. The recombinant bacterial p21 has frequently been produced as an insoluble protein requiring the presence of strong denaturants during purification[1,2] or as a fusion protein[3-5]. Even when soluble it most likely retains the N-terminal methionine[6] (or formyl-methionine residue) and always lacks the post-translational addition of palimitic acid which is essential for its biological activity and its translocation to the membrane[7,8]. Furthermore, soluble E.coli expressed p21 protein can be heterogeneous due to C-terminal proteolysis and possibly incorrect protein-folding. For example, we observe that recombinant-N-ras p21 can elute as several peaks on ion exchange chromatography and can chromatograph anomolously during gel filtration. Palmitoylated p21 has been produced in several eukaryotic cell lines but generally at much lower levels[9,10].

An alternative eukaryotic expression procedure is the insect/baculovirus system which has been used successfully for the production of a wide range of post-translationally modified and processed recombinant proteins[11]. In this system the gene is placed under the control of the strong polyhedrin promoter of the baculovirus Autographa californica. We describe here the use of this system for the high level expression of both palmitoylated and non-palmitoylated forms of the c-Ha-ras protein.

METHODS

Cloned ras c-DNA genes are inserted into the transfer vector p36C, under the control of the polyhedrin promoter of the baculovirus Autographa californica[11]. The genes are then transferred to wild-type virus by homologous recombination. The virus is grown in Spodoptera frugiperda cells in liquid culture to cell densities up to 5 x 10^7/litre.

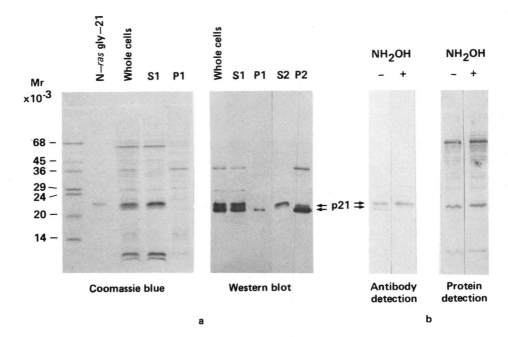

Fig. 1. Expression of Ha-ras p21 in the baculovirus system.
(a) S.Frugiperda cells infected with baculovirus expressing c-Ha-ras
were harvested, briefly sonicated and fractionated to form a 10,000g
supernatant (S1) and pellet (P1). The supernatant was then centrifuged
at 100,000g to yield supernatant (S2) and pellet (P2) fractions.
Samples were run on SDS gels and protein detected by Coomassie blue
staining or by reaction on a blot with polyclonal antibody raised
against N-ras p21; (b) The unfractionated sonicated cells were incubated
with 0.8M-NH$_2$OH, pH8.8, for 1h at 22°C and then electrophoresed and
proteins detected as above.

RESULTS AND CONCLUSIONS

   c-Ha-ras p21 was routinely expressed in the baculovirus system at
levels around 20% of the total cell protein (Fig 1a), i.e. about 100mg
of p21 protein/litre of culture. Expression at even higher levels, up
to 50% of total cell protein could be obtained in small cultures
optimally infected.

   As judged both by Coomassie blue staining of SDS-polyacrylamide
gels, and by measuring the distribution of GDP-binding activity, around
90% of the expressed p21 was soluble and found in a 100,000g supernatant
(Fig 1a). However the remaining p21 was associated with the 10,000g
(comprising cell fragments, organelles and large membrane fragments) and
100,000g (a crude membrane fraction) pellets.

   Western blots revealed that in unfractionated cells two major
antibody reactive bands are present (Fig 1). the low mobility one was
associated with the soluble fraction, the other with the membrane
fraction. Incubation of broken cells with hydroxylamine resulted in the
conversion of the higher mobility band to that of the former suggesting
that the membrane associated protein was acylated (Fig 1b).

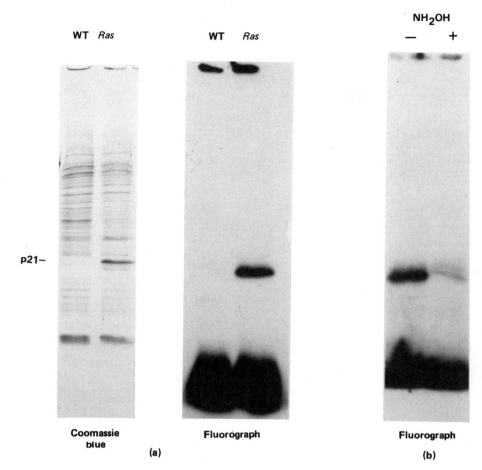

WT *Ras*   WT *Ras*

p21—

Coommassie
blue
(a)

Fluorograph

NH₂OH
— +

Fluorograph
(b)

Fig 2 Labelling of p21 in the baculovirus system by [3H] palmitate.
(a) S.frugiperda cells, infected with either wild type or ras-bearing
baculovirus, were grown in the presence of [3H]-palmitate. The cell
proteins were dissolved in SDS and analysed after electrophoresis by
staining for protein or by fluorography; (b) The ras-expressing cells
grown in the presence of [³H]palmitate were incubated with hydroxylamine
as in Fig 2(b) and after electrophoresis in SDS the radioactive protein
was located by fluorography.

    Growth of cells in the presence of [³H]palmitate resulted in the
specific labelling of p21 protein associated with the membrane fraction
(Fig 2). The bound radioactivity could be removed by incubation with
hydroxylamine (Fig 2b) suggesting that this p21 was palmitoylated.
Using cells grown in the presence of [³⁵S]methionine we have shown that
it is exclusively the faster mobility band which is labelled with
palmitate.

The soluble p21 was purified to homogeneity by chromatography on Mono Q and Superose 12(Prep) with a 25% yield, i.e. around 20 mg of p21/litre. The membrane associated protein can be solubilized with n-octyl-glucoside and we have purified small amounts of it by chromatography in the presence of detergent. Scaling up of the insect cell culture should allow production of sufficient palmitoylated p21 to allow its more detailed characterization.

We are currently working on the expression of K-ras and N-ras genes in this system, to compare the proteins produced with those made in E.coli, since in the latter system we have experienced problems with C-terminal proteolysis and insolubility, respectively, with the products of these two genes.

REFERENCES

1. J.C. Lacal, E. Santos, V. Notario, M. Barbacid, S. Yamazaki, H.-F. Kung, C. Seamans, S. McAndrew and R. Crowl, Expression of normal and transforming H-ras genes in Escherichia coli and purification of their encoded p21 proteins, Proc. Natl. Acad. Sci 81: 5305 (1984).
2. J.P. McGrath, D.J. Capon, D.V. Goeddel and A.D. Lewinson, Comparative biochemical aspects of normal and activated human ras p21 protein, Nature 310: 640 (1984).
3. J.A. Lautenberger, L. Ulsh, T.Y. Shih and T.S. Papas, High level expression in Escherichia coli of enzymatically active Harvey murine sarcoma virus p21 protein, Science 221: 858 (1983).
4. M.P. Poe, E.M. Scolnick and R.B. Stein, Viral Harvey ras p21 expressed in Escherichia coli purifies as a binary one-to-one complex with GDP, J. Biol. Chem. 260: 3906 (1985).
5. E.T. Nakano, M.N. Rao, M. Perucho and M. Inouye, Expression of the Kirsten ras viral and human proteins in Escherichia coli, J. Virol. 61: 302 (1987).
6. J. Tucker, G. Sczakiel, J. Feuerstein, J. John, R.S. Goody and A. Wittinghofer, Expression of p21 proteins in Escherichia coli and stereochemistry of the nucleotide-binding site, EMBO J. 5: 1351 (1986).
7. J.E. Buss and B.M. Sefton, Direct identification of palmitic acid as the lipid attached to p21, Mol. Cell. Biol. 6: 116 (1986).
8. Z.-Q. Chen, L.S. Ulsch, G. DuBois and T.Y. Shih, Post-translational processing of p21 ras proteins involves palmitylation of the C-terminal tetrapeptide containing cysteine-186, J. Virol. 56: 607 (1985).
9. R.J.A. Grand, K.J. Smith and P.H. Gallimore, Purification and characterisation of the protein encoded by the activated human N-ras gene and its membrane localisation, Oncogene 1: 305 (1987).
10. A.I. Magee, L. Gutierrez, I.A. McKay, C. Marshall and A. Hall, Dynamic fatty acylation of p21, EMBO J. 6: 3353 (1987).
11. V.A. Luckow and M.D. Summers, Trends in the development of baculovirus expression vectors, Biotechnology 6: 47 (1988).

PHENOTYPIC AND CYTOGENETIC ALTERATIONS IN HUMAN SKIN KERATINOCYTES

AFTER TRANSFECTION WITH THE CELLULAR HARVEY-ras (c-Ha-ras) ONCOGENE

P. Boukamp*, R.T. Dzarlieva*, Andrea Hülsen*,
E.J. Stanbridge**, and N.E. Fusenig*

*DKFZ, Institut für Biochemie, Im Neuenheimer Feld 280
 6900 Heidelberg
**University of California Irvine, Irvine, CA 92717 USA

Since activated ras oncogenes have been detected in a variety of human tumors, experiments were performed to determine the causal role of the oncogene in the development of neoplasia. So far transformation experiments with the ras oncogene using human epithelial cells was only described twice. The obvious difficulty to transform these cells, although beeing the major target cell of cancer in men, prompted us to study the oncogene action in human epidermal cells in more detail.

So far transfection of normal keratinocytes was unsuccessful and therefore we used spontaneously immortalized human skin keratinocytes designated HaCaT (Boukamp et al., 1988). Despite some in vitro growth characteristics of transformed cells and an aneuploid karyotype these cells remained non-tumorigenic and maintained a nearly normal degree of epidermal differentiation and therefore at present approximates normal cells at the highest degree.

Transfection of these cells with the plasmid containing the EJ-cellular-Harvey-ras oncogene and the neomycin resistance gene resulted in clones with a morphology very similar to the parental cells. Integration of ras-/ and neo and the individuality of the clones was verified by Southern blot analyses. Furthermore, we found a positive correlation between the level of ras expression (mRNA) and integration of oncogene DNA.

To determine which biological functions were altered in the ras transfected cells several growth parameters in vitro and in vivo were tested. The proliferative activity did not increase and similarly the colony forming efficiency on plastic remained uneffected. Interestingly, however, the ability to form colonies in agar was lost in all ras transfected clones, while the parental cells as well as those transfected with the neo-gene only exhibited anchorage independance.

In vivo after subcutaneous injection all control cells as well as clones which only expressed little ras RNA remained non-tumorigenic. Those which exhibited an elevated level of expression formed slowly or rapidly progressing tumors which by histology had to be divided into two groups: a) those which formed large encapsulated epidermal cysts and which were classified as benign and b) those which grew to a malinant squamouse cell carcinomas.

This differential growth behaviour was also maintained in another in vivo test system where the cells are transplanted as suspension onto the muscle fascia of nude mice (Boukamp et al., 1985). Here too, the benign cells formed cysts encapsulated by the mouse mesenchyme while the malignant cells infiltrated the mouse tissue and formed tumor-like nodules. These data strongly indicate that the respective in vivo growth potential was genetically determined.

In respect to differentiation both cell types maintained the capability to stratify and to form epidermal tissues. Moreover, also the differentiation specific proteins were expressed (such as involucrin, filaggrin and the differentiation specific suprabasal keratins K1 and K10). Thus, our data strongly contradict findings showing a positive correlation between ras transformation and inhibition of differentiation (Weissmann and Aaronson, 1985).

The differential in vivo growth pattern could not be explained by molecular biological criteria. Both, the benign and the malignant cells exhibited an elevated level of ras expression obviously sufficient to enable the cells to prolonged growth but insufficient to malignantly convert the cells. For this additional step other mechanisms are required. These could either be i) additional chromosomal rearrangements (translocations, deletions) or ii) specific integration sites of the ras oncogene in the human genom. Preliminary data show arguments for both alternatives.

Firstly, cytogenetic analysis revealed new stable translocations (marker chromosmes) predominently found in the malignant clones. Although new markers also occured in the parental cells during prolonged culture time without obviously leading to malignancy, it cannot be excluded that in combination with the ras oncogene they show a different spectrum of activity.

To identify the chromosomes in which the ras oncogene integrated we used an approach called "micronuclei transfer" (Saxon et al., 1986). The principal of this technique is to transfer single intact chromosomes via formation of micronuclei, isolation and fusion with recipient cells (mouse cells). Southern blots have shown that the ras pattern characteristic for the HaCaT-ras clones was also found in the respective mouse cells. Cytogenetic analysis of these clones revealed that predominantly two chromosomes were found, chromsome No. 12 and marker chromosome 4 [t(4p18q)] of the HaCaT cells. Although these are preliminary experiments, the frequency of these two chromosomes involved in the integration of the tranfected plasmid strongly argues against a random event.

In a second step we isolated the human chromosome out of the mouse cells via micronuclei transfer and transferred it back into the HaCaT cells. So far the chromosome from a malignant HaCaT-ras clone led to malignant conversion of the parental HaCaT cells, while after transfer the chromosome from a non-tumorigenic HaCaT-ras clone the cells remained non-tumorigenic.

Thus, our data have shown that transfection with the cellular Harvey-ras oncogene does not per se lead to malignant conversion of the HaCaT cells. A correlation, however, exists between integration and expression of the ras-oncogene and altered growth regulation in vivo (prolonged growth). Malignancy, on the other hand, seems to require additional events. These either can be structural and/or numerical aberrations with under-/over expression of genes envolved in growth regulation or the requironment for specific integration sites.

Literature

1. Boukamp, P., Petrusevska, R.T., Breitkreutz, D., Hornung, J., Markham, A. (1988) J. Cell Biol. 106, 761-771

2. Boukamp, P., Rupniak, H.T.R., Fusenig, N.E. (1985) Cancer Res. 45, 5582-5592

3. Weissmann, B., Aaronson, S.A. (1985) Mol. Cell. Biol. 5, 3386-3396

4. Saxon, P.J., Srivatsan, E.S., Leipzig, G.V., Sameshima, J.H., Stanbridge, E.J. (1985) Mol. Cell. Biol. 5, 140-146

REGULATION OF CELL DEATH BY RAS AND MYC ONCOGENES

A.H. Wyllie, I. Evans, R.G. Morris and D.A. Spandidos*

Department of Pathology, University Medical
Edinburgh
*Beatson Institute for Cancer Research, Glasgow
*National Hellenic Research Foundation, Athens

INTRODUCTION

Cells within tumours proliferate, differentiate and die. The genes which control their proliferation are becoming better known, and this Workshop has shown the importance of ras genes in this process. The point has also been made within the Workshop that ras expression may influence the movement out of the proliferative state into a differentiation pathway in certain tumour cells. In this paper I wish to discuss the importance of cell death in tumour biology, and to describe some experiments which indicate that expression of ras and other oncogenes may play a role in determining the death of cells also.

Kinetic data have long shown that cell loss plays a quantitatively important part in the overall growth rate of animal and human tumours (reviewed in 1, 2). The "cell loss factor" is a measure of the difference between the observed expansion of tumour volume and the growth expected from direct measurements of cell proliferation alone. When this factor is zero, all cells gained by proliferation contribute to size expansion; when it is 1.0, cells are lost at the same rate as they are generated. Most human tumours for which measurements have been made have cell loss factors close to 1.0. This high rate of cell loss has three non-intuitive implications. First, minor percentage changes in cell loss alone could have major influence on tumour cell number and indeed might determine whether the tumour is growing or regressing. The same argument may be equally important in pre-neoplastic populations, where the size of the target population available for sequential carcinogenic events is directly influenced by the cell loss rate. A notable formulation of this idea appears in Knudson's calculation of the age incidence of cancer, in developing the two-hit model of retinoblastoma development[3]. Second, losses on this scale - were they selective - would be capable of materially changing the quality of the surviving population of tumour cells and so influencing the rate of tumour progression. Third, were entry to the loss pathway to be a control point, there is no a priori reason why such

TABLE 1

CONSTITUTION OF RAT FIBROBLAST LINES

Cell Line	Oncogene	Enhancer
208F	-	-
RFMCMG1	c-myc	MoLTR
RFH06N1	Ha-ras	MoLTR + SV40
RFH05T1	T24-ras	SV40
RF110	Ha-ras	hMTIIA + endogenous 5'
RF108	Ha-ras	hMTIIA

TABLE 2

BEHAVIOUR OF 208F FIBROBLASTS AND THEIR ONCOGENE-EXPRESSING DERIVATIVES IN IMMUNE-SUPPRESSED MICE.

Oncogene	Animals injected	Take (%)	Lung	Node	Total (%)
Nil	7	0(0%)	0	0	0 (0%)
Ha ras	15	14(93%)	0	2	2 (13%)
T24 ras	31	29(94%)	6	5	8 (26%)
C myc	25	13(52%)*	0	0	0(0%)*[+]

*Differs significantly from ras-expressing groups.
+Two metastases appeared later in this group.

150

regulation should be tightly linked to the processes controlling proliferation or differentiation. With these considerations in mind, we asked whether we could demonstrate a role for oncogenes, and the ras oncogene in particular, in influencing tumour cell death.

METHODS

The rodent fibroblast line 208F, and several derivatives of it, obtained by transfection with Ha-ras and c-myc genes in a variety of expression vectors (Table 1), were studied in vitro and in vivo. Cells were maintained in vitro and injected in relatively high numbers subcutaneously to mice rendered immunologically incompetent by thymectomy and irradiation (with cytosine arabinoside protection) as described previously[4]. Quantitative dot blots, using SP6-generated RNA as standards, were used to measure the proportion of oncogene message in total cellular RNA, and expression of mutated or non-mutated Ha-ras was confirmed by Western blotting, using the antibody Y13-259.

RESULTS

Cell Lines Transfected with Different Oncogenes have Differing Aggressiveness In Vivo

In conformity with several other reports[5,6,7], we showed that 208F fibroblasts acquire aggressive growth behaviour, together with the ability to generate frequent metastasis, on transfection with the mutated form of Ha-ras (T24 ras) (Table 2). Cell lines expressing the non-mutated gene also grew aggressively at the injection site, although a smaller number metastasised. The c-myc transfected line, in contrast, showed a lower primary take rate, generated fewer aggressive tumours, and only 2 metastases in 25 injected animals, both at long periods after injection.

Cell Lines Transfected with Different Oncogenes Show Different Rates of Mitosis and Apoptosis

Two types of cell death, necrosis and apoptosis, are commmonly found within tumours[2,8]. Necrosis is character-ised by cell swelling and membrane rupture, and usually occurs in sheets which adopt topographical patterns indicative of vascular insufficiency. Apoptosis occurs in single cells, and is associated with cell volume loss, characteristic surface alterations[9] and activation of endogenous endonucleases which cleave chromatin at internucleosomal sites[10] and appear to be calcium-plus-magnesium dependent[11]. It is widely observed in biology in circumstances where death is "programmed" but can also follow minor toxic stimuli. Even in these latter, its initiation may be dependent upon protein synthesis[12,13].

Histologically, all the aggressive tumours had zones of necrosis, but there were differences in the frequency of

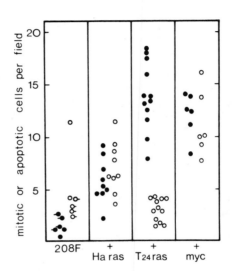

Figure 1. Mitosis and apoptosis within apparently viable portions of animal tumours. The oncogene-expressing tumours were all harvested around the same time after inoculation. The indolent 208F nodules, which were all small, were harvested later, if marked ‑o. At least 10 high power fields were scored for each tumour.

mitosis and apoptosis in the viable parts of the primary tumours. Whereas the indolent nodules generated by the parental fibroblasts contained few mitoses and a similar low number of apoptotic bodies, both apoptosis and mitosis were more frequent in the tumours expressing unmutated Ha-ras, and both were yet more frequent in the tumours derived from myc-expressing cells. T24 ras-expressing cells differed qualitatively from the others in combining a high rate of mitosis with a low rate of apoptosis (figure 1).

Apoptosis Occurs In Vitro As Well As In Vivo

To establish that these differences in apoptosis and mitosis were intrinsic to the cell lines, rather than phenomena imposed by the environment in vivo, the cells were studied in vitro in terms of monolayer density, thymidine incorporation and apoptotic rate. In these experiments near confluent cultures were established at high serum levels, and 48 hours later switched to low serum, in order to minimise the effects of exogenous growth factors. The cell lines containing constitutively expressed human oncogenes were compared with two lines (RF110 and RF108) in which the Ha-ras proto-oncogene is variously truncated in the 5' endogenous promoter region, and placed under control of the human metallothionein IIa promoter[14]. It was hoped that study of these cells, with and without induction of the Ha-ras gene by glucocorticoid, would allow more definitive appraisal of the role of the ras gene in proliferation and death.

The results showed that apoptosis occurred in vitro in the cell lines to different extents. Apoptotic cells floated into the supernatant medium, or were phagocytosed by their neighbours in the monolayer, and showed the morphology and endonuclease activation characteristic of this process. As in vivo, the c-myc expressing lines produced several fold more apoptosis than the others. This was sustained by monolayers which continued to show thymidine uptake, so it appeared to be more a result of increased cell turnover than a "toxic" effect of the c-myc product. The cell line expressing T24-ras was difficult to evaluate for apoptosis, as many of the viable cells in this highly transformed line floated free in the supernatant. However, repeated differential counts of these floating cells revealed less than 1% to be apoptotic. Thus the general pattern of incidence of apoptosis in vitro was similar to that observed in vivo in these cell lines.

Apoptosis in the two lines containing the metallothionein promoter was strongly inhibited by dexamethasone. The inhibition was reversible. It remains to be proven, however, that this was due to induction of the linked ras gene, as the cells containing either of the two inducible constructs showed substantial ras mRNA expression in the absence of the inducing stimulus. Further transfectants are being examined to clarify the relationship between dexamethasone and <u>ras</u> gene expression in inhibiting apoptosis.

FUTURE DIRECTIONS

The data support the view that oncogene expression may have a major role in influencing the relative rates of cell gain and loss within tissue populations. It is possible that in human tumours <u>ras</u> expression, and its interaction with other oncogenes, may expand the population susceptible to "progression" events through inhibiting cell death as well as by inducing proliferation. Such a scenario would fit well with the observation that <u>ras</u> expression is frequently observed in human pre-malignant states[15], but shows less correlation with tumour behaviour once the fully malignant state is achieved[16].

REFERENCES

1. Steel, G.G. Growth Kinetics of Tumours. Clarendon Press, Oxford. 1977.

2. Wyllie, A.H. The biology of cell death in tumours. Anticancer Res. 1985; <u>5</u> 131-136.

3. Knudson, A.G. A two-mutation model for human cancer. In Advances in Viral Oncology (ed Klein, G.) Raven, New York, 1987; <u>7</u>:1-17.

4. Wyllie, A.H., Rose, K.A., Morris, R.G., Steel, C.M., Foster, E., Spandidos, D.A. Rodent fibroblast tumours expressing human <u>myc</u> & <u>ras</u> genes: growth, metastasis and endogenous oncogene expression. Br. J. Cancer 1987; <u>56</u>: 251-259.

5. Vousden , K.H., Eccles, S.A., Purvies, H., Marshall, C.J., Enhanced spontaneous metastasis of mouse carcinoma cells transfected with an activated c-Ha-ras-1 gene. Int. J. Cancer 1986; <u>37</u>: 425-433.

6. Pozzatti, R., Muschel, R., Williams J., Padmanabhan, R., Howard, B., Liotta, L., Khoury, G. Primary rat embryo cells transformed by one or two oncogenes show different metastatic potentials. Science 1986; <u>232</u>: 223-227.

7. Bradley, M.O., Kraynak, A.R., Storer, R.D., Gibbs, J.B. Experimental metastasis in nude mice of NIH3T3 cells containing various <u>ras</u> genes. Proc. Natl. Acad. Sci. 1986; <u>83</u>: 5277-5281.

8. Wyllie, A.H.   Cell Death.   Int. Rev. Cytol. 1987; Suppl. <u>17</u>: 755-785.

9. Duvall, E., Wyllie, A.H., Morris, R.G.   Macrophage recognition of cells undergoing programmed cell death (apoptosis). Immunology 1985; <u>56</u>: 351-358.

10. Wyllie, A.H.   Glucocorticoid-induced thymocyte apoptosis is associated with endogenous endonuclease activation. Nature 1980; <u>284</u>: 555-556.

11. Cohen, J.J., Duke, R.C.   Glucocorticoid activation of a calcium-dependent endonuclease in thymocyte nuclei leads to cell death. J. Immunol. 1984; <u>132</u>: 38-42.

12. Wyllie, A.H., Morris, R.G., Smith, A.L., Dunlop, D. Chromatin cleavage in apoptosis: association with condensed chromatin morphology and dependence on macro-molecular synthesis. J. Path. 1984; <u>142</u>: 67-77.

13. McConkey, D.J., Hartzell, P., Duddy, S.K., Hakansson, H. Orrenius, S.   2,3,7,8 - Tetrachlorodibenzo-p-dioxin kills immature thymocytes by Ca^{2+}-mediated endonuclease activation. Science 1988; <u>242</u>: 256-259.

14. Spandidos, D.A.   The structure and function of eukaryotic enhancer elements and their role in oncogenesis. Anticancer Res. 1986; <u>6</u>: 437-450.

15. Williams, A.R.W., Piris, J., Spandidos, D.A., Wyllie, A.H.   Immunohistochemical detection of the <u>ras</u> oncogene p21 product in an experimental tumour and in human colorectal neoplasms. Br. J. Cancer 1985; <u>52</u>: 687-693.

16. Gallick, G.E., Kurzrock, R., Kloetzer, W.S., Alpinghaus, R.B., Gutterman, J.K.   Expression of p21 <u>ras</u> in fresh primary and metastatic human colorectal tumours. Proc. Natl. Acad. Sci. 1985; <u>82</u>: 1795-1799.

RAS-INDUCED PHENOTYPIC CHANGES IN HUMAN FIBROBLASTS

L Fiszer-Maliszewska and A R Kinsella

Paterson Institute for Cancer Research, Christie Hospital
and Holt Radium Institute, Manchester, M20 9BX, UK

INTRODUCTION

DNA transfection studies have shown activated ras genes to transform
established rodent cell lines in a dominant fashion and to co-operate with
so-called nuclear oncogenes in primary cultures to give transformation
(Land et al, 1983; Ruley, 1983; Reviewed Balmain, 1985). Activated ras
genes have been identified in a wide variety of human tumour and human
tumour cell line DNAs (Santos et al, 1982; Hall et al, 1983; Barbacid,
1986; Lowy and Willumsen, 1986; Bos et al, 1987) suggesting that the ras
oncogene family (Ha-; Ki-; N) might have an important role to play in
human tumourigenesis and human cell transformation. This is in part
confirmed by the limited success of ras-induced in vitro human epithelial
cell transformation (Yoakum et al, 1985; Rhim et al, 1985; Boukamp et al,
1986). However, when compared with rodent fibroblast cell transformation,
very little is known about human cell transformation other than that it is
very difficult to achieve either by the use of conventional carcinogenic
agents (Kakunaga, 1978; Milo and DiPaolo, 1978; Milo et al, 1981) or by
the introduction of activated ras genes. Several groups have reported the
failure of an activated ras gene to transform normal human fibroblasts
(Sager et al, 1983; Spandidos, 1985; Namba et al, 1988), whilst Sutherland
et al (1985) showed pT24 Ha-ras to confer a high incidence of anchorage
independence, but not tumourigenicity on neonate foreskin fibroblasts.
Namba et al (1988) have transformed to tumourigenicity only two out of
seven Ha-ras transfected, immortalised human embryo fibroblast, cell
clones resistant to G418 selection. Hurlin et al (1987) transformed
neonate foreskin fibroblasts to achorage independence, but not
tumourigenicity using the pH06T1 Ha-ras containing plasmid with 3' and 5'
enhancer sequences (Spandidos and Wilkie, 1984) that had previously failed
to transform normal human fibroblasts (Spandidos 1985). Thus, the
evidence provided by these studies again suggests that ras has a role to
play in human cell transformation, but that the factors controlling human
fibroblast cell transformation are outwardly more complex than for rodent
cell transformation.

EXPERIMENTAL

As we commenced the present study we were largely unaware of the
quasi-successful transformation studies of Hurlin et al (1987) and Namba
et al (1988). The aim of the study was to see if we could use oncogenes
to push skin fibroblasts from patients with Familial Adenomatous Polyposis

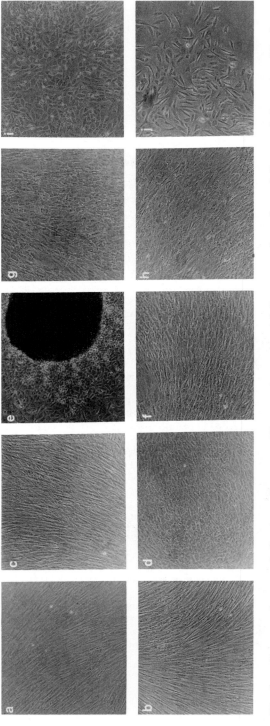

Fig 1　Comparative morphologies of parent FF$_1$, myc-, activated N-ras- and activated Ha-ras-infected cell clones

a) FF$_1$ normal foreskin fibroblasts
b) FF$_1$ neo infected cells
c) myc-infected cells
d) N-ras infected cells
e) dense piled up focus of FF$_1$ N-ras infected cells
f) the same FF N-ras colony after isolation, no longer exhibiting a dense piled up morphology
g) and h) FF$_1$ Ha-ras infected cultures
i) and j) FF^1 cells infected with both myc and Ha-ras and selected in both hygromycin and G418

(FAP/FPC) towards a transformed phenotype as a continuation of previous work conducted in this laboratory (Gainer et al, 1984; Kinsella, 1987).

Unexpectedly, the more interesting results came from observing the effects the introduction of the oncogenes had on the normal foreskin fibroblasts which were being used as control cells. The genes were introduced into the cells using the defective retroviral vector PZIPneo into which had been inserted N-ras cDNA with a codon 12 mutation (Alan Hall, personal communication) and the c-myc cDNA from HL60 (Miyamoto et al, 1985) and a MPSV vector containing a mutated Ha-ras cDNA (Stocking et al, 1986). Infection with pZIPneo alone as a control resulted in no obvious phenotypic changes, Fig 1b, however infection with the myc gene resulted in some clones that exhibited an atypical morphology which was concomitant with a certain degree of loss of contact inhibition of growth in that the cultures formed multilayers, Fig 1c. Infection with the N-ras gene resulted in a marked change in morphology, Fig 1d, and in two of the eight N-ras infected clones studied in detail (clones 4 and 5) dense piled up colonies of cells were identified, Fig 1e. On isolation of these dense piled up colonies, loss of contact inhibition of growth was not maintained, Fig 1f, although patches of cells exhibiting the classical "criss cross" morphology associated with morphological transformation in rodent cells, were observed.

In contrast, two other N-ras infected clones entered premature crisis eight subcultures after infection, passage 15 of the parent FF_1 cell line. In fact, in all of the N-ras infected subclones, Table 1, the presence of the oncogene appeared to increase the cell doubling time and reduce the number of population doublings. The data for the growth rate in 10% versus 1% serum were more dependent on the age of the culture than on the presence of the oncogene and were essentially equivocal. None of the clones became immortalised or produced tumours in nude mice. Infection of the FPC skin fibroblast cell lines with activated N-ras resulted in premature crisis or senescence three passages after infection. Infection of the FF_1 parent cell line with the Ha-ras oncogene using the MPSV vector again produced clones with a spectrum of morphologies which in some cases were distinct from those of the N-ras infected clones, Fig 1g and h. Infection of the FF_1 normal skin fibroblasts with pZIP hygromycin myc resulted in good growing cells with an altered morphology which were infected in turn, two passages later with Ha-ras which produced a morphologically altered cell population resistant to both hygromycin and G418, Fig 1i and j, but in which there was considerable sloughing of cells from the dish and cell loss.

The induction of chromosomal abnormalities did not appear to be associated with the retroviral infections involving any of the oncogenes, although in N-ras infected subclones 5 and 4d, chromosomally altered clones were seen to develop about passage 20.

DISCUSSION

All these data are very preliminary. We know that the oncogenes are stably integrated into the genome, but we have no information about expression.

We do know however that introduction of the activated N-ras and Ha-ras oncogenes alone into normal human skin fibroblasts is obviously not sufficient to induce transformation. Both genes are associated with a spectrum of distinct changes in morphology and in the case of the introduction of the mutated N-ras gene the rare induction of dense piled up foci of cells. Generally the changes in morphology induced by

TABLE 1

Cell Strain	Passage No at infection	Passage No* at isolation	Morphology+	Population Doublings	Cell Doubling Time (Days)	% Growth 1%/10% serum	Senesced Passage No	Chromosomes	Tumours
FF₁ parent	-	-	N	73	1.1 (SC 6) 2.0 (SC 22)	20	23	46 XY	0
FF₁ neo 2	7	10/3	N	42	ND	ND	18	46 XY	0
FF₁ neo 3	7	10/3	N	ND	ND	ND	-	46 XY	0
FF₁ neo 13	7	10/3	N	54	ND	ND	19	46 XY	0
FF₁ neo 7	7	10/3	N	ND	ND	ND	-	46 XY	0
FF₁ N*ras 1	7	10/3	"N"	51	3.1 (18/11)	13.0	22	46 XY, 18 p- polyploidy	0
FF₁ N*ras 2	7	10/3	"N"	44	ND	ND	16	polyploidy	0
FF₁ N*ras 3	7	10/3	"N"	37	ND	ND	15	marker + dicentric	0
FF₁ N*ras 4	7	10/3	T	42	1.9 (16/9)	29.2	23	46 XY	0
FF₁ N*ras 4d	7	10/3	T	62	2.1 (20/13)	45.0	24	46 XY**	0
FF₁ N*ras 5	7	10/3	T	63	1.6 (16/9)	33.9	21	46 XY**	0
FF₁ N*ras 5a	7	10/3	T	70	2.8 (21/14)	40.6	25	46 XY	0
FF₁ myc 1	7	10/3	N	47	ND	ND	17	46 XY	0
FF₁ myc 2	7	10/3	N	49	ND	ND	19	46 XY + aneuploidy	0
FF₁ myc 4	7	10/3	N	42	ND	ND	17	46 XY	0
FF₁ myc 5	7	10/3	Atypical	41	ND	ND	17	46 XY	0
FF₁ myc 9	7	10/3	Atypical	50	2.7 (16/9)	ND	20	46 XY	0
FF₁ Ha*ras (pool)	5	8/3	-	ND	1.4 (8/3)	22.0	ND	46 XY	0
FF₁ Ha*ras 11	5	8/3	T	ND	ND	ND	ND	ND	ND
FF₁ Ha*ras 17	5	8/3	T	ND	ND	ND	ND	ND	ND
FF₁ Ha*ras 18	5	8/3	T	ND	ND	ND	ND	ND	ND
FF₁ Ha*ras 41	5	8/3	T	ND	ND	ND	ND	ND	ND
FF₁ Ha*ras F9	5	6/1	T	ND	ND	ND	ND	ND	ND
FF₁ ZHM myc	6	-	Atypical	ND	ND	ND	ND	ND	ND
FF₁ ZHM Ha*ras	9/3	-	T	ND	ND	ND	13--->	ND	ND

+, N denotes normal morphology and T denotes transformed morphology. * passage no, 10/3 denotes passage no 10 of parent and 3 after infection. **, denotes clone of chromosomally altered cells coming through SC 19/12.

160

activated N-ras were more marked than for activated Ha-ras. Infection of certain cells with mutated N-ras induced the cells into premature crisis several subcultures after infection. Similarly, infection of all the FPC cultures with activated N-ras induced the cultures to crisis and/or senesce. The spectrum of changes induced by activated N-ras or Ha-ras may reflect the infection of cells at different stages in a fibroblast differentiation pathway or different sites for integration of the virus.

The failure to transform normal skin fibroblasts with activated N-ras and Ha-ras genes alone is not surprising and is not inconsistent with previous observations (Sager et al, 1983; Spandidos, 1983; Namba et al, 1988). The positive reports of Ha-ras induced transformation to anchorage independence by Sutherland et al (1985) and Hurlin et al (1987) were both seen in neonatal skin fibroblasts, whilst Namba's observations (1988) of tumourigenicity were made in carcinogen immortalised embryonic skin fibroblasts. Our skin fibroblasts from an eight year old donor might be considered to have the properties of adult skin fibroblasts (Kopelovich, 1982) and be more resistant to transformation. Hurlin's observations might however differ due to the presence of the enhancer sequences in the vector as discussed in the paper (Hurlin et al, 1987), Spandidos et al (1985) possibly only failing using the same construct in human embryonic cells because he looked at fewer independent G418 resistant clones and used a different method for introducing the gene. However, Namba's group succeeded with a pSV2 construct that had failed in earlier studies of Hurlin et al (unpublished).

This apparent abundance of contradictory data regarding the ability of the ras oncogene to induce cellular transformation following its introduction into human fibroblast cell lines probably reflects, more than anything else, the underlying differences in status of the recipient cells used in the different studies. For example, Namba's group (1988) obtained transformation following the introduction of an activated ras gene into immortalised human fibroblasts consistent with both previous observations made in rodent cells (Newbold and Overell, 1983) and theories of multi-stage carcinogenesis. Hurlin's unpublished data regarding the failure of the pSV2ras construct to induce transformation in normal skin fibroblasts is therefore not surprising. The positive report of transformation by Sutherland et al (1985) suggests that their cells although not actively immortalised are more "established" in culture than those of other workers in the field. The important studies now are those just commenced in this laboratory to study the role of co-operating oncogenes in human fibroblast cell transformation.

ACKNOWLEDGEMENTS

The authors would like to thank Geoff Clark for preparation of the pZIP vectors and Dr Carol Stocking for the gift of the MPSV vector. This work was funded by the Cancer Research Campaign.

REFERENCES

Balmain, A., 1985, Transforming ras oncogenes and multi-stage carcinogenesis. Br. J. Cancer, 5:1
Barbacid, M., 1986, Oncogenes and human cancer. Carcinogenesis, 7: 1037
Bos, J. L., Fearon, E. R., Hamilton, S. R., de Vries, V., van Boom, J. H., Van der Eb, A. J. and Vogelstein, B., 1987, Prevalence of ras gene mutations in human colorectal cancers. Nature, 327: 293
Boukamp, P., Stanbridge, E. J., Cerutti, P. A. and Fusenig, N. E., 1986, Malignant transformation of 2 human skin keratinocyte lines by Harvey-ras oncogene. J. Invest. Dermatol, 87: 131
Gainer, H. StC., Schor, S. and Kinsella, A. R., 1984, Susceptibility of

skin fibroblasts from individuals genetically pre-disposed to cancer to transformation by the tumour promoter 12-0-tetradecanoylphorbol-13-acetate. Int. J. Cancer, 34: 349

Hall, A., Marshall, C., Spurr, N. and Weiss, R., 1983, Identification of the transforming gene in two human sarcoma cell lines as a new member of the ras gene family located on chromosome 1. Nature, 303: 396

Hurlin, P. J., Fry, D. G., Maher, V. M. and McCormick, J. J., 1987, Morhplogical transformation, focus formation and anchorage independence induced in diploid human fibroblasts by expression of a transfected Ha-ras oncogene. Cancer Res, 47: 5752

Kakunaga, T., 1978, Neoplastic transformation of human diploid fibroblast cells by chemical carcinogens. Proc. Natl. Acad. Sci, 75: 1334

Kinsella, A. R., 1987, The study of multi-stage carcinogenesis in retinoblastoma and familial polyposis coli patient derived skin fibroblast cell culture systems. Mutation Res, 199: 353

Kopelovich, L., 1982, Are all diploid human strains alike? Relevance to carcinogenic mechanisms in vitro. Exp. Cell Biol, 50: 266

Land, H., Parada, L. and Weinberg, R. A., 1983, Tumourigenic conversion of primary embryo fibroblasts requires at least two co-operating oncogenes. Nature, 304: 596

Lowy, D. R. and Willumsen, 1986, The ras gene family. Cancer Surveys, 5: 273

Milo, G. E. and DiPaolo, J. A., 1978, Neoplastic transformation of human diploid cells in vitro after chemical carcinogen treatment. Nature, 275: 130

Milo, G. E., Noyes, I., Donahoe, J. and Weisbrode, S., 1981, Neoplastic transformation of human epithelial cells in vitro after exposure to chemical carcinogens. Cancer Res, 41: 5096

Miyamoto, C., Chizzonite, R., Crowl, R., Rupprecht, K., Kramer, R., Schaber, M., Kumar, A., Poonain, M. and Ju, G., 1985, Molecular cloning and regulated expression of the human c-myc gene in E.coli and Saccharomyces cerevisiae: comparison of the protein products. Proc. Natl. Acad. Sci, 82: 7232

Namba, M., Nishitani, K., Fukushima, F., Kimoto, T. and Yuasa, Y., 1988, Multi-step neoplastic transformation of normal human fibroblasts by <co-60 gamma rays and Ha-ras oncogenes. Mutation Res, 199: 415

Newbold, R. E. and Overell, R. W., 1983, Fibroblast immortality is a pre-requisite for transformation by EJ c-Ha-ras oncogene. Nature, 304: 648

Rhim, J. S., Jay, G., Arnstein, P., Price, F. M., Sanford, K. K. and Aaronson, S. A., 1985, Neoplastic transformation of human epidermal keratinocytes by Ad12-SV40 and Kirsten sarcoma viruses. Science, 227: 1250

Ruley, H. E., 1983, Adenovirus early region 1A enables viral and cellular transforming genes to transform primary cells in culture. Nature, 304: 602

Sager, R., Lanaka, K., Lau, C. C., Ebina, Y. and Anisowicz, A., 1983, Resistance of human cells to tumourigenesis induced by cloned transforming genes. Proc. Natl. Acad. Sci, 80: 7601

Santos, E., Tronick, S. R., Aaronson, S. A., Pulciani, S. and Barbacid, M., 1982, T24 human bladder carcinoma oncogene is an activated form of the normal human homologue of BALB- and Harvey-MSV transforming genes. Nature, 298: 343

Spandidos, D. A. and Wilkie, N. M., 1984, Malignant transformation of early passage rodent cells by a single mutated human oncogene. Nature. 310: 469

Spandidos, D. A., 1985, Mechanism of carcinogenesis: The role of oncogenes transcriptional enhancers and growth factors. Anticancer Res, 5: 485

Stocking, C., Kollek, R., Bergholz, U. and Ostertag, W., 1986, Point mutations in the U3 region of the LTR of Moloney Murine Leukaemia virus determine disease specificity of the myeloproliferative sarcoma virus. Virology, 153: 145

Sutherland, B., Bennett, P. V., Freeman, A. G., Moore, S. P. and Strickland, P. T., 1985, Transformation of human cells by DNA transfection. Proc. Natl. Acad. Sci, 82: 2399

Yoakum, G. H., Lechner, J. F., Gabrielson, E. W., Korba, B. E., Malan-Shibley, L., Willey, J. C., Valerio, M. G., Shumsudin, A. M., Trump, B. F. and Harris, C. C., 1985, Transformation of human bronchial epithelial cells transfected by Ha-ras oncogene. Science, 227: 1174

SUPPRESSION OF <u>RAS</u> ONCOGENE EXPRESSION USING SEQUENCE SPECIFIC

OLIGODEOXYNUCLEOTIDES

P. Hawley and I. Gibson

School of Biological Sciences
University of East Anglia
Norwich NR4 7TJ

INTRODUCTION

The activation of <u>ras</u> oncogenes has been clearly implicated in the tumorigenic process, being the oncogene most frequently identified in human cancers. (Barbacid 1985). <u>Ras</u> genes containing point mutations as well as amplified genes leading to over-expression of the protein product p21 have been found in human tumours (Der et al 1982, Bos et al 1986). There is, however, no reason to assume that we have yet identified all of the genetic changes involved or indeed that they are necessarily the primary events in tumorigenesis.

We have been involved with the role of the <u>ras</u> family of oncogenes in the cell transformation process <u>in</u> <u>vitro</u> studying changes occurring in inducible cell systems when mutant and normal <u>ras</u> genes are switched on and off. One aspect of this work is to use a technique often referred to as "anti-sense technology" in which oligodeoxynucleotides are directed against nucleotide sequences in the target gene. Success has been claimed with this technique in inhibiting mRNA expression <u>in</u> <u>vivo</u> and <u>in</u> <u>vitro</u> in the replication of certain viruses including H1V (Zamecnik et al., 1986; Maher and Dolnick, 1987; Smith et al., 1986, Cazenave, 1987).

In the area of cancer research Wickstrom and his collaborators have succeeded in preventing HL60 cells entering the S phase and shown inhibition of proliferation of the cells. This was achieved by directing 15-mer oligonucleotides against a critical segment of the <u>myc</u> oncogene (Heikkila et al., 1987; Wickstrom 1986). Indeed others using the same cells have claimed to have induced cell differentiation (Holt et al., 1988) using oligodeoxynucleotides directed against the <u>myc</u> oncogene.

In our own early experiments we used the T15 cell line (McKay et al., 1986) which contains several copies of the normal human N-<u>ras</u> gene, and is induced in the presence of dexamethasone to produce the p21 product and subsequently to exhibit morphological transformation (Tidd et al., 1988, Linstead et al., 1988). Our attempts to delay or prevent transformation using 9-mer methylphosphonate oligodeoxynucleotides at 80mM which are more resistant to nuclease breakdown than phosphodiester molecules (Hawley unpublished), had limited success. We concluded that 1) using longer oligonucleotides, 2) using affinity purified molecules and 3) using cell systems expressing lower levels of target in RNA might facilitate partial or total inhibition of gene expression.

In this paper we describe our initial experiments using another inducible cell line MR4, containing only a single copy of the mutant c-Ha-<u>ras</u> gene, and longer, 15-mer oligodeoxynucleotides directed to both the 5' and 3' end of the translated mRNA.

As phosphodiester oligonucleotides are used in this part of our study we have attempted to effect a more direct entry into the cells using electroporation, thereby minimising the breakdown of the molecules in the tissue culture medium, and allowing the use of lower concentrations of oligonucleotides.

MATERIALS AND METHODS

<u>Cell lines</u>

1)   <u>MR4 cells</u> (Reynolds et al.,1987) are transfected Rat-1 fibroblasts containing a fusion gene of the mouse metallothionein I 5' region and the coding region of the mutated human c-Ha-<u>ras</u> gene (c-<u>ras</u> T24). They were cultured in Dulbeccos MEM containing 10% foetal calf serum and induced to produce mutant p21 protein and to transform by the addition of 100mM $2nSO_4$. They were the gift of Dr. M.W. Lieberman, Fox Chase Cancer Centre, Philadelphia. They are shown in Fig. I, the zinc induced cells (b) and (d) forming foci.

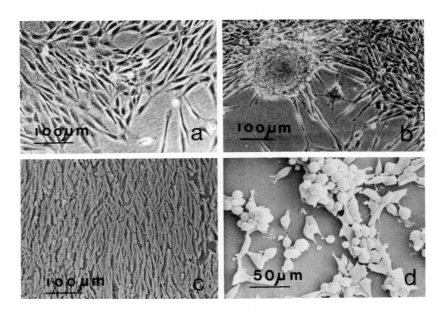

<u>Fig. 1</u>.   Phase micrographs (a) and (b) and scanning electron micrographs (c) and (d) of MR4 cells. Foci can be seen only in the transformed cells (b) and (d).

2)   EC816 cells are mouse NIH 3T3 fibroblasts transfected with
the normal human c-Ha-ras gene which produce p21 constitutively.
They were the gift of Dr. C. Marshall, Chester Beatty Institute,
London. They are used in this study to compare the effects of
oligonucleotides on cells producing large amounts of protein
with the MR4 cells described above.

Treatment of cells by electroporation

The cells were electroporated in a BIO RAD gene pulser using
a method based on a study by Winterbourne et al 1988.
Trypsinized sub-confluent cells were washed once in 280mM
mannitol, 20mM hepes buffer pH 7.2 and resuspended in the same
buffer containing the dextran or oligonucleotide under study to
give 2.5×10^6 cells/ml. 0.8ml volumes of cells were given single
pulses of 350V at 25 MFD capacitance to give a high proportion
(90-100%) of transiently permeabilized cells as shown by
erythrosin B vital staining and approx. 80% cell death. The
cells were centrifuged and resuspended in growth medium in 6 cm
tissue culture dishes. Concentrations of oligonucleotides used
were 0, 5, 10 and 20 mM.   Zinc was added to MR4 cells to induce
p21 protein production at this time. Cell extracts were made
after 24 or 48 hr periods.

Detection of p21

The cell extracts were made and analysed by 'western'
blotting techniques as described previously (Tidd et al., 1988).
30mg total protein was loaded per well on the gels prior to
blotting.

Scanning electron microscopy

The method used has been described elsewhere (Linstead et
al., 1988).

Preparation of oligonucleotides

Oligodeoxynucleotides were prepared in a Cyclone BIO-search
DNA synthesizer (New Brunswick) and purified on C18 Sep-Pak
cartridges (Wates Chromatography Division, Millipore (UK) Ltd.)
by the method described by D.M. Tidd and H. Warenius (in press).

RESULTS

1.   Electroporation of MR4 cells with fluorescent dextran

Cells were electroporated as described previously with
FITC-linked dextran of molecular weight 7,000 Kd at 5 mg/ml in
mannitol-hepes buffer. Control cells were incubated for the
same length of time with the dextran. Figures 2 (a) and (b)
show that electroporated cells show greater fluorescence and
exhibit characteristic 'blebbing' as described by Winterbourne
(1987). A small proportion of unporated cells show fluorescence
due to endocytosis. As this dextran is of greater molecular
weight than the oligonucleotides (approximately 4,500 Kd) it can
be assumed that the cells should be at least equally permeable
to the smaller oligonucleotides.

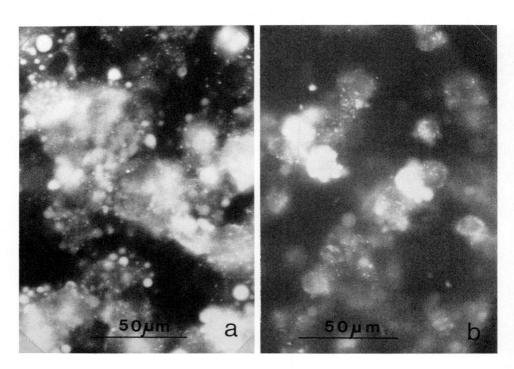

Fig. 2 (a). Cells electroporated with FITC-dextran as described. The cells were then centrifuged washed and resuspended in tissue culture medium and allowed to attach to cover-slips for 2.5 hours. They were fixed with 3.7% paraformaldehyde and mounted.

Fig. 2 (b). Cells were incubated without poration and treated as in (a)

168

2. Oligonucleotides used and their effects on p21 production

These were synthesised on a Cyclone Bio-Search DNA synthesiser to correspond to areas at both the 5' and 3' ends of the mRNA molecule. The 5' end molecules 1. 2. and 3. are antisense, sense and nonsense 15 mers respectively which correspond to the region down-stream from the G of the AUG initiation codon, i.e. bases 3-17 of the message.

```
c-H-ras mRNA   5' AUG-------------------//--------------------- 3'
   5'     bases 3-17
            1.   5'  AGCTTATATTCCGTC      antisense
            2.   5'  GACGGAATATAAGCT      sense
            3.   5'  GATACAGTACGATAG      nonsense
```

The 3' end molecules 4. and 5. respectively antisense and sense, were synthesised to match the 15 bases at the extreme 3' end of the messenger RNA including the termination codon UGA, i.e. bases 560-575.

```
c-H-ras mRNA     5'---------------------//--------------------UGA 3'

                              3'    bases 560-575

               antisense      4.   5' TCAGGAGAGGACACA
                   sense      5.   5' TGTGTGCTCTCCTGA
```

(a) 5' oligonucleotides

A preliminary experiment indicated that 5' antisense oligonucleotides showed no effect on p21 production at concentrations of 2 and 5 mM but were inhibitory at 10 and 20 mM. Figure 3 shows an experiment using oligonucleotides (1) and (3) at 10 and 20mM respectively.

It can be seen that both concentrations of antisense oligonucleotide (1) inhibit the production of p21 for 24 but not 48 hours, nonsense oligonucleotide (3) gives some reduction after a 24 hour incubation but none after 48 hours. The controls show that after 24 hours porated and unporated cells in the presence of Zn^{++} produce similar amounts of p21 and that some protein is produced in the absence of added Zn^{++} possibly as a result of low Zn^{++} levels in the tissue culture medium (Reynolds et al., 1987). An identical blot was prepared (not shown) and stained for total protein with amido black to verify the equal loading of the wells.

A further experiment (results not shown) showed that the sense oligonucleotide (2) slightly decreased the amount of p21 after 24 but not 48 hours at 20mM. These results are being further verified by the method of immunoprecipitation, but are not reported here.

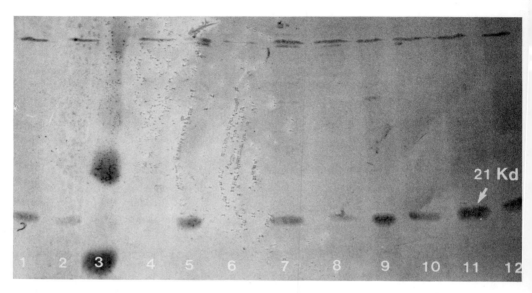

Fig. 3. Western blot of proteins from MR4 cells treated with antisense oligonucleotide (1) and sense oligonuclotide (3).

Lanes
1. Porated cells. $+Zn^{++}$.
2. Porated cells. $-ZN^{++}$.
3. Molecular weight markers.
4. Oligonucleotide (1) 10mM+Zn^{++}, incubated 24 hrs.
5. Oligonucleotide (1) 10mM+Zn^{++}, incubated 48 hrs.
6. Oligonucleotide (1) 20mM+Zn^{++}, incubated 24 hrs.
7. Oligonucleotide (1) 20mM+Zn^{++}, incubated 48 hrs.
8. Oligonucleotide (3) 10mM+Zn^{++}, incubated 24 hrs.
9. Oligonucleotide (3) 10mM+Zn^{++}, incubated 48 hrs.
10. Oligonucleotide (3) 20mM+Zn^{++}, incubated 24 hrs.
11. Oligonucleotide (3) 20mM+Zn^{++}, incubated 48 hrs.
12. Cells unporated +Zn^{++}

(b) 3' oligonucleotides

     No reduction in p21 levels as shown by western blotting
methods has been found using the 3' oligonucleotides 4 and 5.

(ii) EC816 cells

     A preliminary experiment using 5' oligonucleotides (1), (2)
and (3) (see Fig. 4) shows that there is no obvious reduction of
protein in this p21 overproducing cell line after 24 hours
incubation.

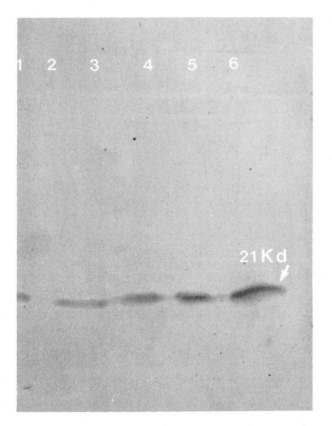

Fig. 4. Western blot of proteins from EC816 cells treated with
antisense oligonucleotide (1) and sense oligonucleotide (2) and
nonsense oligonucleotide (3).

Lanes (1) unporated control.

           (2)   blank.
           (3)   unporated control.
           (4)   porated cells with 20mM oligo. (1).
           (5)   Porated cells with 20mM oligo. (2).
           (6)   Porated cells with 20mM oligo. (3).

Further experiments are planned using 3' oligonucleotides, varying times of incubation after treatment, and utilisation of immunoprecipitation techniques.

DISCUSSION

The use of oligonucleotides as inhibitors of gene expressions has been the subject of an interesting review (Stein & Cohen 1988). The potential value of such compounds for gene therapies, although a long way off, holds some promise. Many problems remain to be solved including selectivity and availability of the specific target sequence, stability and, of course, solubility and passage through the cell membrane. Despite these problems, however, dramatic effects have been recorded on viral replications in vitro. Our own initial work with anti-sense methylphosphonate oligonucleotide analogues, directed against the N-ras oncogene sequences has been less successful (Tidd et al., 1988). It did, however, suggest new approaches including the development of new biological test systems. This has been reported in the current paper.

Although it is our intention to use various oligodeoxynucleotides including phosphorothioates and chimaeras containing methylphosphonates, our initial work has been with oligonucleotides with phosphodiester linkages. These have been introduced into the cells by electroporation, as we have shown (unpublished data) that such compounds are substantially broken down within 4 hours in tissue culture medium. We are currently undertaking microinjection and transfection with plasmids as means of introducing anti-sense molecules into cells.

The results described here show a dramatic decrease in p21 levels 24 hours after induction of the H-ras gene with zinc. The relevant antisense oligonucleotide was directed against a sequence following the initiation codon. Nonsense and sense sequences surprisingly, however, caused small reductions in protein levels but did not cause complete disappearance. The effect was not seen at 48 hours possibly because of the instability of the oligonucleotide, but this effect might be prolonged by the use of methylphosphonate and other stable molecules. The possibility that there might be complete inhibition of protein synthesis during this time is being investigated by radio labelling studies. The same antisense sequence has also been shown to delay the formation of clones of MR4 cells in soft agar for five days when compared with electroporated controls. Matching sense and nonsense molecules gave a shorter delay of two days (data not shown).

If these effects on the ras system can be prolonged it will be interesting to determine what further cell functions are disturbed, e.g. whether elevated levels of g-glutamyltransforase and glutathione transferase as described (Li et al., 1988) are reduced in the MR4 cells. Of even greater interest would be the secondary effects on ras expressing established tumour lines which are non-inducible. This might be expected to be most successful in cell-lines with mutated ras genes rather than cells which over-express ras protein as we have shown that there

is little, if any, effect on the proteins produced by EC816
cells. Recent evidence of the involvement of an untranslated
region of the c-H-ras gene in the genes function has been
described (Cohen et al., 1988). The use of antisense
oligonucleotides directed to these regions might increase our
knowledge of the functioning of the gene. Our laboratory is
also engaged in studies of the mechanism of action of
oligonucleotides in vitro, such studies will, we hope, aid us in
selecting active sequences against cellular targets.

The role of ras oncogenes in the development and maintenance
of the transformed phenotype at the biochemical and
physiological level is still unknown. Combined with studies on
human tumours, molecular biological studies in vitro using
ribozymes and cell studies as described above will, we believe,
contribute to this understanding.

Acknowledgements

We thank the Norfolk & Norwich Big C charity and the Cancer
Research Campaign for financial support. We also wish to thank
the following individuals:

Dr. M. Lieberman who supplied the MR4 cell line, Dr. C.J.
Marshall for the EC816 cell line and Paul Linstead for the
electron microscopy.

References

Barbacid, M. (1985). In "Important Advances in Oncology 1986"
  ed. V de Vita, S. Hellman, S. Rosenberg, pp 3-22.
  Philadelphia Lippincot.
Bos, J.L., Toksoz, D., Marshall, C., J and 6 others (1986).
  Amino acid substitutions at codon 13 in the N-ras oncogene in
  human acute mycloid leukaemia. Nature 315, 726.
Cazenave, C., Loreau, N., Thuong, N.T., Toulemé, J.J. and
  Hélène, C. (1987). Enzymatic amplification of translation
  inhibition of rabbit b-globin in RNA mediated by anti mesenger
  oligodeoxynucleotides covalently linked to intercalting
  agents. Nucleic Acids Res. 15, 4717.
Cohen, J.B. and Levinson, A.D. (1988). A point mutation in the
  last intron responsible for increased expression and
  transforming activity of the c-Ha-ras oncogenes. Nature 334,
  119.
Der, C.J., Krontiris, T.G. and Cooper, G.M. (1982).
  Transforming genes of human bladder and lung carcinoma cell
  lines are homologous to the ras genes of Harvey & Kirsten
  carcinoma viruses. Proc. Natl. Acad. Sci. USA 79, 3637-3640.
Heikkila, R., Schwab G., Wickstrom, E. and 4 others. (1987). A
  c-myc antisense oligodeoxynucleotide inhibits entry in S phase
  but not progress from G0-G1. Nature 328, 445.
Holt, J.T., Redner, R.L. and Neinhuis, A.W.(1987). An oligomer
  complementary to c-myc inhibits proliferation of HL60
  promyclocytic cells and induces differentiation. Mol. Cell
  Biol. 8, 963.
Li, Y., Seyama, T., Godison, A.K., Winokun, T.S., Lebovitz, R.M.
  and Lieberman, M.W. (1988). MTras T24, a metallothionein-ras
  fusion gene, modulates expression in cultured rat liver cells

of two genes associated with *in vivo* liver cancer. Proc. Natl. Acad. Sci. USA 85, 344-348.

Linstead, P., Jennings, B., Prescott, A., Hawley, P., Warn, R., Gibson, I. (1988). Scanning electron microscopy and the transformed phenotype. Micron and Microsc. Acta 19, 155-162.

Maher, L.J. ad Dolnick, B.J. (1987). Specific hybridization arrest of dihydrofolate reductase in RNA *in vitro* using anti-sense RNA or anti-sense oligonucleotides. Arch. Biochem and Biophys. 252(1), 214-220.

McKay, I.A., Marshall, C.J., Cales, C. and Hall, A. (1986). Transformation and stimulation of DNA synthesis in NIH 3T3 cells are a titratable function of normal p21 N-ras expression. EMBO Journal 5, 157.

Reynolds V.L., Lebovitz, R.M., Warren, S., Hawley, T.S., Godwin, A.K., Lieberman, M.W. (1987). Regulation of a metallothionein-ras T24 fusion gene by zinc results in graded alterations in cell morphology and growth.

Smith, C.C., Aurelian, L., Reddy, M.P., Miller, P.S. and Ts'O, P.O.P. (1986). Antiviral effect of an oligo(nucleoside methylphosphonate) complementary to the splice junction of herpes simplex virus type 1 immediate early pre mRNAs 4 and 5. Proc. Natl. Acad. Sci. USA 83, 2787-2791.

Stein, C.A. and Cohen, J.S. (1988). Oligodeoxynucleotides as inhibitors of gene expression: A review. Cancer Res. 48, 2659-2668.

Tidd, D.M., Hawley, P., Warenius, H.M. and Gibson, I. (1988). Evaluation of N-ras oncogene anti-sense, sense and nonsense sequence methylphosphonate oligonucleotide analogues. Anti Cancer Drug Design 3, 117-127.

Wickstrom, E.L., Wickstrom, E., Lyman, G.H. and Freeman, D.L. (1986). HL60 cell proliferation inhibited by an anti-c-myc pentadecadeoxynucleotide. Federation proceedings 45, 1708.

Winterbourne, D.J., Thomas, S., Herman-Taylor, J., Hussain, I., Johnston, Alan P. (1988). Electric shock mediated transfection of cells. Biochem. J. 251, 427-434.

Zamecnik, P.C., Goodchild, J., Taguchi, Y. and Sarin, P.S. (1986). Inhibition of replication and expression of human T-cell lymphotropic virus type III in cultured cells by exogenous synthetic oligonucleotides complementary to viral RNA. Proc. Natl. Acad. Sci. USA, 83, 4143.

EXPRESSION OF RAS, MYC AND FOS ONCOGENES DURING THE
DMSO-INDUCED DETRANSFORMATION OF A COLON CARCINOMA LINE
(HT29)

Susan A. Newbould and I. Gibson

School of Biological Sciences
University of East Anglia
Norwich NR4 7TJ

INTRODUCTION

The effects of differentiating agents on cancer cells
are of considerable interest, both from a therapeutic
viewpoint and as a means of elucidating the underlying
mechanisms of cell transformation. A number of cell systems
have been used to study the phenomenon of malignant cell
detransformation and differentiation, in particular the
HL-60 human promyelocytic leukemia line which can mature
along either a granulocytic or monocytic pathway (1). Other
workers have investigated differentiation of murine
neuroblastoma cells (2), murine rhabdomyosarcoma cells (3),
human melanocytes (4) and human colon carcinoma cells
(5-14).

The HT29 human colorectal carcinoma cell line has been
shown to exhibit a reversible pattern of differentiation in
which the cells assume an enterocytic phenotype under
certain conditions including i) the replacement of glucose
in the culture media by other sugars/nucleosides such as
galactose, uridine or inositol (15,16), ii) the total
absence of glucose in the media (17,18) or iii) treatment
with sodium butyrate or polar solvents such as
dimethylformamide (DMF) or dimethylsulphoxide (DMSO)
(19-21). They have also been reported to exhibit permanent
differentiation after subculture in sodium butyrate, the
subsequent clones having heritable phenotypes of either
mucous-secreting or absorptive intestinal epithelial cells
(i.e. enterocytes) (22,23). The former can be identified by
positive staining of mucoproteins (12,23,24) whilst the
latter are characterised by the presence of a brush border
with associated enzymes such as sucrase-isomaltase (SI) and
alkaline phosphatase (ALP) (18,25-27). HT29 cells appear to
be multipotential stem cells with the potential to mature
into either absorptive or mucin-producing enterocytes. It
has been suggested that there may also be other possible
pathways of differentiation for HT29 cells to produce Paneth
or goblet cells (23).

Since oncogenes appear to be intimately involved in the control of cell growth and development (28,29) it is of interest to know how their expression may be affected during the detransformation and differentiation of HT29 cells. Previous studies have investigated the expression of the ras and myc gene families in HT29 cells induced to differentiate by exposure to sodium butyrate and found considerably higher levels of the p21 ras gene product in certain of the mucin-secreting clones than in the parent HT29 line. In this study it was chosen to assess the effect of another differentiation-inducer, DMSO, on the expression of the oncogenes ras, myc and fos in the HT29 cell line. It was shown that the levels of ras and myc decrease during DMSO treatment of the cells which produces a partial, reversible differentiation pattern. Expression of c-fos remains at a low level both before and after exposure to DMSO.

MATERIALS AND METHODS

Cell Culture

HT29 cells (30) were grown as monolayers in $25cm^2$ tissue-culture flasks or 56mm diameter petri-dishes. The cells were cultured in Dulbeccos modified Eagles minimum essential medium (DMEM) supplemented with 10% foetal bovine serum at $37^{\circ}C$ in an atmosphere of 90% air, 10% CO_2.

Growth Curves

The cells were cultured in triplicate $25cm^2$ flasks, harvested using trypsin:EDTA (0.1%:0.04%) and samples counted on a Coulter Electronic Cell-Counter.

Cloning in Soft Agar

Cells were seeded at a density of $0.5 \times 10^1 - 0.5 \times 10^5$ per ml in 0.3% Noble agar according to standard techniques for the clonogenicity assay (31).

Immunocytochemistry

The cells were grown on sterile glass coverslips, washed in PBS pH 7.6, fixed in 3.7% paraformaldehyde for 10-15 mins at room temperature, permeabilised in ice-cold acetone for 1 minute, blocked with normal rabbit serum (1:20) before exposure to antibody. The following antibody-detection systems were used :
Actin: Phalloidin-Rhodamine at 2ug/ml (32)
Carcinoembryonic Antigen (CEA): A5B7 antibody (kindly donated by Dr. G.T. Rogers) used at a dilution of 1:100 (33)
Ras p21: Cetus p21 pan-reactive mouse monoclonal (1:100)
SI: anti-SI antibody (kindly supplied by Dr. H.-P. Hauri (34)) used at a dilution of 1:100.

Immunoassays

Enzyme-linked immunosorbent assays for c-myc and c-fos were performed using antibodies and protocols kindly supplied by Dr. G. Evan (35,36).

CEA was detected using the CEA Enzyme Immunoassay Duomab 60 <Roche> system (Roche Products Ltd., Welwyn Garden City, HERTS)

Western Blotting

Cell extracts were prepared, electrophoresed and blotted as described previously (37). Ras p21 protein was visualised using the anti-ras p21 pan-reactive monoclonal antibody and detection system (Cetus Corporation, Emeryville, CA 94608).

Southern Blotting

Standard methods of DNA extraction and Southern blotting were used. Filters were hybridised with probes for H-ras (38), N-ras (39), K-ras (40), c-myc (41) and c-fos (42) and also with an n-globin gene probe (43) to control for DNA loading. Relative intensities of bands were ascertained by densitometry.

Scanning Electron Microscopy

Cells were grown on sterile glass coverslips and processed by standard techniques (44). They were examined in a CamScan MK4 scanning electron microscope.

ALP Assay

Brush-border membrane fractions from normal and treated HT29 cells and from the small intestine and colon of rat were prepared (45). Levels of alkaline phosphatase were determined using the method of Garen and Levinthal (46).

RESULTS

DMSO treatment of HT29 cells produces a decrease in both growth rate and saturation density (Fig. 1).The doubling time increases from 2 days to 5.5 days and the saturation density decreases from $3X10^6$/ml to $8X10^5$/ml. The cloning efficiency of the cells when grown in soft agar falls from >90% for the untreated cells to <10% for the DMSO-treated cells.

The morphology of the cells appears altered within 24-48 hours after addition of DMSO (Fig. 2), the cells exhibiting a more flattened and elongated morphology with concomitant changes in the cytoskeletal arrangement within the cells (Fig. 3).

Scanning E. M. reveals no significant surface structural differences (Fig. 4). Although the concentration of microvilli appears less on the DMSO-treated cells this may be due solely to their increased surface area. The presence of the brush-border associated enzymes sucrase-isomaltase and alkaline phosphatase could not be detected either by immunocytochemical means or traditional enzyme assay methods.

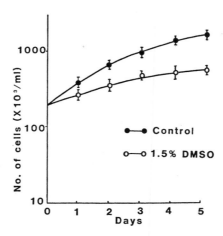

Figure 1. Effect of 1.5% DMSO on the growth of HT29
cells. (Standard error bars shown).

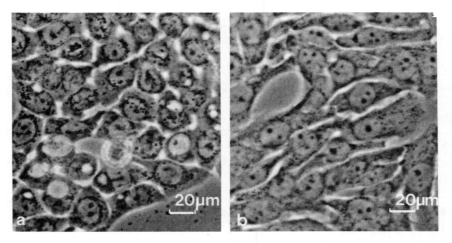

Figure 2. Effect of 1.5% DMSO on the morphology of HT29
cells: a) HT29 cells before addition of DMSO,
b) HT29 cells after 48 hours exposure to 1.5%
DMSO.

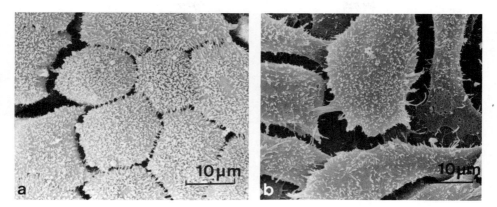

Figure 4. The surface structures of HT29 cells (a)
before and (b) after exposure to DMSO appear
similar under scanning electron microscopy.

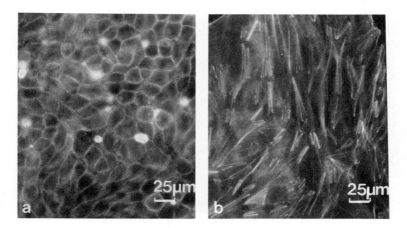

Figure 3. The arrangement of actin filaments
changes from:
a) untreated HT29 cells to b) DMSO-treated
cells as visualised by phalloidin-rhodamine
fluorescence.

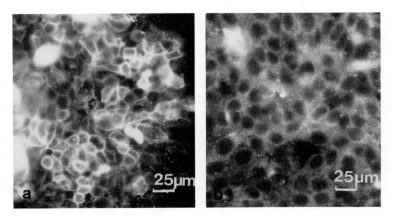

Figure 5. Immunofluorescent staining of CEA shows a
decrease in intensity between   a) untreated
and   b) DMSO-treated HT29 cells.

The expression of CEA is significantly decreased during the
first 5 - 6 days of DMSO exposure after which it remains at
a low level in the DMSO-treated cells. This effect was
observed by immunofluorescence studies and an immunoassay
(Figs. 5 and 6). All of the above changes were reversible,
the cells resuming their previous characteristics upon the
withdrawal of DMSO.

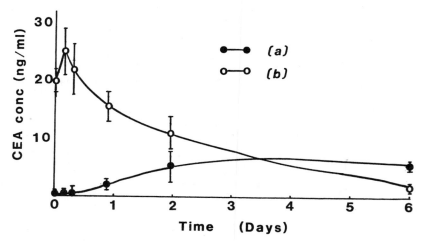

Figure 6. Effect of 1.5% DMSO on the levels of CEA in
HT29 cell lysates: a) HT29 cells pretreated
with 1.5% DMSO for 2 weeks, the DMSO being
removed at time 0 indicated on the graph
b) 1.5% DMSO added to cells at time 0
(Bars indicate S.E.M).

HT29 cells	ng p62^{c-myc} / mg total cell protein
Control	27.2 ($\pm$ 2.3)
1.5% DMSO	5.0 ($\pm$ 0.15)

Table 1. Concentrations of c-myc protein present in HT29 cells before and after exposure to DMSO.

Southern blotting revealed 3-5 fold amplification of the c-myc gene and single copies of the three members of the ras gene family and c-fos. The c-myc protein is expressed at 4-6 fold higher levels in the untreated HT29 cells than in the treated, as demonstrated by an ELISA method (Table 1). Levels of the ras p21 protein appear also to be lower in the treated cells as demonstrated by Western blotting (Fig. 7) Levels of c-fos appeared low at all times as shown by ELISA.

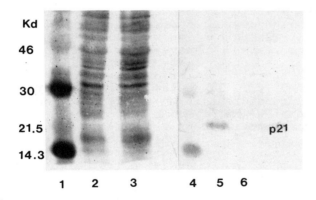

Figure 7. Western blot of the levels of p21 in HT29 cells with and without exposure to 1.5% DMSO. Lanes 1 and 4: molecular weight markers, lanes 2 and 5: HT29 cells untreated, lanes 3 and 6: HT29 cells treated with 1.5% DMSO. Lanes 1-3 were stained with Amido-black to indicate total protein, lanes 4-6 immunostained for ras p21 protein.

181

DISCUSSION

The apparent changes in the growth and morphology of the HT29 cells observed in this study correspond to partial detransformation and loss of the malignant phenotype similar to that reported previously with other colon cell lines (5,6,8,47). Certain aspects of this detransformation resemble the effects of sodium butyrate, in particular the morphology of the DMSO treated cells seems similar to the spindle-shaped cells reported to occur in flat foci after exposure to sodium butyrate (22). However sodium butyrate is capable of producing more marked characteristics of differentiation in colon cells, such as the appearance of brush borders and significant increases in some of the brush border associated enzymes such as alkaline phosphatase (8,9,19). These changes resemble closely those produced by altering the nutritional status of HT29 cells.

It is probable that DMSO exerts an effect via the plasma membrane, since it is known that DMSO decreases membrane fluidity (48,49). Sodium butyrate acts by inhibiting the deacetylation of nuclear proteins (50,51) and removal of glucose in the medium has been reported to result in alterations in intracellular nucleotide levels which are suggested to affect glycosylation of membrane proteins (16). Since oncogenes are implicated in the control of differentiation they are likely to be involved in these pathways of maturation produced by the above-mentioned compounds. It is not known at present whether these alternative methods of action bring about a common pattern of oncogene expression.

In general tissue CEA contents have been reported to correlate with the degree of differentiation of human colorectal tumours (52,53); poorly differentiated or anaplastic tumours tend to have a lower CEA content than do well-differentiated colon tumours. Higher CEA content also seems to correlate with better differentiated human colorectal tumour cell lines (54,55). However polar solvents appear to be inconsistent in their effects; increases in membrane-associated CEA of human colon cancer cell lines were reported after treatment with DMF (6,7) whilst in this study DMSO treatment reduced the CEA content as has been observed previously with another colon carcinoma line (10). It is possible that the progression from a malignant to a normal cell line indicated by the change in growth parameters and morphology may not be closely related to the pathways of differentiation. Alternatively it may be that different cell lines express characteristic patterns of phenotypic markers during differentiation. Whichever conclusion is valid it is obvious that CEA cannot be taken as a general indicator of degree of differentiation in colon cells although within a specified system it may be a useful marker.

182

It has previously been shown that HT29 cells have normal but overexpressed H-ras p21 (56). This is in keep with reports which suggest that frequent overexpression not activating point mutations of ras genes occurs in primary human gastric cancers (57). Although experiments on rodent fibroblasts suggest that p21 may be an important signal transducer in the regulation of proliferation of these cells (58-61), it has been shown that modulation of p21 did not affect cell cycle progression of HT29 cells (21) which suggests that it is not involved in primary control of HT29 cell proliferation. However the decrease in p21 observed in this study may be a secondary effect of the altered growth rate. It has been suggested elsewhere that since high levels of H-ras expression at the level of transcription were observed in mucin-secreting differentiated clonal cell lines derived from HT29 that ras gene expression is a marker for a particular differentiated state in colon cells rather than being directly related to tumorigenicity or transformation (62). Further support for this suggestion comes from the findings of Dexter (62) that expression of H-ras and K-ras increases in certain colon carcinoma cell lines when induced to differentiate with N-methylformamide (NMF) or sodium butyrate. Since ras expression has been shown by immunohistochemistry to be highest in the most differentiated cells at the top of the colonic crypt (63) it seems possible that ras expression is also related to differentiation in vivo. However in this study, no change in mucin-secreting status was detected in the DMSO-treated HT29 cells though a decrease in p21 was observed. A previous study also noted decreases in H-ras p21 expression in HT29 cells grown in hyperosmolar medium with no concomitant signs of differentiation (21) so it would appear that p21 expression is not only related to differentiation.

The c-myc gene has been shown to be involved in cell proliferation and differentiation; analysis of mRNA from human placentae show that there is a time-dependent variation in the level of c-myc expression, the peak transcriptional activity occurring between four and five weeks after conception (64). Induction of differentiation of the HL60 human promyelocytic leukemia line results in a marked decrease in transcription and expression of c-myc (65,66), the decrease being related to differentiation rather than to cellular proliferation (65). However no differences in expression of the myc family were observed in HT29 cells and differentiated clones derived from them (62). It is possible that the differences in the level of c-myc protein observed in this study are partly due to the differences in growth rate of the cells although since the decrease in c-myc expression is somewhat greater than the decrease in the growth rate there must be other contributory factors involved.

Amplification of the c-myc oncogene has been reported in primary human colon tumours (67) and in human colon carcinoma lines (68). Since it appears to play a significant role in cell proliferation and transformation it is probable that the amplification observed in this study is partly responsible for the malignant phenotype of the HT29 cell line.

It appears from this study that expression of both the ras and myc oncogenes is reduced by DMSO treatment of HT29 cells though further work is required to ascertain to what extent this is related to the decrease in growth rate of the cells. At present there are encouraging reports of the effects of anti-sense oligonucleotides directed against the c-myc oncogene on the inhibition of proliferation and induction of differentiation of HL-60 cells (69,70). It is planned to investigate the effects of myc and ras antisense oligonucleotides in the HT29 system to determine the importance of these oncogenes in the differentiation of the cells.

ACKNOWLEDGEMENTS

We wish to thank Paul Linstead for help with the S.E.M., Dr. Alan Prescott for assisting in the immunofluorescence work and Ruth Magrath for producing the photographs. S. N. is supported by an XNI grant from the Department of Education for Northern Ireland.

REFERENCES

(1)   S. J. Collins, F. W. Ruscetti, R. E. Gallagher, and R. C. Gallo, Terminal differentiation of human promyelocytic   leukemia cells induced by dimethylsulfoxide and other polar compounds, Proc. Natl. Acad. Sci. USA 75:2458 (1978).

(2)   Y. Kimhi, C. Palfrey, I. Spector, Y. Barak and U. Z. Littauer, Maturation of neuroblastoma cells in the presence of dimethylsulphoxide, Proc. Natl. Acad. Sci. USA 73: 462 (1976).

(3)   D. L. Dexter, N,N-Dimethylformamide-induced morphological differentiation and reduction of tumorigenicity in cultured mouse rhabdomyosarcoma cells, Cancer Res. 37:3136 (1977).

(4)   E. Huberman, C. Heckman, and R. Langenbach, Stimulation of differentiated functions in human melanoma cells by tumor-promoting agents and dimethyl sulphoxide, Cancer Res. 39:2618 (1979).

(5)   D. L. Dexter, J. A. Barbosa and P. Calabresi, N,N-Dimethylformamide-induced alteration of cell culture characteristics and loss of tumorigenicity in cultured human colon carcinoma cells, Cancer Res. 39:1020 (1979).

(6) D. L. Dexter and J. C. Hager, Maturation-induction of tumor cells using a human colon carcinoma model, Cancer 45:1178 (1980).

(7) J. C. Hager, D. V. Gold, J. A. Barbosa, Z. Fligiel, F. Miller and D. L. Dexter, N,N-Dimethylformamide-induced modulation of organ- and tumor-associated markers in cultured human colon carcinoma cells, J. Natl. Cancer Inst. 64:439 (1980).

(8) Y. S. Kim, D. Tsao, B. Siddiqui, J. S. Whitehead, P. Arnstein, J. Bennett and J. Hicks, Effects of sodium butyrate and dimethyl-sulphoxide on biochemical properties of human colon cancer cells, Cancer 45:1185 (1980).

(9) A. Morita, D. Tsao and Y. S. Kim, Effect of sodium butyrate on alkaline phosphatase in HRT-18, a human rectal cancer cell line, Cancer Res. 42:4540 (1982).

(10) D. Tsao, A. Morita, A. Bella, P. Luu and Y. S. Kim, Differential effects of sodium butyrate, dimethyl sulfoxide and retinoic acid on membrane-associated antigen, enzymes and glycoproteins of human rectal adenocarcinoma cells, Cancer Res. 42:1052 (1982).

(11) M. Pinto, S. Robine-Leon, M-D. Appay, M. Kedinger, N. Triadou, E. Dussaulx, B. Lacroix, P. Simon-Assmann, K. Haffen, J. Fogh and A. Zweibaum, Enterocyte-like differentiation and polarization of the human colon carcinoma cell line CaCO-2 in culture, Biol. Cell 47:323 (1983).

(12) Y. S. Chung, I. S. Song, R. H. Erickson, M. H. Sleisenger and Y. S. Kim, Effect of growth and sodium butyrate on brush border membrane-associated hydrolases in human colorectal cancer cell lines, Cancer Res. 45:2976 (1985).

(13) M. Rousset, The human colon carcinoma cell lines HT29 and CaCO-2 : two in vitro models for the study of intestinal differentiation, Biochimie 68:1035 (1986).

(14) I. Chantret, A. Barbat, E. Dussaulx, M. Brattain, and A. Zweibaum, Epithelial polarity, villin expression and enterocytic differentiation of cultured human colon carcinoma cells: a survey of twenty cell lines, Cancer Res. 48:1936 (1988).

(15) M. Pinto, M-D. Appay, P. Simon-Assmann, G. Chevalier, N. Dracopoli, J. Fogh and A. Zweibaum, Enterocytic differentiation of cultured human colon cancer cells by replacement of glucose by galactose in the medium, Biol. Cell 44:193 (1982).

(16) B. M. Wice, G. Trugman, M. Pinto, M. Rousset, G. Chevalier, E. Dussaulx, B. Lacroix and A. Zweibaum, The intracellular accumulation of UDP-N-acetylhexosamines is concomitant with the inability of human colon cancer cells to differentiate, J. Biol. Chem. 260:139 (1985).

(17) E. Chastre, S. Emami, G. Rosselin and C. Gespach, Vasoactive intestinal peptide receptor activity and specificity during enterocyte-like differentiation and retrodifferentiation of the human colonic cancerous subclone HT29-18, FEBS Lett 188:197 (1985).

(18) A. Zweibaum, M. Pinto, G. Chevalier, E. Dussaulx, N. Triadou, B. Lacroix, K. Haffen, J.-L. Brun and M. Rousset, Enterocytic differentiation of a subpopulation of the human colon tumor cell line HT-29 selected for growth in sugar-free medium and its inhibition by glucose, J. Cell. Physiol. 122:21 (1985).

(19) F. Herz, A. Schermer, M. Halwer and L. H. Bogart, Alkaline phosphatase in HT-29, a human colon cancer cell line: influence of sodium butyrate and hyperosmolality, Arch. Biochem. Biophys. 210:581 (1981).

(20) F. Herz and M. Halwer, Synergistic induction of alkaline phosphatase in colonic carcinoma cells by sodium butyrate and hyperosmolality, Biochim. Biophys. Acta 718:220 (1982).

(21) B. Czerniak, F. Herz, R. P. Wersto and L. G. Koss, Modification of Ha-ras oncogene p21 expression and cell cycle progression in the human colonic cancer cell line HT-29, Cancer Res. 47:2826 (1987).

(22) C. Augeron and C. L. Laboisse, Emergence of permanently differentiated cell clones in a human colonic cancer cell line in culture after treatment with sodium butyrate, Cancer Res. 44:3961 (1984).

(23) C. Huet, C. Sahuquillo-Merino, E. Coudrier and D. Louvard, Absorptive and mucus-secreting subclones isolated from a multipotent intestinal cell line (HT29) provide new models for cell polarity and terminal differentiation, J. Cell. Biol. 105:345 (1987).

(24) B. Dudouet, S. Robine, C. Huet, C. Sahuquillo-Merino, L. Blair, E. Coudrier and D. Louvard, Changes in villin synthesis and subcellular distribution during intestinal differentiation of HT29-18 clones, J. Cell Biol. 105:359 (1987).

(25) A. Zweibaum, N. Triadou, M. Kedinger, C. Augeron, S. Robine-Leon, M. Pinto, M. Rousset and K. Haffen, Sucrase-isomaltase: a marker of foetal and malignant epithelial cells of the human colon, Int. J. Cancer 32:407 (1983).

(26) A. Zweibaum, H.-P. Hauri, E. Sterchi, I. Chantret, K. Haffen, J. Bamat and B. Sordat, Immunohistological evidence, obtained with monoclonal antibodies, of small intestinal brush border hydrolases in human colon cancers and foetal colons, Int. J. Cancer 34:591 (1984).

(27) A. Zweibaum, Enterocytic differentiation of cultured human colon cancer cell lines: negative modulation by D-glucose, in: "Ion Gradient-Coupled Transport" INSERM Symposium No.26 F. Alvarado and C. H. van Os, eds. Elsevier Science Publishers B.V.(1986).

(28) C. J. Marshall, Oncogenes, J. Cell. Sci. Suppl. 4:417 (1986).

(29) S. Nishimura and T. Sekiya, Human cancer and cellular oncogenes, Biochem. J. 243:313 (1987).

(30) J. Fogh and G. Trempe, New human tumor cell lines, in:

"Human Tumor cells in Vitro," J. Fogh, ed., Plenum
Press, New York (1975).

(31) R. I. Freshney, "Culture of animal cells," Alan R.
Liss, Inc., New York (1983).

(32) H. Faulstich, H. Trischmann and D. Mayer, Preparation
of Tetramethylrhodaminyl-phalloidin and uptake of the
toxin into short-term cultured hepatocytes by
endocytosis, Exp. Cell Res. 144:73 (1983).

(33) P. J. Harwood, D. W. Britton, P. J. Southall, G. M.
Boxer, G. Rawlins and G. T. Rogers, Mapping epitope
characteristics on carcinoembryonic antigen, Br. J.
Cancer 54:75 (1986).

(34) H. P. Hauri, E. E. Sterchi, D. Bienz, J. A. Fransen and
A. Marxer, Expression and intracellular transport of
microvillus membrane hydrolases in human intestinal
epithelial cells, J. Cell Biol. 101:838 (1985).

(35) J. P. Moore, D. C. Hancock, T. D. Littlewood and G. I.
Evan, A sensitive and quantitative enzyme-linked
immunosorbence assay for the c-myc and N-myc
oncoproteins, Oncogene Res. 2:65 (1987).

(36) J. P. Moore and G. I. Evan, Immunoassays for
oncoproteins, Nature 327:733 (1987).

(37) D. M. Tidd, P. Hawley, H. M. Warenius and I. Gibson,
Evaluation of N-ras oncogene antisense, sense and
nonsense sequence methylphosphonate oligonucleotide
analogues, Anti-Cancer Drug Design 3:117 (1988).

(38) D. J. Capon, E. Y. Chen, A. D. Levinson, P H. Seeburg
and D. V. Goeddel, Complete nucleotide sequences of the
T24 human bladder carcinoma oncogene and its normal
homologue, Nature 302:33 (1983).

(39) M. J. Murray, J. M. Cunningham, L. F. Parada, F.
Dautry, P. Lebowitz and R. A. Weinberg, The HL-60
transforming sequence: A ras oncogene coexisting with
altered myc genes in hematopoietic tumors, Cell 33:749
(1983).

(40) M. S. McCoy, C. I. Bargmann and R. A. Weinberg,
Human-colon carcinoma Ki-ras2 oncogene and its
corresponding proto-oncogene, Mol.Cell.Biol. 4:1577
(1984).

(41) T. H. Rabbitts, A. Forster, P. Hamlyn and R. Baer,
Effect of somatic mutation within translocated c-myc
genes in Burkitt's lymphoma, Nature 309:592 (1984).

(42) R. Treisman, Transient accumulation of c-fos RNA
following serum stimulation requires a conserved 5'
element and c-fos 3' sequences, Cell 42:889 (1985).

(43) L. W. Coggins, G. J. Grindlay, J. K. Vass, A. A.
Slater, P. Montague, M. A. Stinson and J. Paul,
Repetitive DNA sequences near 3 human beta-type globin
genes, Nuc. Acids Res. 8:3319 (1980).

(44) P. Linstead, B. Jennings, A. Prescott, P. Hawley, R.
Warn and I. Gibson, Scanning electron microscopy and
the transformed phenotype, Micron et Microscopica Acta
19:155 (1988).

(45) J. Schmitz, H. Preiser, D. Maestracci, B. K. Ghosh, J.
J. Cerda and R. K. Crane, Purification of the human

intestinal brush border membrane, Biochim. Biophys. Acta 323:98 (1973).

(46) A. Garen and C. Levinthal, A fine-structure genetic and chemical study of the enzyme alkaline phosphatase of E. Coli. I. Purification and characterization of alkaline phosphatase, Biochim. Biophys. Acta 38:470 (1960).

(47) D. L. Dexter, G. W. Crabtree, J. P. Stoeckler, T. M. Savarese, L. Y. Ghoda, T. L. Rogler-Brown, R. E. Parks and P. Calabresi, N,N-dimethylformamide and sodium butyrate modulation of the activities of purine-metabolizing enzymes in cultured human colon carcinoma cells, Cancer Res. 41:808 (1981).

(48) G. H. Lyman, H. D. Preisler and D. Papahadjopoulos, Membrane action of DMSO and other chemical inducers of Friend leukaemic cell differentiation, Nature 262:360 (1976).

(49) S. H. C. Ip and R. A. Cooper, Decreased membrane fluidity during differentiation of human promyelocytic leukemia cells in culture, Blood 56:227 (1980).

(50) G. Vidali L. C. Boffa, E. M. Bradbury and V. G. Allfrey, Butyrate suppression of histone deacetylation leads to accumulation of multiacetylated forms of histones H3 and H4 and increased DNase I sensitivity of the associated DNA sequences, Proc. Natl. Acad. Sci. USA 75:2239 (1978).

(51) R. Reeves and P. Cserjesi, Sodium butyrate induces new gene expression in Friend erythroleukemic cells, J. Biol. Chem. 254:4283 (1979).

(52) H. Denk, G. Tappeiner, R. Eckerstorfer and J. H. Holzner, Carcinoembryonic antigen (CEA) in gastrointestinal and extragastrointestinal tumors and its relationship to tumor-cell differentiation, Int. J. Cancer 10:262 (1972).

(53) J. Breborowicz, G. C. Easty and A. M. Neville, The production of carcinoembryonic antigen (CEA) by human colonic carcinomas and normal colonic mucosa in monolayer and organ culture, Ann. Immunol. 124:613 (1973).

(54) A. Leibovitz, J. C. Stinson, W. B. McCombs, C. E. McCoy, K. C. Mazur and N. D. Mabry, Classification of human colorectal adenocarcinoma cell lines, Cancer Res. 36:4562 (1976).

(55) Z. R. Shi, D. Tsao and Y. S. Kim, Subcellular distribution, synthesis, and release of carcinoembryonic antigen in cultured human colon adenocarcinoma cell lines, Cancer Res. 43:4045 (1983).

(56) M. Perucho, M. Goldfarb, K. Shimizu, C. Lama, J. Fogh and M. Wigler, Human-Tumor-Derived cell lines contain common and different transforming genes, Cell 27:467 (1981).

(57) K. Fujita, N. Ohuchi, T. Yao, M. Okumara, Y. Fukushima, Y. Kanakura, Y. Kitamura and J. Fujita, Frequent overexpression, but not activation by point mutation, of ras genes in primary human gastric cancers, Gastroenterology 93:1339 (1987).

(58) J. R. Feramisco, M. Gross, T. Kamata, M. Rosenberg and R. W. Sweet, Microinjection of the oncogene form of the human H-ras (T-24) protein results in rapid proliferation of quiescent cells, Cell 38:109 (1984).

(59) D. W. Stacey and H-F Kung, Transformation of NIH-3T3 cells by microinjection of Ha-ras p21 protein, Nature 310:508 (1984).

(60) J. R. Feramisco, R. Clark, G. Wong, N. Arnheim, R. Milley and F. McCormick, Transient reversion of ras oncogene-induced cell transformation by antibodies specific for amino acid 12 of ras protein, Nature 314:639 (1985).

(61) L. S. Mulcahy, M. R. Smith and D. W. Stacey, Requirement for ras proto-oncogene function during serum-stimulated growth of NIH 3T3 cells, Nature 313:241 (1985)

(62) L. H. Augenlicht, C. Augeron, G. Yander and C. Laboisse, Overexpression of ras in mucus-secreting human colon carcinoma cells of low tumorigenicity, Cancer Res. 47:3763 (1987).

(63) P. Garin-Chesa, W. J. Rettig, M. R. Melamed, L. J. Old and H. L. Niman, Expression of p21ras in normal and malignant human tissues: Lack of association with proliferation and malignancy, Proc. Natl. Acad. Sci, USA 84:3234 (1987).

(64) S. Pfeifer-Ohlsson, A. S. Gronsten, J. Rydnert and 4 others, Spatial and temporal pattern of cellular myc oncogene expression in developing human placenta: Implications for embryonic cell proliferation, Cell 38:585 (1984).

(65) J. Filmus and R. N. Buick, Relationship of c-myc expression to differentiation and proliferation of HL-60 cells, Cancer Res. 45:822 (1985).

(66) L. E. Grosso and H. C. Pitot, Transcriptional regulation of c-myc during chemically induced differentiation of HL-60 cultures, Cancer Res. 45:847 (1985).

(67) R. J. Alexander, J. N. Buxbaum and R. F. Raicht, Oncogene alterations in primary human colon tumours, Gastroenterology 91:1503 (1986)

(68) K. Alitalo, M. Schwab, C. C. Lin, H. E. Varmus and J. M. Bishop, Homogeneously staining chromosomal regions contain amplified copies of an abundantly expressed cellular oncogene (c-myc) in malignant neuroendocrine cells from a human colon carcinoma, Proc. Natl. Acad. Sci. USA 80:1707 (1983).

(69) E. L. Wickstrom, E. Wickstrom, G.H. Lyman and D. L. Freeman, HL-60 cell proliferation inhibited by an anti-c-myc pentadecadeoxynucleotide, Fed. Proc. 45:1708 (1986).

(70) J. T. Holt, R. L. Redner and A. W. Neinhuis, An oligomer complementary to c-myc inhibits proliferation of HL60 promyelocytic cells and induces differentiation, Mol.Cell.Biol. 8:963 (1987).

THE *ras* ONCOGENE AND MYOGENIC COMMITMENT AND DIFFERENTIATION

Terry P. Yamaguchi, Helen H. Tai, David J. Kelvin, Gilles Simard, Andrew Sue-A-Quan, Michael J. Shin and Joe A. Connolly

Department of Anatomy, University of Toronto, Toronto, Canada M5S 1A8

ABSTRACT   When proliferating BC3H1 muscle cells are shifted to low serum conditions, they exit from the cell cycle and differentiate, activating a family of muscle-specific genes. Addition of the purified growth factors, fibroblast growth factor (FGF) or thrombin reverses this process and stimulates these cells to reenter the cell cycle. Pertussis toxin (PT) blocks thrombin's, but not FGF's, effects on muscle proliferation/differentiation. Thrombin, therefore, requires a G protein to transduce its signal. In addition, PT promoted differentiation in the presence of high concentrations of serum. Serum then contains a mitogen that signals through a PT-sensitive pathway in order to promote proliferation and inhibit muscle gene transcription. Transfection of the activated Ha-*ras* oncogene into BC3H1 and 10T1/2 cells blocked muscle differentiation in both of these lines. PT could not rescue the *ras*-mediated inhibition of differentiation. These results suggest that G protein-like molecules play important roles in transducing growth factor signals that control myogenesis.

INTRODUCTION

The development of muscle cells provides an excellent model system for experimental analysis of the processes of commitment, proliferation and differentiation. Myoblasts are cells that arise from the conversion of multipotential mesodermal progenitor cells into commited muscle precursors. They are proliferating, mononucleated stem cells which, upon withdrawal from the cell cycle, differentiate into myotubes. Muscle cell differentiation is characterized morphologically by the fusion of myoblasts into multinucleated myotubes, and biochemically by the coordinate activation of a family of muscle-specific genes including a-actin, and several myosins (Merlie et al, 1977).

Growth factor-like molecules may have important roles in development acting as morphogenetic and differentiation signals (for review, see Mercola and Stiles, 1988). Indeed, growth factors appear to play a significant role in muscle development. Several *in vitro* muscle cell systems (both cell lines and primary muscle cell cultures) have provided evidence that growth factors, including FGF and TGF-B, (both of which induce the formation of mesoderm tissue from *Xenopus* animal pole ectoderm explants (Slack et al., 1987; Kimelman and Kirschner, 1987; Rosa et al., 1988)) regulate myoblast proliferation and differentiation (for review, see Florini and Magri, 1988).

Our approach has been to utilize two mouse embryonic cell lines, C3H10T1/2 and BC3H1. C3H10T1/2 cells (clone 8) are multipotential mesodermal precursor cells to muscle, adipocyte and chondrocyte cell types (Taylor and Jones, 1982) that we are using in order to understand the signals that specify commitment to these lineages. The role of growth factors in the subsequent processes of myoblast proliferation and differentiation can be uniquely studied using the non-fusing mouse muscle cell line BC3H1 (Schubert et al., 1974). These cells stop proliferating when they are serum or growth factor deprived, and begin to differentiate, expressing muscle-specific proteins such as creatine kinase (Munson et al., 1982). However readdition of serum or purified growth factors to these differentiated cells shuts down muscle protein synthesis and stimulates them to reenter the cell cycle (Lathrop et al., 1985).

Relatively little is known about these growth factors and their signal transduction mechanisms in muscle development. G proteins are membrane-associated, GTP-binding proteins which are thought to transduce signals from extracellular receptors to intracellular effector molecules (for review, see Gilman, 1987). A subset of G proteins have been extensively characterized through the use of pertussis toxin (PT), a protein that catalyzes the ADP-ribosylation of them and thereby interferes with their function. Recent evidence suggests that G proteins may act to link growth factor receptors and inositol phosphate hydrolysis (Taylor and Merritt, 1986). In addition, the *ras* oncogene, a membrane-associated GTP-binding protein with high homology to the alpha subunit of the G protein family (Hurley et al., 1984), could be involved in signalling from cell surface receptors to the nucleus (Smith et al., 1986).

We show that the growth factors FGF and thrombin both inhibit muscle differentiation in BC3H1 cells and that thrombin's effect, but not FGF's, is sensitive to PT. PT alone will induce differentiation, even in the presence of high concentrations of serum. Introduction of an activated Ha-*ras* oncogene completely blocked muscle differentiation in BC3H1 cells. In addition, stable incorporation of this oncogene into 10T1/2 cells, also prevented the formation of muscle colonies.

MATERIALS AND METHODS

Cells: BC3H1 cells (Schubert et al., 1974), obtained from the American Type Culture Collection, were cultured as described (Kelvin et al., 1989a). 10T1/2 cells and the cell line 245 (Trimble et al., 1986) was provided by T. Haliotis and N. Hozumi (Mount Sinai Hospital Research Institute, Toronto). These cells were maintained in Dulbecco's modified Eagle's medium (DME) with 4.5 g/l glucose, 10mM HEPES supplemented with 10% FBS, penicillin (100 U/ml) and streptomycin (100 ug/ml). To induce differentiation cells were switched to the same medium containing 2% horse serum (HS).

Differentiation and Northern Analysis: CK activity measurements and Northern blot analysis were performed as previously described (Kelvin et al., 1989a).

Muscle Colony Immunostaining: 10T1/2 and 245 cells were plated at 2000 cells/100 mm collagenized dish in 10% FBS medium. Cells were treated with 3 uM 5-azacytidine for 24 hours. Cells were then grown in 10% FBS medium until confluency (7-10 days) whereupon cells were switched to 2% HS to induce differentiation. Dishes were fixed in AFA (Davis et al., 1987) 4-5 days later. Fixed colonies were incubated in a 1/500 dilution of QBM-2 (a mouse monoclonal against myosin light chain provided by Dr. P.A. Merrifield) in Tris-buffered saline/Tween (TBST) overnight. Immunostaining was performed using the Protoblot alkaline phosphatase detection kit (Promega Biotec). Colonies were scored for myogenesis by examining dishes on a grid under brightfield and phase contrast optics.

RESULTS

BC3H1 cells proliferate in 20% FBS (growth medium) and do not express detectable amounts of CK mRNA or CK activity (Fig. 1). Shifting the cells to 1/2% FBS (differentiation medium) results in cessation of proliferation and a turn-on of muscle genes exemplified by the increase in CK mRNA over four days (Fig. 1b) and a corresponding increase in CK activity (Fig. 1a). Readdition of serum to these cells results in the shutdown of these muscle genes and reentry into the cell cycle (Munson et al., 1982). The same effect is achieved if purified FGF or thrombin is added to differentiated cells. Both FGF and thrombin in 1/2% FBS inhibit the activity of CK (Fig. 2) in a dose-dependent manner and stimulate the incorporation of [3H]-thymidine (Kelvin et al., 1989b).

In an effort to elucidate the mechanisms through which serum exerts its effects, we used PT to identify G proteins that could play a role in transducing growth signals. We found that PT, in a dose-dependent fashion, reversibly blocked the normal proliferative BC3H1 cell response to 20% FBS and induced differentiation, even though these cells were still in growth medium (Kelvin et al., 1989a). PT in 20% FBS induced the turn-on of at least two muscle-specific markers: myosin light chain protein, and CK mRNA (Kelvin et al., 1989a). PT also enhanced CK activity seen when BC3H1 cells were shifted to 1/2% FBS (Fig. 2). It appears then that PT is interfering with a mitogenic signal present in serum which normally represses muscle gene activity.

Could PT be blocking signals from the FGF or thrombin receptor? FGF inhibited CK activity in the presence or absence of PT (Fig. 2). In addition, PT did not block the FGF induced incorporation of [3H]-thymidine (Kelvin et al., 1989b) suggesting that FGF acts to stimulate proliferation and inhibit differentiation in BC3H1 cells through a PT-insensitive pathway. PT, however, completely blocked thrombin's effects on muscle differentiation (Fig. 2) and proliferation (Kelvin et al., 1989b) suggesting that thrombin signals through a PT-sensitive pathway. PT therefore distinguishes at least two distinct growth factor pathways involved in controlling muscle differentiation. In addition, PT's effects suggest that G proteins may play an important role in transducing signals controlling this development.

We attempted to further elucidate the role of G proteins in muscle differentiation by transfecting cells with the val-12 Harvey *ras* oncogene, an oncogenic G protein homologue. Stable incorporation of this oncogene into BC3H1 cells (cell line designated BCT9; see Kelvin et al., 1989b) resulted in the complete inhibition of muscle differentiation, in agreement with previously published reports (Olson et al., 1987; Payne et al., 1987). Figure 3 demonstrates that the *ras* oncogene blocks the CK activity normally seen in differentiation medium. In addition, it shows that PT did not induce any detectable levels of CK activity in *ras*-transfected cells. Thus the inhibition of differentiation by the activated *ras* oncogene could not be rescued by PT.

In an attempt to try and uncover factors involved in controlling developmental lineages, we obtained the cell line 245, a 10T1/2 cell line containing a stably incorporated Ha-*ras* oncogene (Trimble et al., 1986). The mesoderm-like cell line10T1/2 can be converted into differentiated muscle, adipocytes and chondrocytes by treatment with 5-azacytidine (Taylor and Jones, 1979). Constitutive expression of the *ras* oncogene in 10T1/2 cells almost completely blocks the formation of muscle colonies induced by 5-azacytidine. Immunostaining with monoclonal antibodies to skeletal muscle myosin revealed that the activated *ras* oncogene blocked both morphological and biochemical muscle differentiation (Fig. 4). Thus *ras* blocks muscle differentiation in very different muscle cell lines, BC3H1 and 10T1/2.

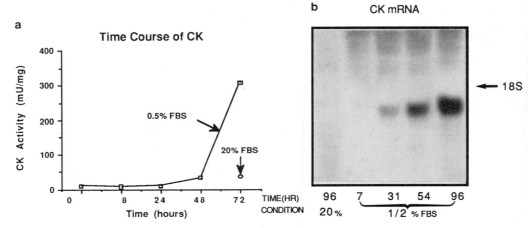

Figure 1. Time course of creatine kinase (CK) turn-on. (a) Graph showing increase in CK activity levels over 72 hours in 1/2% FBS (squares). The circle indicates CK activity in cultures maintained in 20% FBS for 72 hours. (b) Northern blot of BC3H1 RNA isolated from cultures shifted to 1/2% FBS for 0, 7, 31, 54, and 96 hours and in 20% FBS for 96 hours. Blots were probed with a CK cDNA probe labelled with [32P].

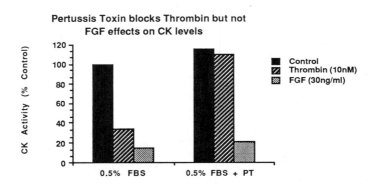

Figure 2. Growth factor and pertussis toxin effects on creatine kinase expression in BC3H1 cells. Control cells were plated in 20% FBS and switched to 1/2% FBS 24 hours later to induce differentiation. CK activity was measured 4 days later and experimental values expressed as percentages of this control level of CK activity. Cells were treated with 10nM thrombin or 30 ng/ml FGF with or without 10 ng/ml pertussis toxin in 1/2% FBS.

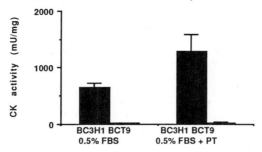

Figure 3. Effect of activated *ras* on differentiation. Control BC3H1 cells (untransfected) and *ras*-transfected cells (BCT9) were grown in 20% or 1/2% FBS for 4 days and CK activity was measured at this time.

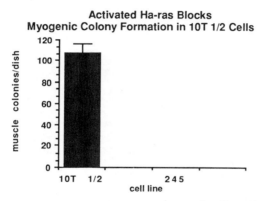

Figure 4. *ras* inhibits muscle colony formation in 10T1/2 cells. Transfection of a Ha-*ras* oncogene and selection of a clonal cell line (245) were as described (Trimble et al., 1986). Cells were treated and assessed for myosin positive colonies as described in Materials and Methods.

DISCUSSION

BC3H1 cells appear to be an excellent model system for the study of growth factor control of myoblast proliferation and differentiation. High concentrations of serum are mitogenic for these cells and appear to negatively control their differentiation. Removal of the serum component from the media creates a condition which allows BC3H1 cells to differentiate (Schubert et al., 1974). Serum-free or low serum conditions then allow for the dissection of serum components in order to establish the growth factors controlling myogenesis. Several growth factors inhibit muscle differentiation including FGF (Lathrop et al., 1985), EGF (Wang and Rubenstein, 1988), and TGF-B (Olson et al., 1986). We have now shown that thrombin inhibits muscle differentiation (Fig. 2) and stimulates the incorporation of [3H]-thymidine (Kelvin et al., 1989b). While TGF-B does not promote proliferation of these cells (Olson et al., 1986), FGF and EGF are mitogenic, acting in a synergistic fashion (Kelvin et al., 1989c). Thus the growth factor regulation of myoblast proliferation and differentiation is complex and appears to involve several signals operating in concert.

How does the muscle cell distinguish these signals? PT, even in 20% FBS, inhibits proliferation and stimulates BC3H1 differentiation. It ADP-ribosylates an intracellular substrate, thought to be a G protein (Kelvin et al., 1989a). PT presumably blocks a signalling pathway that is controlled by a serum component which normally stimulates proliferation and inhibits differentiation. In addition, cholera toxin (CT) which interferes with the function of a G protein distinct from the PT substrate, inhibits proliferation and differentiation, possibly by elevating cAMP levels (Kelvin et al., 1989a). Thus G proteins play an important role in transducing signals controlling myogenesis.

Do any of the previously mentioned growth factors signal through the PT-sensitive pathway? Our results show that FGF (and EGF- Kelvin et al., 1989b) stimulate thymidine incorporation and inhibit differentiation in the presence of PT. However, PT completely blocks the effects of thrombin suggesting that thrombin signals via a PT-sensitive G protein. Unfortunately, it is unclear whether thrombin is the component of serum that is blocked by PT. It appears then that growth factors signal through at least two different pathways to control myogenesis, one of which involves G proteins.

We investigated the possibility that the G protein homologue, p21*ras*, could also function in transducing signals that control myogenesis. In agreement with previous reports (Payne et al., 1987) we found that expression of the activated *ras* oncogene blocks BC3H1 differentiation (Kelvin et al., 1989b). In what signal pathway could *ras* be acting in? When *ras*-transfected cells were treated with PT, they continued to proliferate. Furthermore, PT could not stimulate these cells to differentiate. *ras* could be exerting its effect through stimulating the release of autocrine factors into the growth medium. However, conditioned media from activated *ras* transfected cells did not block the differentiation of control BC3H1 cells (Tai et al., 1989). Alternatively, *ras* could be exerting an effect upon growth factor receptors, by changing their number or affinity for ligand for instance. This seems unlikely since no significant difference was detected when the binding of iodinated EGF to BC3H1 and BCT31 (*ras*-transfected clonal cell line) cells was assessed (data not shown). Therefore, *ras* presumably functions "downstream" of PT's site of action, or it acts by transmitting growth factor signals through an intracellular PT-insensitive pathway.

The activated *ras* oncogene also blocks the azacytidine induced formation of muscle colonies from mesodermal 10T1/2 cells. The nuclear phosphoprotein myoD1 is expressed in myoblasts derived from 10T1/2 cells after azacytidine treatment but not in 10T1/2 cells themselves (Tapscott et al., 1988). Expression of a complementary DNA (cDNA) encoding MyoD1 in 10T1/2 cells (as well as in a number of other fibroblast and adipoblast cell lines) converts them into myoblasts (Davis et al., 1987). Thus myoD1 appears to be a gene that specifies commitment to the muscle lineage. We are currently testing for the possibility that *ras* could be preventing muscle colony formation by preventing the expression of the myoD1 gene and thereby blocking the commitment event. We are also interested in the effects that *ras* may have on the development of other mesodermal lineages.

If *ras* acts as a transducer of growth factor receptors, then it is plausible to reason that growth factors may act to signal turn-on of myoD1, and hence commitment to the muscle lineage. These signals, if they exist, have not yet been identified. It has been reported recently that the growth factors FGF and TGF-B can induce mesoderm tissue from explants of Xenopus animal pole (Kimelman and Kirschner, 1987; Rosa et al, 1988). We and others have shown that FGF and other growth factors stimulate myoblast proliferation and inhibit their differentiation. In addition, Seed and Hauschka (1988) have observed that FGF can also induce differentiation in chick wingbud. Thus it appears likely that growth factors will play a role in specifying developmental lineages. It is of interest then that preliminary experiments reveal that FGF alone could not induce the formation of muscle colonies from 10T1/2 cells, but enhanced the number formed after azacytidine treatment. Figure 5 summarizes the stages of muscle development where we think growth factors and *ras* may play regulatory roles.

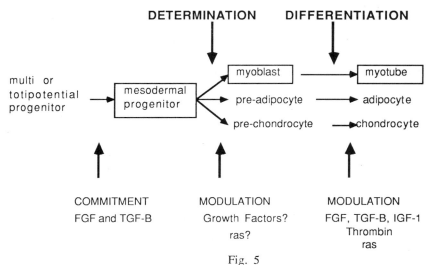

Fig. 5

REFERENCES

Davis, R.L., Weintraub, H., and Lassar, A. B.,1987, Expression of a single transfected cDNA converts fibroblasts to myoblasts, Cell, 51:987.

Florini, J.R. and Magri, K.A.,1988, Effects of growth factors on myogenic differentiation, Amer. J. Physiol. (in press).

Gilman, A.G.,1987, G proteins: tranducers of receptor-generated signals, Ann. Rev. Biochem., 56:615.

Hurley, J.B., Simon, M.I., Teplow, D.B., Robishaw, J.D., Gilman, A.G., 1984, Homologies between signal transducing G proteins and *ras* gene products, Science, 226:860.

Kelvin, D.J., Simard, G., Tai, H.H., Yamaguchi, T.P., and Connolly, J.A., 1989a, Growth factors, signalling pathways and the regulation of proliferation and differentiation in BC3H1 muscle cells. I. A pertussis toxin sensitive pathway is involved, J. Cell Biol., 108:159.

Kelvin, D.J., Simard, G., Sue-A-Quan, A., and Connolly, J.A., 1989b, Growth factors, signalling pathways and the regulation of proliferation and differentiation in BC3H1 muscle cells. II. Two signalling pathways distinguished by pertussis toxin and a potential role for the *ras* oncogene, J. Cell Biol., 108:169.

Kelvin, D.J., Simard, G., and Connolly, J.A., 1989c, FGF and EGF act synergistically to induce proliferation in BC3H1 myblasts, J. Cell. Physiol. (in press).

Kimelman, D. and Kirschner, M., 1987, Synergistic induction of mesoderm by FGF and TGF-B and the identification of an mRNA coding for FGF in the early X *Xenopus* embryo, <u>Cell</u>, 51:869.

Lathrop, B., Olson, E., and Glaser, L,1985, Control by fibroblast growth factor of differentiation in the BC3H1 muscle cell line, <u>J. Cell Biol.</u>, 100:1540.

Mercola, M., and Stiles, C.D., 1988, Growth factor superfamilies and mammalian embryogenesis, <u>Development</u>, 102:451.

Merlie, J. P., Buckingham, M.E., and Whalen, R.G., 1977, Molecular aspects of myogenesis, <u>Curr. Top. Dev. Biol.</u>, 11:61.

Munson, R., Caldwell, K.L., and Glaser, L., 1982, Multiple controls for the synthesis of muscle-specific proteins in BC3H1 cells, <u>J. Cell Biol.</u>, 92:350.

Olson, E.N., Sternberg, E., Hu, J.S., Spizz, G., and Wilcox, C., 1986, Regulation of myogenic differentiation by type-B transforming growth factor. <u>J. Cell Biol.</u>, 103:1799.

Olson, E.N., Spizz, G., and Tainsky, M.A.,1987, The oncogenic form of N-*ras* or H-*ras* prevent skeletal myoblast differentiation, <u>Mol. Cell. Biol.</u>,7:2104-2111.

Payne, P.A., Olson, E.N., Hsiau, P., Roberts, M.B., Perryman, M.B., and Schneider, M.D., 1987, An activated c-Ha-*ras* allele block the induction of muscle-specific genes whose expression is contingent upon mitogen withdrawal, <u>Proc. Natl. Acad. Sci. U.S.A.</u>, 84:8956.

Rosa, F., Roberts, A.B., Danielpour, D., Dart, L.L., Sporn, M.B., and Dawid, I.B., 1988, Mesoderm induction in amphibians: the role of TGF-B2-like factors, <u>Science</u>, 239:783.

Schubert, D., Harris, A.J., Devine, and C.E., Heinemann, S., 1974, Characterization of a unique muscle cell line, <u>J. Cell Biol.</u>, 61:398.

Seed, J., and Hauschka, S.D.,1988, Clonal analysis of vertebrate myogenesis. VIII. Fibroblast growth factor (FGF)-dependent and FGF-independent muscle colony types during chick wing development, <u>Dev. Biol.</u>, 128:40.

Slack, J.M.W., Darlington, B.G., Heatth, J.K., and Godsave, S.F., 1987, Mesoderm induction in early *Xenopus* embryos by heparin-binding growth factors, <u>Nature</u>, 326:197.

Smith, M.R., DeGudicibus, S.J., and Stacey, D.J.,1986, Requirement for c-ras proteins during viral oncogene transformation, <u>Nature</u>, 320:540.

Tai, H.H., Shin, M.J., Simard, G., Strauch, A.R., Connolly, J.A., 1989, Signalling pathways in the inhibition of myogenic differentiation by the *ras* oncogene, <u>J. Cell Biol.</u>, 107: abstr. 2738.

Tapscott, S.J., Davis, R.L., Thayer, M.J., Cheng, P.-F., Weintraub, H., and Lassar, A.B., 1988, MyoD1: A nuclear phosphoprotein requiring a *myc* homology region to convert fibroblasts to myoblasts, <u>Science</u>, 242:405.

Taylor, C.W., and Merritt, J.E.,1986, Receptor coupling to polyphosphoinositide turnover: a parallel with the adenylate cyclase system, <u>Trends Pharmacol. Sci.</u>, 7:238.

Taylor, S.M., and Jones, P.A., 1979, Multiple new phenotypes induced in 10T1/2 and 3T3 cells treated with 5-azacytidine, <u>Cell</u>, 17:771.

Trimble, W.S., Johnson, P.W., Hozumi, N., and Roder, J.C., 1986, Inducible cellular transformation by a metallothionein-*ras* hybrid oncogene leads to natural killer cell susceptibility, <u>Nature,</u> 321:782.

Wang, Y.-C. and Rubenstein, P.A.,1988, Epidermal growth factor controls smooth muscle a-isoactin expression in BC3H1 cells, <u>J. Cell Biol.</u>, 106:797.

Yamaguchi, T.P., Sue-A-Quan, G.T. and Connolly, J.A.,1989, An activated c-Ha-*ras* allele inhibits the formation of muscle colonies from 10T1/2 cells, <u>J. Cell Biol.,</u> 107:abstr.2739.

TUMORIGENIC CONVERSION OF <u>IN VIVO</u> DIFFERENTIATION COMPETENT MAMMARY

CELLS BY INTRODUCTION AND EXPRESSION OF <u>RAS</u> OR <u>MIL</u>(<u>RAF</u>) BUT NOT <u>MYC</u>

Walter H. Günzburg, Brian Salmons[*] and Robert Ullrich[*]

Abteilung für Molekulare Zellpathologie, Gesellschaft für Strahlen
und Umweltforschung mbH (GSF), D-8042 Neuherberg, FRG
[*]Lehrstuhl für Molekulare Tierzucht an der Ludwig-Maximillians
Universität, D-8000 München, FRG
[*]Oak Ridge National Laboratory, Biology Division, Oak Ridge,
TN 37831, USA

INTRODUCTION

Mammary cancer affects around 9% of women and is the most frequent
cause of death in women aged between 40 to 54 years old (Miller and
Bulbrook, 1986). Whilst the mechanisms involved in mammary tumorigenesis
are not understood, there is evidence for the existence of a human mammary
tumour virus that is related to the Mouse Mammary Tumour Virus (MMTV)
(Schlom et al., 1971; Crepin et al., 1984; Keydar et al., 1984; Ono et al.,
1986; Ono et al., 1987; Franklin et al., 1988). Additionally, a number of
oncogenes have been implicated in the tumorigenic process. Elevated
expression and/or amplification of the oncogenes <u>ras</u>, <u>myc</u> and <u>neu</u> have been
correlated with human mammary tumorigenesis (Horan-Hand et al., 1984;
DeBortoli et al., 1985; Ohuchi et al., 1986; Kraus et al., 1984; King et
al., 1985; Slamon et al., 1987; Kozbor and Croces, 1984; Escot et al.,
1986; Varley et al., 1987).

The mouse has been extensively studied as a model system for mammary
tumorigenesis. In addition to the <u>ras</u>, <u>myc</u> and <u>neu</u> oncogenes, a series of
genes (<u>int</u>-1, <u>int</u>-2, <u>int</u>-3, <u>int</u>-H) have been identified that become
transcriptionally activated upon integration of an MMTV provirus in their
vicinity (Nusse and Varmus, 1982; Peters et al., 1983; Gray et al., 1986;
Gallahan and Callahan, 1987). Recent evidence suggests that the human
homologue of <u>int</u>-2 is rearranged and amplified in primary breast carcinomas
(Varley et al., 1988; Zhou et al., 1988). However introduction and expression of cloned <u>int</u> genes in normal mouse mammary epithelial cells, although
subtly altering morphology and normal growth arrest mechanisms, does not
result in frank transformation (Brown et al., 1986), leaving the role of
these genes in the tumorigenic process still open to speculation (Salmons
and Günzburg, 1987).

Transgenic mice, in which transgenes have been expressionally targeted
to the mammary gland by the use of mammary gland specific promoters such as
that of MMTV or Whey Acidic Protein (WAP), have been created in an attempt
to directly demonstrate the involvement of the <u>ras</u>, <u>myc</u> and <u>neu</u> oncogenes
in the neoplastic conversion of the mammary gland. Although expression of
the <u>ras</u> and <u>myc</u> oncogenes **predisposed** such transgenic animals to mammary
cancer, other, as yet uncharacterised, events were also required for
complete mammary transformation (Stewart et al., 1984; Andres et al., 1987;
Sinn et al., 1987; Schoenenberger et al., 1988). These experiments leave

unanswered the question of whether the <u>ras</u> and <u>myc</u> oncogenes are directly involved in the genesis of mammary cancer.

A straightforward approach to determine the role of these oncogenes in mammary neoplasia is the direct introduction and expression of these oncogenes in mammary epithelial cells <u>in vitro</u>.

There are a number of difficulties associated with this approach. The mammary gland consists of a number of, as yet, poorly characterized cell types. Two major epithelial cell components, myoepithelial and alveolar, and one major stromal fibroblast cell type have been recognized. Attempts have been made to define biochemical markers for the major mammary gland components. These markers have been used to characterize the different-iation pathway involved from the presumed progenitor stem cell to the mature differentiated types of mammary epithelial cell. Although a number of independent workers have defined maturation lineages both for the human and rodent mammary gland (Rudland et al., 1979; Asch and Asch, 1985; Sonnenberg et al., 1986a; Sonnenberg et al., 1986b), little progress has been made in the overall understanding of these pathways. Primary mammary epithelial cells are thus heterogeneous mixtures of different cell types each at a different stage of differentiation.

Introduction and expression of genes into primary mammary cells there-fore is tantamount to introducing a gene into a mixture of cells, observing an effect, but not being able to identify (and thus characterize) the initial cell into which the gene was introduced. This approach thus does not allow the investigation of the more subtle, potentially differen-tiation-specific, effects of oncogene expression.

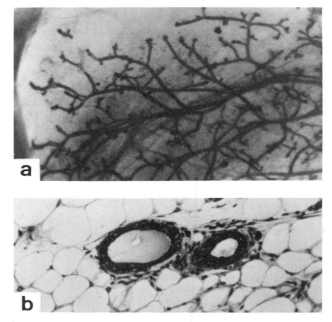

Fig. 1.(a) Cleared mammary fat pad after repopulation with EF43 cells
(b) Histology of alveolar differentiation after pituitary isograft

A number of established and characterized mammary cell lines are available which could serve as recipient cells for testing oncogene effects. Most of these however, whilst suited for investigations regarding the gross changes, brought about by oncogene expression, in cell growth and tumorigenicity, do not allow investigation of differentiation-related changes which are more likely to be relevant in the highly differentiation competent mammary gland.

We have established a unique, non-tumorigenic, mammary gland cell line, EF43, which although able to grow indefinitely in vitro, retains the ability to participate in normal differentiation when cultured in vivo. If the cell line is injected as a single cell suspension into the cleared mammary fat pad of a syngeneic mouse, the cells will organize themselves and grow out giving rise to a branching ductile system typical of the mammary gland (Fig. 1.a). Additionally, if the recipient mice receive pituitary isografts these duct systems differentiate further to give rise morphologically to both epithelial and myoepithelial components (Fig. 1.b) (Günzburg et al., 1988). This cell line is a perfect candidate recipient for testing oncogene effects upon mammary gland differentiation.

RESULTS AND DISCUSSION

Introduction and Expression of Oncogenes in EF43 cells

The clonal cell line, EF43, is amenable to both calcium-phosphate mediated DNA transfection and retroviral infection (Günzburg and Salmons, 1986; Günzburg et al., 1988). The ras oncogene was introduced into the EF43 cells by either transfection of the activated cellular Ha-ras in pEJ (Shih and Weinberg, 1982) or by infection with a retroviral vector construct (Fig. 2A) based on Moloney Murine Leukaemia Virus (MoMLV)(Cepko et al., 1984) in which the viral ras oncogene is expressed from the MoMLV promoter. Southern blot analysis of single cell transfected clones (J series) or a population of recombinant retrovirus infected cells (pZIP-ras-pop) revealed that these cells had acquired the ras oncogene (Fig. 3).

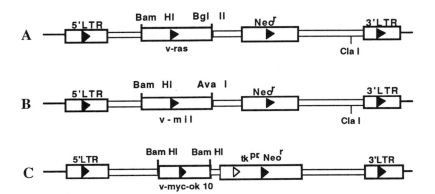

Fig. 2. Retroviral constructs used for introduction of oncogenes (A) pZIP-ras, (B) pZIP-mil, (C) pMMCV-neo (v-myc) (Wagner et al., 1985).

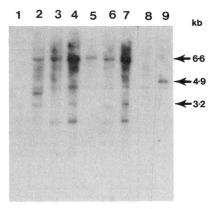

Fig. 3. Southern Blot analysis of ras transfected and
infected EF43 cells. 10 μg of cellular DNA was digested
with either Bam HI; lane 1, non-transfected EF43 cells;
lanes 2-4 transfected EF43 cell clones J1, J2 and J3;
lane 5, tumour outgrowth from J3; lanes 6-7, transfected
EF43 cell clones J6 and J7; or with Bam HI and Cla I;
lane 8, EF43 cells infected with "empty" virus; lane 9,
EF43 cells infected with ras virus (Fig. 2A). Hybridization probe is the first exon of the activated c-Ha-
ras in pEJ (Shih and Weinberg, 1982). Indicated are the
indicative 6.6kb c-ras and 3.2kb pBR322 fragments from
pEJ, and the 4.9kb v-ras fragment (Fig. 2A).

Both the transfected cell clones, and the infected population expressed
authentic p21ʳᵃˢ as determined by western blotting analysis using a polyclonal rabbit anti-ras serum (Fig. 4A). The levels of endogenous murine c-
ras were virtually undetectable. All of the ras expressing cells, regard-
less of whether the oncogene had been introduced by transfection or retroviral infection, showed identical properties, indicating that the method of
introduction does not affect the results obtained in such analyses.

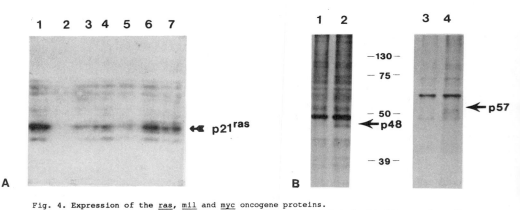

Fig. 4. Expression of the ras, mil and myc oncogene proteins.
(A) Membrane proteins from EF43 cells (lane 2); EF43 cells infected with ZIP-ras virus (lane
1) and five clones of EF43 cells transfected with the activated c-Ha-ras, J1 (lane 3); J2
(lane 4); J3 (lane 5); J6 (lane 6) and J7 (lane 7) were separated on a 14% polyacrylamide--
SDS gel. After transfer to nitrocellulose, ras specific proteins were visualized by
treatment of the filter with anti-p21ʳᵃˢ followed by incubation with iodinated protein A. The
p21ʳᵃˢ protein is marked.
(B) EF43 cells infected with empty virus (lanes 1 and 3), ZIP-mil (lane 2) or ZIP-myc (lane
4) virus were cultured in DMEM with 10⁻⁵M methionine plus 200μCi ³⁵S methionine for 3 hours.
Total cell proteins were extracted and immunoprecipitated with anti-mil (lanes 1 and 2) or
anti-myc (lanes 3 and 4) serum. The anti-mil serum specifically precipitates a protein of
48K. The anti-myc serum specifically precipitates p57ᵐʸᶜ. The positions of reference standard
proteins are marked.

EF43 cells expressing _ras_ or _mil(raf)_ but not _myc_ are transformed

EF43 cells containing and expressing the _ras_ oncogene, exhibit properties typically associated with transformed cells such as growth in semisolid medium and tumorigenicity in nude mice (Table 1). Similar transformation markers (Table 1) are displayed by EF43 cells expressing the _mil-
(raf)_ oncogene (Fig. 4B) introduced by retroviral infection with a recombinant _mil_ carrying virus (Fig. 3). This is in sharp contrast to EF43 cells
infected with, and expressing (Fig. 4B), a _myc_ containing retrovirus (Fig.
3) which do **not** display a classical transformed phenotype (Table 1).

The gross morphology of the _ras_ expressing EF43 cells was examined both
on plastic and in collagen matrices. The _ras_ transfected or infected EF43
cells are a little more refractile than either normal EF43 cells or EF43
cells infected with 'empty' retroviral vector (the recombinant viral construct lacking an oncogene) (Fig. 5). These morphological changes are even
less apparent when the cells reach confluence. The use of altered morphology as a marker for transformation thus is of less significance in these
studies with epithelial cells than in classical transformation studies
using fibroblasts where foci formation and morphological changes are
commonly used parameters.

EF43 cells, like other mammary epithelial cells, form branching ductile
structures when they are grown in collagen matrices (Fig. 5c). The introduction and expression of the _ras_ oncogene into these cells does not abolish the branching growth obtained in collagen (Fig. 5d).

EF43 cells expressing the _ras_ oncogene were more metabolically active
and showed a slightly increased rate of cell division as compared to parental EF43 cells or EF43 cells that had acquired and express the _mil(raf)_

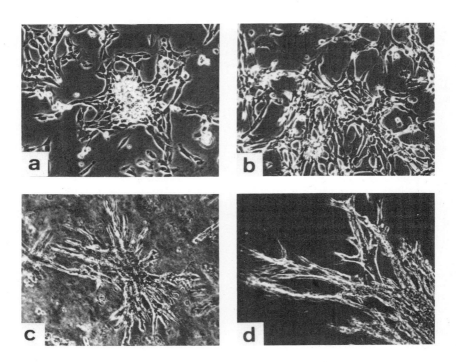

Fig. 5. Morphology of "empty" virus infected EF43 cells (a and c) or _ras_ expressing EF43
cells (b and d) grown on plastic (a and b) or in collagen matrices (c and d).

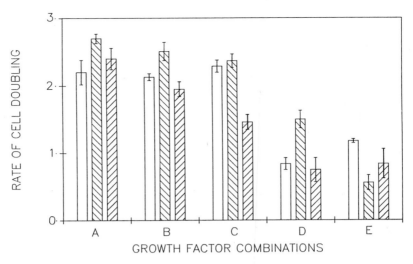

Fig. 6. Graphic representation of growth factor dependence of EF43 cells. The rate of growth from duplicate points of cells grown logarithmically over a period of six days was determined in medium containing (A) 0.5% foetal calf serum, 1μg/ml Insulin and 5ng/ml EGF; (B) 0.5% foetal calf serum and 5ng/ml EGF; (C) 0.5% foetal calf serum and 1μg/ml insulin; (D) 0.5% foetal calf serum; (E) 5ng/ml EGF and 1μg/ml insulin. Growth rate of EF43 cells is shown by open blocks, of EF43 cells infected with the ZIP-ras virus, EF43zip-ras-pop by descending shading (left to right) and of EF43 cells infected by the ZIP-mil virus, EF43zip-mil-pop by ascending shading (left to right). The error bars indicate the deviation of the actual data from the plotted mean.

oncogene (Fig. 6). Additionally expression of the ras oncogene rendered the cells less dependent upon exogenous growth factors such as Epidermal Growth Factor (EGF) or insulin, supplied in the culture medium. This is clearly a specific effect of the ras oncogene, since the mil(raf) oncogene, although also transforming EF43 cells (as determined by anchorage independent growth and tumorigenicity), does not alleviate growth factor dependence of EF43 cells. Another mammary derived cell line transformed by a ras oncogene has recently been shown to be similarly unresponsive to EGF, possibly due to a reduction in the number and affinity of EGF receptors (Salomon et al., 1987).

Interestingly, the ras expressing EF43 cells do not grow as well as the parental cells in defined (serum-free) culture conditions of Dulbecco's Modified Eagle Medium containing EGF and insulin. Presumably another factor, present in serum, is required for optimal growth of the ras expressing mammary cells. Recently, a serum component, that is not a known growth factor, has been found to enhance the emergence of tumorigenic fibroblasts upon transfection of the Ha-ras oncogene (Hsiao et al., 1987). Taken together, these observations suggest that the ras expressing EF43 cells show an **altered** rather than a **reduced** requirement for growth factors.

Expression of the ras oncogene may modify the pattern of secreted proteins synthesized by mammary epithelial cells. EF43 cells and the ras transfected clone, J3, were metabolically labeled in medium lacking foetal bovine serum but containing variously insulin, EGF, insulin and EGF, or no additives. In all growth conditions, the major secreted protein from the parental EF43 cells was a 42 KDa protein. The expression of this protein was much reduced in the ras expressing J3 cell clone and instead a protein of 34 KDa was the major secreted protein from these cells. Again the expression of this protein was not affected by any of the growth conditions

Table 1. EFFECTS OF ONCOGENE EXPRESSION UPON EF43 MAMMARY CELLS

ONCOGENE	GROWTH IN SOFT AGAR	NUMBER OF TUMOURS (LATENCY)		REPOPULATION OF CLEARED FAT PAD
		NUDE MICE	SYNGENEIC FAT PAD	
none	0	0/2 (49 days)	0/16 (54 days)	7/16
ras (transfected)	24%	9/10 (21 days)	63/64 (24 days)	0/64
ras (infected)	22%	2/2 (35 days)	16/16 (38 days)	0/16
mil (infected)	33%	n.t.	9/16 (54 days)	0/16
myc (infected)	0	n.t.	0/16 (54 days)	4/16

N.B. ras (transfected) refers to the average values obtained from 5 individual cell clones. n.t. = not tested.

used. Experiments are currently in progress to identify and characterize these proteins. The 42 KDa protein may be associated with the higher rate of cell doubling observed for EF43 cells when grown in serum-free medium.

Oncogene expression and the differentiation state of EF43 cells

EF43 cells transformed by the ras oncogene are no longer able to participate in the repopulation of cleared mammary fat pads upon injection into syngeneic mice. In contrast, such injected ras expressing EF43 cells form tumours very quickly. Since all five ras transfected cell lines and the ras virus infected population resulted in tumour formation, and 79 of the 80 mice tested had mammary tumours with an average latency of around 26 days (Table 1), it is unlikely that a subset of the injected cells were selected out on the basis of tumour formation. Thus, these experiments directly demonstrate that expression of the Ha-ras oncogene causes mammary tumorigenesis in this system.

One potential caveat to extending this conclusion to mammary cells in general is the fact that the EF43 cells are not normal mammary cells in that they grow indefinitely when cultured in vitro. However, the observation that both parental and empty virus infected EF43 cells repopulate the cleared mammary fat pad suggests that indefinite growth (immortality) is not sufficient, per se, to cause the development of mammary cancer. This would be consistent with the observation that expression of so called immortalizing genes in transgenic animals predispose them to mammary cancer, but that other events are also required for tumorigenic conversion (Stewart et al., 1984; Schoenenberger et al., 1988). It is still not clear whether high level expression of the ras oncogene alone (Spandidos and Wilkie, 1984) is sufficient for the development of mammary tumours. The existing data from transgenic animal studies suggest that other events are also required, alternatively high enough levels of p21^{ras} in the cell type that is destined to proliferate into a mammary tumour may not have been reached in these studies (Andres et al., 1987).

The mil(raf) oncogene, when expressed in EF43 cells, also results in tumour formation upon injection of the cells into cleared mammary fat pads (Table 1). This, together with the differences in in vitro growth factor dependence between ras and mil(raf) transformed EF43 cells, suggests that

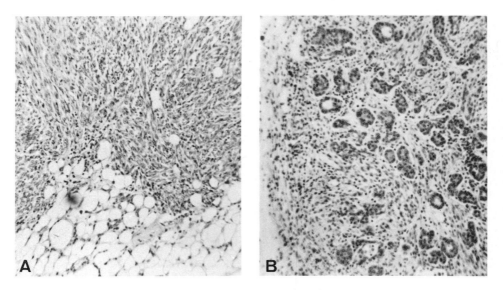

Fig. 7. Histology of tumours induced by ras (A) or mil (B). Cleared mammary fat pads were injected with 2x10⁻⁵ cells. EF43 cells expressing ras give rise to poorly differentiated anaplastic tumours whereas the mil expressing cells form relatively well differentiated adenocarcinomas.

alternative tumorigenic pathways exist. This oncogene is clearly not as potent as the ras oncogene, inducing tumours with a efficiency of 56% and requiring a slightly longer latency period (Table 1). These tumours are also relatively well differentiated, the ductal structures still being visible, as compared to the undifferentiated ras induced tumours (Fig. 7).

It is unlikely that all of the known oncogenes would confer tumorigenicity upon the EF43 cell line (indeed the myc oncogene does not - see below) and the transforming ability of the mil(raf) oncogene in EF43 cells may be explained by the previously shown involvement of this oncogene in carcinoma formation (Keski-Oja et al., 1982; Ishikawa et al., 1985).

The in vivo repopulation of the cleared mammary fat pad by EF43 cells was not abrogated by expression of v-myc (Table 1). It is possible that a critical threshold concentration of v-myc protein required for transformation was not reached in these cells. However this seems unlikely since the same retroviral construct has been shown to transform other cell types (Wagner et al., 1985), demonstrating the biological activity of the v-myc in this retrovirus. It has recently been reported that the myc oncogene causes enhanced growth of mammary structures that are histologically normal (Edwards et al., 1988). This is reminiscent of the preneoplastic stage of mammary gland transformation, a generalized hyperplasia (reviewed by Medina, 1988). Indeed, the immortalized nature of the EF43 cells could be regarded as hyperplasia, making transformation of these cells by myc, generally regarded as an immortalizing oncogene, unlikely. This is consistent with previous studies using established (immortalized) cell lines (Palmieri et al., 1983; Falcone et al., 1987; Zerlin et al., 1987).

Our results strongly suggest that the transformation of a single mammary epithelial cell leads to mammary tumorigenesis and thus argues

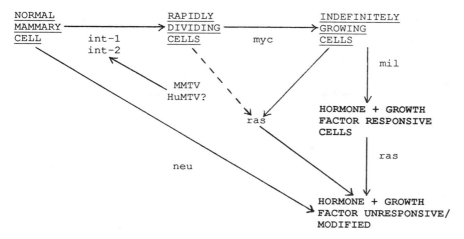

Fig. 8. Scheme of potential steps in mammary tumorigenesis
Underlined are hypothetical cells that are able to differentiate and are non-tumorigenic. In
bold are cells that are tumorigenic and show limited or no differentiation.

against one current hypothesis that mammary transformation is initiated in
a stromal cell and that the mitogenic signal is transferred to epithelial
cells as a result of the known interaction of different cell types in the
mammary gland. It is likely that there is more than one pathway that leads
to the uncontrolled growth of mammary cells. The frequency with which a
given pathway leads to mammary cancer depends on the number of events
required for overt tumour formation (Fig. 8).

Mammary gland targeted expression of the activated c-neu has been
recently shown to directly induce mammary adenocarcinomas (Muller et al.,
1988). In humans, amplification of the human homologue of this gene is
generally accepted to be associated with more aggressive tumours with poor
prognosis (Slamon et al., 1987) although this has been challenged (Ali et
al., 1988). It will be of interest to introduce the neu oncogene into EF43
cells and to compare the effects of the expression of this oncogene upon
the mammary gland specific properties. The availability of such a defined
cell line and the extraordinary ability of these cells to re-create the
mammary gland should allow both the elucidation of the biochemical pathways
that exist in normal mammary epithelial cells and how these pathways become
subverted by the action of oncogenes.

ACKNOWLEDGEMENTS

We would like to thank Alex Schlaeffli and Sylvianne Moritz-Legrand for
excellent technical assistance.

REFERENCES

Ali, I.U., Campbell, G., Lidereau, R. and Callahan, R., 1988, Amplifi-
    cation of c-erbB-2 and aggressive human breast tumors? Science 240,-
    1795-1796.

Andres, A.C., Schoenenberger, C.A., Groner, B., Henninghausen, L., LeMeur,
    M. and Gerlinger, P., 1987, Ha-ras oncogene expression directed by a
    milk protein gene promoter: Tissue specificity, hormonal regulation,
    and tumor induction in transgenic mice. Proc. Natl. Acad. Sci. USA 84,-
    1299-303.

Asch, H.L. and Asch, B.B., 1985, Heterogeneity of keratin expression in mouse mammary hyperplastic alveolar nodules and adenocarcinomas. Cancer Res. **45**,2760-2768.

Brown, A.M.C., Wildin, R.S., Prendergast, T.J. and Varmus, H.E., 1986, A retrovirus vector expressing the putative mammary oncogene int-1 causes partial transformation of a mammary epithelial cell line. Cell **46**,1001--1009.

Cepko, C.L., Roberts, B.E. and Mulligan, R.C., 1984, Construction and applications of a highly transmissible murine retrovirus shuttle vector. Cell **37**,1053-1062.

Crepin, M., Lidereau, R., Chermann, J.C., Pouillart, P., Magdamenat, H. and Montagnier, L., 1984, Sequences related to mouse mammary tumor virus genome in tumor cells and lymphocytes from patients with breast cancer. Biochem. Biophys. Res. Commun. **118**,324-331.

DeBortoli, M.E., Abou-Issa, H., Haley, B.E. and Cho-Chung, Y.S., 1985, Amplified expression of p21 ras protein in hormone-dependent mammary carcinomas of humans and rodents. Biochem. Biophys. Res. Commun. **127**,699-706.

Edwards, P.A.W., Ward, J.L. and Bradbury, J.M., 1988, Alteration of morphogenesis by the v-myc oncogene in transplants of mammary gland. Oncogene **2**,407-412.

Escot, C., Theillet, C., Lidereau, R., Spyratos, F., Champeme, M.H., Gest, J. and Callahan, R., 1986, Genetic alteration of the c-myc protooncogene (MYC) in human primary breast carcinomas. Proc. Natl. Acad. Sci. USA **83**,4834-4838.

Falcone, G., Summerhayes, I.C., Paterson, H., Marshall, C.J. and Hall, A., 1987, Partial transformation of mouse fibroblastic and epithelial cell lines with the v-myc oncogene. Exp. Cell Res. **168**,273-284.

Franklin, G.C., Chreitien, S., Hanson, I.M., Rochefort, H., May, F.E.B. and Westley, B., 1988, Expression of human sequences related to those of mouse mammary tumor virus. J. Virol. **62**,1203-1210.

Gallahan, D. and Callahan, R., 1987, Mammary tumorigenesis in feral mice: Identification of a new int locus in mouse mammary tumor virus (Czech II)-induced mammary tumors. J. Virol. **61**,66-74.

Gray, D.A., McGrath, C.M., Jones, R.F. and Morris, V.L., 1986, A common mouse mammary tumor virus integration site in chemically induced precancerous mammary hyperplasias. Virology **148**,360-368.

Günzburg, W.H. and Salmons, B., 1986, Mouse mammary tumour virus mediated transfer and expression of neomycin resistance to infected cultured cells. Virology **155**,236-248.

Günzburg, W.H., Salmons, B., Sclaeffli, A., Moritz-Legrand, S., Jones, W., Sarkar, N.H. and Ullrich, R., 1988, Expression of the oncogenes mil and ras abolishes abolishes the in vivo differentiation of mammary epithelial cells. Carcinogenesis **9**, 1849-1856.

Horan-Hand, P., Thor, A., Wunderlich, D., Muraro, R., Caruso, A. and Schlom, J., 1984, Monoclonal antibodies of predefined specificity detect activated ras gene expression in human mammary and colon carcinomas. Proc. Natl. Acad. Sci. USA. **81**,5227-5231.

Hsiao, W-L.W., Lopez, C.A., Wu, T. and Weinstein, I.B., 1987, A factor present in fetal calf serum enhances oncogene-induced transformation of rodent fibroblasts. Mol. Cell. Biol. **7**,3380-3385.

Ishikawa, F., Takaku, F., Ochiai, M., Hayashi, K., Hirohashi, S., Terada, M., Takayama, S., Nagao, M. and Sugimura, T., 1985, Activated c-raf gene in a rat hepatocellular carcinoma induced by 2-amino-3-methyl-imidazo[4,5-f]quinoline. Biochem. Biophys. Res. Commun. 132,186-192.

Keski-Oja, J., Rapp, U.R. and Vaheri, A., 1982, Transformation of mouse epithelial cells by acute murine leukaemia virus 3611: inhibition of collagen synthesis and induction of novel polypeptides. J. Cell. Biochem. 20,139-148.

Keydar, I., Ohno, T., Nayak, R., Sweet, R., Simoni, F., Weiss, F., Karby, S., Mesa-Tejada, R. and Spiegelman, S., 1984, Properties of retro-virus-like particles produced by a human breast carcinoma cell line: immunological relationship with mouse mammary tumor virus proteins. Proc. Natl. Acad. Sci. USA. 81,4188-4192.

King, C.R., Kraus, M.H. and Aaronson, S.A., 1985, Amplification of a novel v-erbB-related gene in a human mammary carcinoma. Science 229,974-976.

Kozbor, D. and Croces, C.M., 1984, Amplification of the c-myc oncogene in one of five human breast carcinoma cell lines. Cancer Res. 44,438-441.

Kraus, M.H., Yuasa, Y. and Aaronson, S.A., 1984, A position 12-activated H-ras oncogene in all HS578T mammary carcinosarcoma cells but not normal mammary cells of the same patient. Proc. Natl. Acad. Sci. USA 81,5384--5388.

Medina, D., 1988, The preneoplastic state in mouse mammary tumorigenesis. Carcinogenesis 9,1113-1119.

Miller, A.B. and Bulbrook, R.D., 1986, UICC multidisciplinary project on breast cancer: the epidemiology, aetiology and prevention of breast cancer. Int. J. Cancer 37, 173-177.

Muller, W. J., Sinn, E., Pattengale, P.K., Wallace, R. and Leder, P., 1988, Single-step induction of mammary adenocarcinoma in transgenic mice bearing the activated c-neu oncogene. Cell 54,105-115.

Nusse, R. and Varmus, H.E., 1982, Many tumors induced by the mouse mammary tumor virus contain a provirus integrated in the same region of the host genome. Cell 31,99-109.

Ohuchi, N., Thor, A., Page, D.L., Horan-Hand, P., Halter, S.A. and Schlom, J., 1986, Expression of the 21,000 molecular weight ras protein in a spectrum of benign and malignant human mammary tissues. Cancer Res. 46,2511-2519.

Ono, M., Yasunaga, T., Miyata, T. and Ushikubo, H., 1986, Nucleotide sequence of human endogenous retrovirus genome related to the mouse mammary tumor virus genome. J. Virol. 60,589-598.

Ono, M., Kawakami, M. and Ushikubo, H., 1987, Stimulation of expression of the human endogenous retrovirus genome by female steroid hormones in human breast cancer cell line T47D. J.Virol. 61,2059-2062.

Palmieri, S., Kahn, P. and Graf, T., 1983, Quail embryo fibroblasts trans-formed by four v-myc-containing virus isolates show enhanced prolifera-tion but are non tumorigenic. The EMBO J. 2,2385-2389.

Peters, G., Brookes, S., Smith, R. and Dickson, C., 1983, Tumorigenesis by mouse mammary tumor virus: evidence for a common region for provirus integration in mammary tumors. Cell 33,369-377.

Rudland, P.S., Bennett, D.C. and Warburton, M.J., 1987, Hormonal control of growth and differentiation of cultured rat mammary gland epithelial cells. Cold Spring Harbor Conf. on Cell Proliferation 6,677-699.

Salmons, B. and Günzburg, W.H., 1987, Current perspectives in the biology of mouse mammary tumour virus. Virus Research 8,81-102.

Salomon, D.S., Perroteau, I., Kidwell, W.R., Tam, J. and Derynck, R., 1987, Loss of growth responsiveness to epidermal growth factor and enhanced production of alpha-transforming growth factors in ras-transformed mouse mammary epithelial cells. J. Cell. Physiol. 130,397-409.

Schlom, J., Spiegelman, S. and Moore, D., 1971, RNA-dependent DNA polymerase activity in virus-like particles isolated from human milk. Nature 14,231-298.

Schoenenberger, C-A., Andres, A-C., Groner, B., van der Valk, M., LeMeur, M. and Gerlinger, P., 1988, Targeted c-myc gene expression in mammary glands of transgenic mice induces mammary tumours with constitutive milk protein gene transcription. The EMBO J. 7,169-175.

Shih, C. and Weinberg, R.A., 1982, Isolation of a transforming sequence from a human bladder carcinoma cell line. Cell 29,161-9.

Sinn, E., Muller, W., Pattengale, P., Tepler, I., Wallace, R. and Leder, P., 1987, Coexpression of MMTV/v-Ha-ras and MMTV/c-myc genes in transgenic mice: synergistic action of oncogenes in vivo. Cell 49,465-475.

Slamon, D.J., Clark, G.M., Wong, S.G., Levin, W.J. Ullrich, A., McGuire, W.L., 1987, Human breast cancer: correlation of relapse and survival with amplification of the HER-2/neu oncogene. Science 235, 177-182

Sonnenberg, A., Daams, H., Calafat, J. and Hilgers, J., 1986a, In vitro differentiation and progression of mouse mammary tumor cells. Cancer Res. 46,5913-5922.

Sonnenberg, A., Daams, H., van der Valk, M.A., Hilkens, J. and Hilgers, J., 1986b, Development of mouse mammary gland: Identification of stages in differentiation of luminal and myoepithelial cells using monoclonal antibodies and polyvalent antiserum against keratin. J. Histochem. Cytochem. 34,1037-46.

Spandidos, D.A. and Wilkie, N.M., 1984, Malignant transformation of early passage rodent cells by a single mutated human oncogene. Nature 310,-469-475

Stewart, T.A., Pattengale, P.K. and Leder, P., 1984, Spontaneous mammary adenocarcinomas in transgenic mice that carry and express MTV/myc fusion genes. Cell 38,627-637.

Varley, J.M., Swallow, J.E., Brammar, W.J., Whittaker, J.L. and Walker, R.A., 1987, Alterations to either c-erbB-2 (neu) or c-myc proto-oncogenes in breast carcinoma correlate with poor short term prognosis. Oncogene 1,423-430.

Varley, J.M., Walker, R.A., Casey, G. and Brammar, W.J., 1988, A common alteration to the int-2 proto-oncogene in DNA from primary breast carcinomas. Oncogene 3,87-91.

Wagner, E.F., Vanek, M. and Vennstrom, B., 1985, Transfer of genes into embryonal carcinoma cells by retrovirus infection: efficient expression from an internal promoter. The EMBO J. 4,663-666.

Zerlin, M., Julius, M.A., Cerni, C. and Marcu, K.B., 1987, Elevated expression of an exogenous c-myc gene is insufficient for transformation and tumorigenic conversion of established fibroblasts. Oncogene 1,19-28.

Zhou, D.J., Casey, G. and Cline, M.J., 1988, Amplification of human int-2 in breast cancers and squamous carcinomas. Oncogene 2, 279-282.

ACTIVATION OF THE HUMAN C-HA-RAS-1 GENE BY

CHEMICAL CARCINOGENS IN VITRO

Peter G. Lord, Terry J. Nolan* and Terry C. Orton

ICI Pharmaceuticals, Mereside, Alderley Park
Nr Macclesfield, Cheshire, U.K.;* School of Health
Sciences, Liverpool Polytechnic, Byrom St.,
Liverpool, U.K.

INTRODUCTION

The study of genetic alterations which are associated with carcinogenesis has progressed rapidly following the discovery of cellular protooncogenes. In the study of chemical carcinogenesis it is perhaps significant that members of the ras gene family are activated in a large number of tumours induced by a variety of chemical compounds. In such tumours the activation of ras genes has been shown to be by point mutation at specific codons, mainly codons 12 and 61, (Barbacid, 1987; Sukumar and Barbacid 1986, Stowers et al. 1987). Direct acting carcinogens are believed to act by modifying bases of DNA, for example by alkylation or bulky adduct formation. Point mutations arise as a consequence of base modifications (Saffhill et al. 1985).

An in vitro procedure for studying the interactions of chemical compounds with the human c-Ha-ras gene was developed by Marshall and colleagues (Marshall et al. 1984, Vousden et al. 1986). In this procedure a compound is reacted directly with a plasmid containing the wild type human c-Ha-ras gene. The reacted plasmid DNA is transfected into NIH 3T3 cells. Foci of transformed cells are then scored, subcultured and analysed for the presence of a mutated (activated) Ha-ras sequence (which will have resulted from the modification of bases in that sequence by the compound).

In the present study we have explored this procedure further with 3 potent mutagens, N-ethyl-N-nitrosourea and β-propiolactone, which are both alkylating agents, and 1-chloromethyl pyrene.

EXPERIMENTAL PROCEDURE

In the case of each compound, several different concentrations of the compound were reacted with 10 μg of the plasmid pbc-N1 which contains a 6.4 Kb fragment spanning the human c-Ha-ras-1 gene. After chloroform or ether extraction to remove unreacted chemical the plasmid DNA was ethanol precipitated and dried. A calcium phosphate precipitate was prepared with resuspended plasmid DNA plus mouse genomic DNA as carrier. The precipitate was layered onto subconfluent monolayers of NIH3T3 cells in petri dishes such that each dish of cells received 1 μg of plasmid DNA. One set of cells was transfected with unreacted pbc-N1. Foci of transformed cells were scored 14-20 days later.

Individual foci were removed using cloning rings and subcultured. Genomic DNA was extracted from each subculture and regions spanning codons 12 and 61 of the Ha-ras gene were selectively amplified by the polymerase chain reaction (PCR) technique (Saiki et al. 1988). PCR amplified DNA was dot-blotted onto Gene Screen Plus nylon membranes and hybridized with Allele Specific Oligonucleotide (ASO) probes for wild type and mutant human c-Ha-ras sequences (7 for codon 12, 8 for codon 61) following published procedures (Verlaan de Vries et al. 1986).

RESULTS

The number of foci obtained with each concentration of the three compounds are presented in Table 1. Each focus was subcultured and DNA from each subculture probed for activating point mutations in codons 12 and 61 using ASO probes after amplifying those regions by the PCR method. Four codon 61 and eight codon 12 mutations were identified from the three experiments (Table 2). 1-chloromethyl pyrene produced an A→T transversion at position 2 of codon 61 and two G→T transversions at position 2 of codon 12. The codon 61 mutation was identified in a subculture which also contained a codon 12 mutation.

N-Ethyl-N-nitrosourea produced two A→C transversions and one A→T transversion at position 2 of codon 61. It also produced three G→T transversions at codon 12 position 2.

Three G→T transversions at position 2 of codon 12 were produced by beta-propiolactone.

Table 1. Transfections with pbc-N1 modified by three chemical compounds

Compound	DNA/compound ratio/(wt/wt)	No. foci/ No. plates	Foci/μg
1-Chloromethyl pyrene	10:1	3/9	0.033
	20:1	3/9	0.033
	100:1	5/9	0.056
	200:1	4/9	0.044
	1000:1	6/9	0.067
	1:0	0/9	0.0
N-ethyl-N-nitrosourea	1:1	5/9	0.056
	2:1	0/9	0.0
	10:1	4/9	0.044
	100:1	6/9	0.067
	1000:1	3/9	0.033
	1:0	1/9	0.011
β-propiolactone	1:10	4/7	0.057
	1:2	8/9	0.089
	1:0	0/9	0.0

Table 2. Base substitutions at codons 12 and 61 of
         c-Ha-ras-1 after modification by three
         chemical mutagens

Modifying agent	Codon	Mutation	Number
1-chloromethyl pyrene	12	GGC→GTC	2
	61	CAG→CTG	1
N-ethyl-N-nitrosourea	12	GGC→GTC	3
	61	CAG→CCG	2
	61	CAG→CTG	1
β-propiolactone	12	GGC→GTC	3

DISCUSSION

Each compound modified the c-Ha-ras sequence to the extent that activatin point mutations were introduced after transfection into NIH 3T3 cells. Several foci of transformed cells were produced although in each case the number of foci produced was not dose dependent, unlike in other similar studies (Marshall et al. 1984; Vousden et al. 1986; Hashimoto et al. 1987). Point mutations in codons 12 and 61 were identified in several of the foci. It is possible that in other foci, the point mutation was in other codons (e.g. 13, 59. 117). However it should be noted that the identification of foci is subjective and several types of foci can be generated in transfection experiments (Perucho et al. 1981), not all of which are "truly" transformed cells. Although few foci were detected upon transfection with unmodified pbc-N1, some of what we identified as foci may not have been produced by the chemically modified plasmid. In support of the latter we noted that the wild type probes did not produce a hybridization signal with c.35% of the foci DNA samples.

We identified two mutations (one at codon 12 and one at codon 61) in one focus. These mutations are not necessarily in the same Ha-ras sequence. More than one plasmid molecule will have been taken up by the cells therefore the mutations may well have been introduced on separate plasmids and are consequently in separate sequences. Secondary transfection of DNA from this focus would separate the sequences and confirm this interpretation.

The most prevalent mutation produced by the three compounds was a G→T transversion at position 2 of codon 12. This was unexpected for N-ethyl-N-nitrosourea since it generates O^6-ethylguanine adducts which primarily lead t G→A transitions (Saffhill et al. 1985).

The system used here is attractive for _in vitro_ screening of potential carcinogens since it uses: 1) a human gene; 2) a gene (c-Ha-ras) implicated i carcinogenesis; and 3) a mammalian cell detection procedure. However it suffers from several drawbacks. 1) It requires much larger amounts of compound in comparison to other short term _in vitro_ mutagenic assays. 2) It works only for direct acting mutagens and because of 1) the inclusion of a metabolizing system for activation of the compound would be impracticable. 3) Identification of foci is very subjective. 4) It is much more complicate than current regulatory _in vitro_ tests and therefore difficult to implement a routine basis.

Possible improvements to the system are: 1) increasing the sensitivity of the NIH 3T3 cell transfection process for example by using electroporation; and 2) cotransfection with the G418 resistance gene to reduce the subjectivity of focus identification.

ACKNOWLEDGEMENTS

We wish to thank Alan Balmain, for providing the plasmid pbc-N1, and John Ashby for providing 1-chloromethylpyrene. We owe much gratitude to Jo Loughlin for excellent technical assistance.

REFERENCES

Barbacid, M., 1987, Involvement of ras oncogenes in the initiation of carcinogen-induced tumours, in: "Oncogenes and Cancer," S.A. Aaronson et al., eds., Japan Sci. Soc. Press, Tokyo, pp. 43-53.

Hashimoto, Y., Kawachi, E., Shudo, K., Sekiya, T., and Sugimura, T., 1987, Transforming activity of Human c-Ha-ras-1 proto-oncogene generated by the binding of 2-amino-6-methyl-dipyrido [1,2-a:3',2'-D]imidazole and 4-nitroquinoline N-oxide: direct evidence of cellular transformation by chemically modified DNA, Jpn. J. Cancer Res., 78:211.

Marshall, C.J., Vousden, K.H., and Phillips, D.H., 1984, Activation of c-Ha-ras-1 proto-oncogene by in vitro modification with a chemical carcinogen, benzo(a)pyrene diol-epixide, Nature, 310,586.

Perucho, M., Goldfarb, M., Shimizu, K., Lama, C., Fogh, J., and Wigler, M., 1981, Human tumour derived cell lines contain common and different transforming genes, Cell, 27:467.

Saffhill, R., Margison, G.P., and O'Connor, P.J., 1985, Mechanisms for carcinogenesis induced by alkylating agents, Biochimica et Biophysica Acta, 823:111.

Saiki, R.K., Gelfand, D.H., Stoffel, S., Scharf, S.J., Higuchi, R., Horn, G.T., Mullis, K.B., and Erlich, H.A., 1988, Primer-directed enzymatic amplification of DNA with a thermostable DNA polymerase, Science, 239:487.

Stowers, S.J., Maronpot, R.R., Reynolds, S.H., and Anderson, M.W., 1987, The role of oncogenes in chemical carcinogenesis, Environmental Health Perspectives, 75:81.

Sukumar, S., and Barbacid, M., 1986, The role of ras oncogenes in chemically-induced tumours, Pontificiae Academiae Scientarum Scripta Varia, 70:35.

Verlaan de Vries, M., Bogaard, M.E., van den Elst, H., van Boom, J.H., van der Eb, A.J. and Bos, J.L., 1986, A dot-blot screening procedure for mutated ras oncogenes using synthetic oligodeoxynucleotides, Gene, 50:313.

Vousden, K.H., Bos, J.L., Marshall, C.J., and Phillips, D.H., 1986, Mutations activating human c-Ha-ras-1 proto-oncogene (HRAS1) induced by chemical carcinogens and depurination, Proc. Nat. Acad. Sci., 83:1222.

ACTIVATION OF RAS ONCOGENES DURING COLONIC CARCINOGENESIS:

DETECTION BY THE POLYMERASE CHAIN REACTION

A. Haliassos, J.C. Chomel, S. Grandjouan*,
J.C. Kaplan and A. Kitzis

Institut de Pathologie Moléculaire and *Service de
Gastroentérologie, C.H.U. Cochin, 24, rue du
Faubourg Saint-Jacques, 75014 PariS, France

INTRODUCTION

Oncogene activations are known to occur during the multiple step process of colonic carcinogenesis. However, no specific activation has been demonstrated that was relevant to colonic carcinogenesis until the detection of point mutations in total genomic DNA became feasible. The new developments in Molecular Biology permitted us to screen for point mutations of ras oncogenes, not only in solid tumor material but also in archeived histological slices and in DNA extracted from a single cell.

In our studies we used the Polymerase Chain Reaction (P.C.R.) which is an in vitro method for the primer directed enzymatic amplification of specific DNA sequences[1]. This technique leads to a very sensitive method for the detection of point mutations in genes already sequenced. The amplified DNA product is analysed by a slot blot hybridization, with synthetic allele specific radiolabeled oligodeoxynucleotide probes (ASO), under high stringency conditions[2]. The probes are specific of normal DNA sequence or of each possible mutation of a given codon. The results of the hybridization are revealed by autoradiography.

MATERIALS AND METHODS

1- DNA samples

DNA from colorectal cancer, adenomatous polyps and adjacent normal mucous was extracted according either to the classical method of phenol chloroform, or by the simplified guanidium method[3].

DNA from histological slides, fixed by formol and embeded in paraffin, was extracted by boiling in a lysis mixture after the dissolution of paraffin in chloroform.

Single cell was prepared from blood samples by a Ficoll gradient under microscopic control. DNA was extracted by the boiling method.

2- Oligonucleotides primers and probes

The oligodeoxynucleotides were synthesised by the solid phase triester method in a Biosearch 8600 synthesiser. The ASO probes were end labeled using $[\gamma^{32}P]ATP$ and T4-Polynucleotide kinase.

3- Polymerase chain reaction

In vitro enzymatic DNA amplification (PCR) was performed on an automated apparatus (DNA thermal Cycler from Perkin Elmer Cetus). The principe of this method is shown schematically in Fig.1.

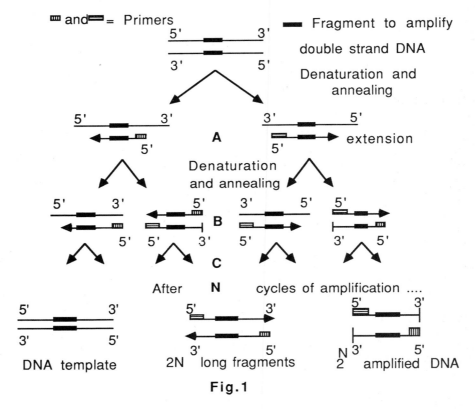

Fig.1

Normaly we perform between 30 (DNA from solid tumors and mucosas) and 50 (DNA from a single cell) cycles of amplification. Each cycle includes 3 steps:
- denaturation of DNA at 94°C for 50 sec.
- annealing of the primers at 37°C for 1 min.
- enzymatic extention at 72°C for 2 min.

The annealing temperature affects a little the specificity of the PCR. In spite of our low annealing temperature, we observed very rarely parasitic amplifications. For the elongation we use the DNA polymerase from thermophilus aquaticus (Perkin Elmer Cetus). An aliquot of the product of the PCR was controlled in a 2-3% Nu Sieve gel for the presence of the amplified fragment. Fig. 2.

A B C D E F G

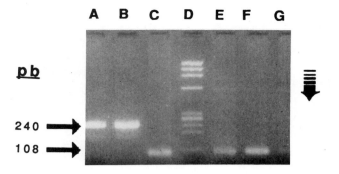

pb

240 ➡

108 ➡

RESULTS AND DISCUSSION

Before the clinical studies, we tested the validity of this method on the EJ cell line[5,6] which carries a known mutation at the second base of codon 12 of the Harvey ras gene. After the enzymatic amplification of a DNA region including the codon 12 of the Ha ras gene we hybridized the PCR product with probes specific of normal DNA sequence or of each possible mutation of the second base of codon 12. Only the probe corresponding at the known mutation (and the normal one) left hybridized after the stringent wash (Fig. 4).

We used the above technique to detect the mutations of codon 12 of the Kirsten ras oncogene in cell populations from adenoma, adenocarcinoma, coexistent gastric tumor and adjacent mucosa in 8 cases of colon neoplasia. Our results are represented in the table 1.

We found three cases of colon carcinomas and one coexistent gastric tumor carrying a codon 12 mutation. At the other hand we have not found any mutation at the same codon in the adjacent mucosas and in the two adenomatous polyps.

Attempts were made to establish correlation between the presence of an activated Kirsten ras allele and Dukes stages in the above cases but our results do not permit this, because our sample was statistically insufficient.

Further studies were designed to detect activated Kirsten ras alleles in benign sporadic adenomas in rapport with other available prognostic features (dysplasia, size, number and recurrence rate of adenomas).

CONCLUSIONS

We proved that our approach was sensitive and specific for the detection of point mutations in already sequenced genes. It is about 1000 times more sensitive than the previous method of transfection assay, and many times faster.

Slot-blots

1- N Cycles of DNA amplification

2- Spotting of the amplified DNA on to a nylon filter

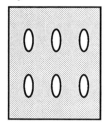

3-Hybridization with specific probes (normal Ras p.ex.), wash at a non stringent temperature and autoradiography

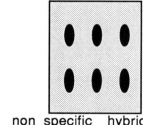

(non specific hybridization)

4- Wash at $T° = Tm-2°C$ where $Tm = (GC \times 4) + (AT \times 2)$ and autoradiography

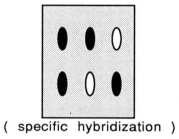

(specific hybridization)

● = homozygous normal/normal

or heterozygous normal/mutated

◯ = homozygous mutated/mutated

Fig.3

SLOT BLOT of Ha ras CODON 12

probes	H12N(Gly)	H12d(Ala)	H12e(Asp)	H12f(Val)
codon	GGC	GCC	GAC	GTC
tm	74°C	74°C	74°C	72°C

Hybridization
and non stringent
wash at 64°C

stringent wash
at 72°C

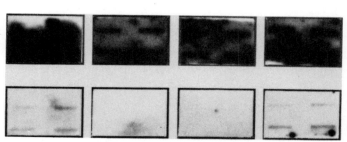

Fig. 4

Table 1

Patient code \ Sample	SG3	SG6	SG7	SG8	SG11	SG12	SG13	SG18
Adjacent mucosa	−	−	−	−	−	−	−	−
Polyp	−	N.A.	N.A.	−	N.A.	N.A.	N.A.	N.A.
Tumor 1	+①	−	−	N.A.	+	−	−	+
Tumor 2	+②	N.A.	N.A.	N.A.	+③	N.A.	N.A.	N.A.

+ Kirsten ras codon 12 mutated   N.A. not available
− Kirsten ras codon 12 not mutated
① Sigmoid ② Ancending colon ③ Gastric

However tumors are consisted of heterogenous cell populations (neoplastic cells, infiltrated inflammatory cells, normal tissue cells). This heterogeneity rises the background "noise" elevating the number of amplified non mutated sequences. Actually we can detect cells carrying a mutation, among a population of non carrying cells, if the mutated cells represent at least 5% of the total cell population. We can avoid this inconvenience using our method of amplification of DNA from a single cell.

Otherwise, we can study conjointly the mutations of a given codon and the histological aspect of a tissue, by using the DNA from a series of fixed slides, which alternatively are controlled by common histological methods. This last approach is applicable also to the study of rare forms of tumors in retrospective.

Finally, this philosophy permits to correlate clinical stages with data from histology and molecular biology.

REFERENCES

1. Saiki, R. K., Scharf, S., Faloona, F., Mullis, K. B., Horn, G. T., Erlich, H. A., Arnheim, N. (1985). Enzymatic amplification of ß-globin genomic sequences and restriction site analysis for diagnosis of sickle cell anemia. *Science*, **230**,1350-1354.
2. Verlan-de Vries, M., Bogaard, M. E., Elst, H., Boom, J. H., Eb, A. J., & Bos, J. L. (1986). A dot-blot screening procedure for mutated ras oncogenes using synthetic oligodeoxy-nucleotides. *Gene*, **50**, 313-320.
3. Jeanpierre, M. (1987). A rapid method for purification of DNA from blood. *Nucleic Acids Research*, **15**, 9611.
4. Melchior, W.B., Von Hippel, P.H. (1973). Alteration of the relative stability of dA.dT and dG.dC base pairs in DNA. *Proc. Natl. Acad. Sci. USA*, **70**, 298-302.
5. Marshall, C.J., Francs, L.M., and Carbonell, A.W. (1977). Markers of neoplastic transformation in epithelial cell lines derived from human carcinomas. *J. Nat. Cancer Inst.*, **58**, 1743-1747.
6. Shih, C. and Weinberg, R.A. (1982). Isolation of a transforming sequence from a human bladder carcinoma cell line. *Cell*, **29**, 161-169.

APPENDIX

Cell lysis solution:

    0,1M NaOH
    2M NaCl

PCR mixture:

    700ng of each primer
    1,5mM of each dNTP
    67mM Tris HCl pH 8,8
    6,7μm EDTA

6,7mM MgCl$_2$

10mM β-mercaptoethanol

16,6mM ammonium sulfate

10% DMSO

DNA to amplify

2,5 units of Taq polymerase

H$_2$O to 100 μl

Prehybridation solution:

3,0M tetramethylammonium Cl

50mM Tris HCl pH 8,0

2mM EDTA

100μg/ml sonicated, denatured herring sperm DNA

0,1% sodium dodecylsulfate

5x Denhart's solution

First and second wash solutions:

20mM sodium phosphate pH 7,0

0,36M NaCl

2mM EDTA

0,1% sodium dodecylsulfate

Third and fourth wash solutions:

Prehybridation solution without Denhart's solution
and herring sperm DNA

POST-TRANSLATIONAL MODIFICATION OF RAS PROTEINS: PALMITOYLATION AND PHOSPHORYLATION OF YEAST RAS PROTEINS

Fuyuhiko Tamanoi, Alexander R. Cobitz, Asao Fujiyama*, Laurie E. Goodman and Charles Perou

Department of Biochemistry and Molecular Biology
The University of Chicago
Chicago, IL. 60637

INTRODUCTION

The ras-genes are conserved during evolution and appear to play an essential role in the growth regulation of cells. Products of these genes are localized in the plasma membrane[1,2] and exhibit well-defined biochemical activities to bind guanine nucleotides and hydrolyze GTP[3-6]. Because a large amount of the purified proteins can be obtained after their expression in E. coli, extensive structural studies have been carried out. The protein has been crystalized and its three dimensional structure has been determined[7]. Furthermore, recent identification of the GAP protein[8-10] raises the possibility that protein-protein interactions involving the ras protein can be elucidated in biochemical terms.

To gain further understanding of the ras proteins, we have investigated post-translational modification events. Whereas most of the above mentioned structural studies utilized proteins purified from E. coli, we have studied ras proteins purified from their natural environment. In particular, we have concentrated on characterizing yeast RAS proteins purified from yeast cells. Two proteins, RAS1 and RAS2, are present in yeast cells[11,12]. These proteins have extensive homology with the mammalian ras protein and they exhibit guanine nucleotide binding as well as GTPase activity[13-15]. Genetic as well as biochemical studies have established that these yeast RAS proteins are involved in the stimulation of adenylate cyclase[16,17]. In this paper, we report that the yeast RAS proteins undergo two different types of post-translational modification; palmitoylation and phosphorylation. The palmitoylation plays an important role in membrane association of the ras protein. The phosphorylation, on the other hand, might be important for the regulation of its activity.

RESULTS AND DISCUSSION

I. Palmitoylation of yeast RAS proteins

The ras proteins are post-translationally modified by the addition of palmitic acid[18,19]. It is believed that the palmitic acid is attached to a cysteine at the C-terminus via thioester linkage. This cysteine is present within a sequence CysAAX (A is an aliphatic amino acid and X is the last amino acid)[20,21]. Because the ras protein itself is not hydrophobic, it is proposed that the palmitic acid provides hydrophobic moiety needed for its association with the membrane. Terms, CAAX box and sticky finger, have recently been used to describe this modification of ras proteins[22]. To further gain insights into this palmitoylation and membrane association of ras protein, we have used the yeast system and analyzed RAS1 and RAS2 proteins.

Palmitic acid attachment. Palmitoylation of yeast RAS proteins was demonstrated by labeling the protein with ³H-palmitic acid[23]. Briefly, yeast cells overproducing RAS1 or RAS2 protein were labeled with ³H-palmitic acid and the incorporation of the radioactivity into RAS1 and RAS2 proteins was detected by the immunoprecipitation of these proteins using an anti-ras antibody Y13-259. Subcellular fractionation experiments demonstrated that the palmitic acid labeled RAS proteins were present exclusively in the membrane fraction[23].

The palmitic acid could be removed from the RAS proteins by the treatment with alkali[23]. Recently, we have shown that virtually all the radioactivity can be released by the treatment with hydroxylamine, suggesting that the palmitic acid attachment involves a thioester linkage. The released radioactivity, when analyzed on a HPLC column, were found to be exclusively palmitic acid. A similar result was obtained even when we labeled the cells with ³H-myristic acid. Presumably, myristic acid was converted to palmitic acid inside the cell and the palmitic acid was incorporated into the RAS proteins. Therefore, palmitic acid appears to be the only fatty acid attached to the RAS proteins.

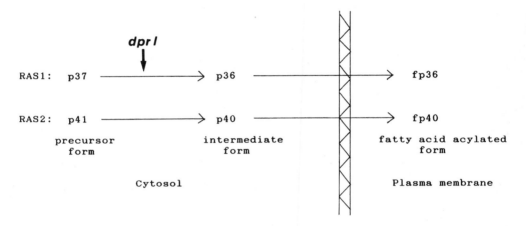

Figure 1. A model for the biosynthesis of yeast RAS proteins.

Palmitic acid addition consists of at least two biochemical reactions.

The addition of palmitic acid to the RAS proteins seems to involve two distinct biochemical steps[23-25]. From the analyses of RAS proteins by ³⁵S-methionine labeling, we have proposed a model for the modification of the RAS proteins at their C-termini[23,24]. As shown in Figure 1, we propose that the RAS proteins are synthesized as precursor forms in the cytosol which are quickly converted to intermediate forms. This intermediate form migrates slightly faster than the precursor form on a SDS polyacrylamide gel (there is a difference of approximately 1000 dalton molecular weight difference between the two proteins). The addition of palmitic acid occurs on the intermediate form and this fatty acid acylated form is localized in the membrane.

Three different forms of the RAS protein described above could be separated by a HPLC column. When the RAS proteins labeled with ³⁵S-methionine were purified by using the antibody Y13-259 covalently coupled to Sepharose beads and ran over a C4 HPLC column, we observed three peaks separated by an acetonitrile gradient. These peaks corresponded to the precursor, intermediate and the fatty acid acylated forms in the order of their appearance during elution. The three forms have identical N-terminal sequences. This was shown by labeling RAS proteins with different amino acids and then determining in

226

which cycle during sequencing we observe the radioactivity. By this experiment, we found that all three forms started with proline which is the second amino acid expected from the sequence. Thus, the first amino acid is removed. Whereas the three forms have identical N-terminal structures, their C-terminal structures appear to be different. Our preliminary results suggest that the C-terminal peptide of the precursor form elutes from HPLC column at a position different from those of the intermediate and the fatty acid acylated forms. Therefore, it appears that the modifications occur only at their C-termini. This idea is in line with the demonstration by Deschenes and Broach[26] that a mutant RAS2 protein which has a cysteine to serine change at the 4 th residue from the C-terminus remains as a precursor protein.

Similar biosynthetic events were observed when H-ras protein was expressed in yeast[25]. Although the conversion of the precursor form to the intermediate form was much slower than that seen with the yeast protein, we did observe three different forms and it was possible to follow the conversion of the precursor form to the intermediate form and then to membrane bound fatty acid acylated form by pulse-chase experiment.

Genetic approach to define biosynthesis steps. The above results suggest that the C-terminal modification of the RAS protein is not just a simple addition of palmitic acid but is rather a complex reaction consisting of at least two biochemical steps. It has recently been reported that there is also a carboxy-methylation event taking place[27]. This could mean that at least three biochemical events are involved in the modification. To further define these events, we have utilized yeast genetics and looked for mutants which are defective in the processing of ras proteins[24].

The rationale for the isolation of mutants is to make use of heat shock sensitivity of yeast cells expressing RAS2[val119] mutant protein. These cells are mutagenized and heat shock resistant cells are recovered. In collaboration with Kunihiro Matsumoto (DNAX Research Institute), we have obtained 52 such isolates, 40 of which were found to have mutations in adenylate cyclase. This was expected, since inactivation of adenylate cyclase can compensate for the overactivation of the cyclase by the RAS2 mutant protein. Three isolates, however, were found not to have mutations in the adenylate cyclase. One of these mutants, termed dpr1, has been further investigated.

Processing of RAS proteins is defective in the dpr1 mutant cells[24]. When RAS2 protein was expressed in the dpr1 mutant cells, the protein remained as a precursor form. Subcellular fractionation experiments revealed that the protein accumulated in the cytosol. Thus, it appears that the defect of the mutant is in the first step of the biosynthesis (conversion of a precursor to an intermediate form). The dpr1 mutant exhibits two phenotypes, temperature sensitive growth and sterility specific to MATa cells. These points are described in more detail below.

Cloning of processing genes. To further gain insights into the steps involved in the processing, we cloned the DPR1 gene from yeast. A genomic library of yeast DNA was introduced into the dpr1 mutant and cells capable of growth at a high temperature were isolated. Plasmid DNAs were recovered from these cells. The DNA thus cloned was capable of complementing deficiency in the processing of ras proteins as well as complementing a-sterile phenotype of the dpr1 cells.

A region responsible for the complementing activity was narrowed down and this DNA fragment was sequenced[28]. A single open reading frame encoding a protein of 431 amino acids was found. This open reading frame was bounded by typical regulatory elements; TATA like sequence 140 nucleotides upstream of ATG and a transcription termination signal (TAG---CAGT---TTT) right after the gene. The gene was actively transcribed producing a polyadenylated message of approximately 1.6 kb.

The gene product is a hydrophilic protein lacking any hydrophobic stretches, suggesting that it is a cytoplasmic protein. The C-terminal half of the protein is unusually rich in cysteines and glycines. The gene product has been identified as a 43 kd protein by cell free translation of DPR1 RNA.

Biosynthesis of a-factor. In addition to being involved in the processing of ras proteins, the DPR1 gene is also involved in the processing of a mating factor a-factor. This was indicated by the phenotype of the dpr1 mutant. The mutant was sterile specific to MATa cells. Furthermore, the dpr1 mutant is found to be allelic to a mutant ram, independently isolated by Powers et al[29], which is allelic to STE16 believed to be involved in the processing of the mating factor, a-factor[30].

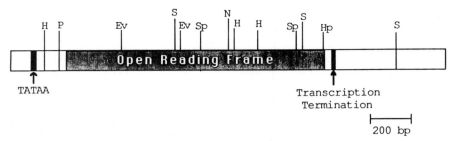

Figure 2.   Structure of <u>DPR1</u> genes.

The biosynthesis of the <u>a</u>-factor has been proposed to involve processing of a 36 (or 38) amino acids precursor protein[31] to a mature protein which is 12 amino acids. The C-terminal sequence of the <u>a</u>-factor precursor, as deduced from its gene sequence, ends with CysAAX[31], a sequence found to be present in all ras proteins. Interestingly, the mature <u>a</u>-factor ends with the cysteine, suggesting that the three amino acids are removed[32]. This cysteine is also known to be carboxymethylated as well as modified by a hydrophobic residue[32].

A working hypothesis for the biosynthesis of ras proteins. Based on the above results, we propose a model for the biosynthesis of ras proteins. As shown in Figure 3, the ras protein is synthesized as a precursor on free

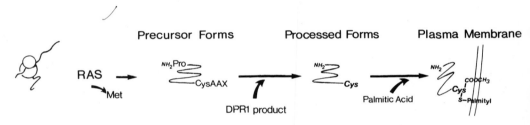

Figure 3.   A working hypothesis for the biosynthesis of ras proteins.

ribosomes. The precursor starts with a second amino acid in the sequence (proline in the case of yeast RAS2) and ends with the CAAX box. The three amino acids at the C-terminus will be removed by a reaction involving DPR1 protein and palmitic acid will be attached to the cysteine exposed to the C-terminus of an intermediate protein. The C-terminus of the mature ras protein is carboxymethylated.

The idea that three amino acids are removed at the C-terminus of ras protein is primarily based on the analogy with the processing of the <u>a</u>-factor. Three amino acids at the C-terminus of the <u>a</u>-factor precursor is clearly removed during its biosynthesis and the <u>DPR1</u> gene plays a role in this processing. In keeping with this idea, our preliminary results suggest that the C-termini of the processed ras proteins are distict from that of the precursor ras protein. Further work, however, is necessary to determine the exact structure of the ras C-terminus. Details concerning the carboxymethylation events such as determining at which step this modification occurs also remain to be clarified.

II. Phosphorylation of RAS proteins

In addition to the palmitoylation, we have recently discovered that the yeast RAS proteins undergo yet another type of post-translational modification; phosphorylation. This modification appears to occur after the proteins are localized in the membrane. Whereas the processing of ras proteins described above concerns events prior to their attachment to the membrane, the phosphorylation seems to involve events after they are membrane localized. Although the physiological roles of the phosphorylation have not been established, it may be involved in the feedback regulation of cAMP production in yeast.

Yeast RAS proteins are phosphorylated. An indication that yeast RAS proteins are phosphorylated was obtained when we purified RAS1 and RAS2 proteins by using the anti-ras monoclonal antibody Y13-259 coupled to Sepharose beads. Characterization of the purified RAS1 protein on a SDS polyacrylamide gel revealed that the protein consisted of heterogeneous species, a protein of an expected molecular weight and two or more species having higher molecular weights. These high molecular weight species could be converted to the expected molecular weight species by the treatment with calf intestinal alkaline phosphatase, suggesting that these forms represent phosphorylated RAS1 proteins. With RAS2 protein, we observed only one band on a SDS polyacrylamide gel. However, on a two-dimensional gel, three major spots were identified, two of which corresponded to phosphorylated RAS2 proteins.

Direct evidence for the phosphorylation of the RAS1 and RAS2 proteins was obtained by the labeling with ^{32}P-orthophosphate. It was found that both RAS1 and RAS2 proteins can be labeled with [32]P. Acid hydrolyses of the phosphorylated proteins indicated that the phosphorylation occurred exclusively on serine residues.

Characterization of phosphorylated yeast RAS proteins. Two kinds of observation have been made regarding the phosphorylation of RAS1 and RAS2 proteins. First, we carried out tryptic fingerprinting of the phosphorylated RAS2 protein. Results clearly showed that there are two major and three minor phosphopeptides. Thus, there seems to be a small number of phosphorylation site on the RAS2 protein. Second, fractionation of the labeled cell extracts into soluble and membrane fractions suggested that the phosphorylated RAS1 proteins are predominantly localized in the membrane. Together with the above demonstration that the phosphorylation occurs exclusively on serine residues, these results suggest that the phosphorylation of the ras proteins is not just a random event but may be a specific event.

Possible physiological roles of the phosphorylation. Our work described above has established that the yeast RAS proteins are phosphorylated in yeast cells. At the moment, we can only speculate about the physiological role of the phosphorylation. As shown in figure 4, yeast RAS proteins are involved in the stimulation of adenylate cyclase. cAMP produced activates cAMP dependent protein kinase (A-kinase)[16]. Recently, Nikawa et al[33] have shown that there is an inverse correlation between the level of cAMP and the activity of the A-kinase; cAMP level was increased in yeast cells with low A-kinase activity whereas a lowered cAMP level was seen in yeast cells with increased A-kinase activity. Although this feedback regulation may involve any proteins in this pathway, it is tempting to speculate that the phosphorylation of RAS proteins plays a role in this feedback regulation. This idea has been proposed by Resnick and Racker[34] who demonstrated that RAS2 protein purified after its expression in E. coli can be phosphorylated in vitro by a purified mammalian A-kinase. We have also shown that RAS2 protein purified from yeast can be phosphorylated by a mammalian A-kinase. Comparison of tryptic fingerprints is being carried out to assess whether the A-kinase is responsible for the phosphorylation of RAS2 protein.

A low but significant level of serine phosphorylation has been observed with mammalian H-ras protein[35]. A serine phosphorylation event was also seen with C-K-ras protein after phorbol ester treatment[36]. Thus, phosphorylation occurs not only on yeast proteins but also on mammalian proteins. Further investigation into biochemical consequences of the phosphorylation should be revealing in understanding the mechanism of the action of ras proteins.

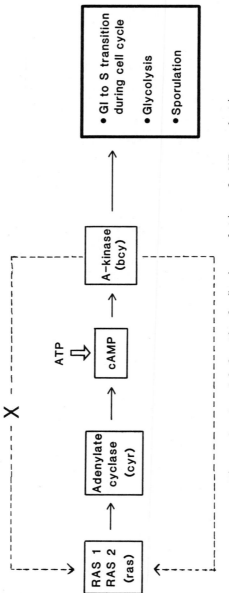

Figure 4. A model for the feedback regulation of cAMP production.

ACKNOWLEDGEMENTS

This investigation was supported by NIH grant CA41996 and in part by Goldblatt Cancer Research Foundation.

*Current address: National Institute of Genetics, Mishima, Shizuoka, 411 Japan.

REFERENCES

1. M. C. Willingham, I. Pastan, T. Y. Shih and E. M. Scolnick, Localization of the src gene product of the Harvey strain of MSV to plasma membrane of transformed cells by Electron Microscopic Immunocytochemistry. Cell 19:1005 (1980).

2. T. Y. Shih, M. O. Weeks, P. Gruss, R. Dhar, S. Oroszlan and E.M. Scolnick, Identification of a precursor in the biosynthesis of the p21 transforming protein of Harvey murine sarcoma virus. J. Virol. 42:253 (1982).

3. E. M. Scolnick, A. G. Papageorge and T. Y. Shih, Guanine nucleotide binding activity as an assay for the src protein of rat-derived murine sarcoma viruses. Proc. Natl. Acad. Sci. USA 76:5355 (1979).

4. J. P. McGrath, D. J. Capon, D. V. Goeddel and A. D. Levinson, Comparative biochemical properties of normal and activated human ras p21 protein. Nature 310:644 (1984).

5. R. W. Sweet, S. Yokoyama, T. Kamata, J. R. Feramisco, M. Rosenburg and M. Gross, The product of ras is a GTPase and the T24 oncogenic mutant is deficient in this activity. Nature 311:273 (1984).

6. J. B. Gibbs, I. S. Sigal, M. Poe and E. M. Scolnick, Intrinsic GTPase activity distinguishes normal and oncogenic ras p21 molecules. Proc. Natl. Acad. Sci. USA 81:5704 (1984).

7. A. M. DeVos, L. Tong, M. V. Milburn, P. M. Matias, J. Jancarik, S. Noguchi, S. Nishimura, K. Miura, E. Ohtsuka and S-H Kim, Three-dimensional structure of an oncogene protein: catalytic domain of human c-H-ras p21. Science 239:888 (1988).

8. M. Trahey and F. McCormick, A cytoplasmic protein stimulates normal N-ras p21 GTPase, but does not affect oncogenic mutants. Science 238:542 (1987).

9. U. S. Vogel, R. A. F. Dixon, M. D. Schaber, R. E. Diehl, M. S. Marshall, E. M. Scolnick, I. S. Sigal and J. B. Gibbs, Cloning of bovine GAP and its interaction with oncogenic ras p21. Nature 335:90 (1988).

10. C. Cales, J. F. Hancock, C. J. Marshall and A. Hall, The cytoplasmic protein GAP is implicated as the target for regulation by the ras gene product. Nature 332:548 (1988).

11. D. DeFeo-Jones, E. M. Scolnick, R. Koller, and R. Dhar, ras-related gene sequences identified and isolated from Saccharomyces cerevisiae, Nature 306:707 (1983).

12. S. Powers, T. Kataoka, O. Fasano, M. Goldfarb, J. Strathern, J. Broach and M. Wigler, Genes in S. cerevisiae encoding proteins with domains homologous to the mammalian ras proteins. Cell 36:607 (1984).

13. F. Tamanoi, M. Walsh, T. Kataoka and M. Wigler, A product of yeast RAS2 gene is a guanine nucleotide binding protein. Proc. Natl. Acad. Sci. USA 81:6924 (1984).

231

14. A. Fujiyama, N. Samiy, M. Rao and F. Tamanoi, Biochemistry of yeast RAS1 and RAS2 proteins. in "Yeast Cell Biol." J. Hicks ed., Alan R. Liss, New York (1986) p125.

15. G. L. Temeles, J. B. Gibbs, J. S. D'Alonzo, I. S. Sigal and E. M. Scolnick, Yeast and mammalian ras proteins have conserved biochemical properties. Nature 313:700 (1985).

16. T. Toda, I. Uno, T. Ishikawa, S. Powers, T. Kataoka, D. Broek, J. Broach, K. Matsumoto and M. Wigler, In RAS proteins are controlling elements of the cyclic AMP pathway. Cell 40:27 (1985).

17. D. Broek, N. Samiy, O. Fasano, A. Fujiyama, F. Tamanoi, J. Northup and M. Wigler, Differential activation of yeast adenylate cyclase by wild-type and mutant RAS proteins. Cell 41:763 (1985).

18. B. M. Sefton, I. S. Trowbridge, J. A. Cooper and E. M. Scolnick, The transforming proteins of Rous sarcoma virus and Abelson virus contain tightly bound lipid. Cell 31:465 (1982).

19. J. E. Buss and B. M. Sefton, Direct identification of palmitic acid as the lipid attached to p21 ras. Mol. Cell. Biol. 6:116 (1986).

20. B. Willumsen, A. Christiansen, N. L. Hubbert, A. G. Papageorge and D. Lowy, The p21 ras C-terminus is required for transformation and membrane association, Nature (London), 311:583 (1984).

21. M. O. Weeks, G. L. Hager, R. Lowe and E. M. Scolnick, Development and analysis of a transformation-defective mutant of Harvey murine sarcoma tk virus and its gene product, J. Virol. 54:586 (1985).

22. A. Magee and M. Hanley, Sticky fingers and CAAX boxes, Nature 335:114 (1988).

23. A. Fujiyama and F. Tamanoi, Processing and fatty acid acylation of RAS1 and RAS2 proteins in Saccharomyces cerevisiae, Proc. Natl. Acad. Sci. USA, 83:1266 (1986).

24. A. Fujiyama, K. Matsumoto and F. Tamanoi, A novel yeast mutant defective in the processing of ras proteins: assessment of the effect of the mutation on processing steps, EMBO J. 6:223 (1987).

25. F. Tamanoi, E. C. Hsueh, L. E. Goodman, A. R. Cobitz, R.J. Detrick, W. R. Brown and A. Fujiyama, Posttranslational modification of as proteins: Detection of a modification prior to fatty acidacylation and cloning of a gene responsible for the modification, J. Cell. Biochem. 36:261 (1988).

26. R. J. Deschenes and J. R. Broach, Fatty acylation is important but not essential for Saccharomyces cerevisiae RAS function, Mol. Cell. Biol. 7, 2344 (1987).

27. S. Clarke, J. P. Vogel, R. J. Deschenes and J. Stock, Posttranslational modification of the Ha-ras oncogene protein: Evidence or a third class of protein carboxyl methyltransferases, Proc. Natl. Acad. Sci. USA 85:4643 (1988).

28. L. E. Goodman, C. M. Perou, A. Fujiyama and F. Tamanoi, Structure and expression of yeast DPR1, a gene essential for the processing and intracellular localization of ras proteins, Yeast in press.

29. S. Powers, S. Michaelis, D. Broek, A. S. Santa-Anna, J. Field, I. Herskowitz and M. Wigler, RAM, a gene of yeast required for a functional modification of RAS proteins and for production of mating pheromone a-factor. Cell 47:413 (1986).

30. K. L. Wilson and I. Herskowitz., STE16, a new gene required for pheromone production by a cells of Saccharomyces cerevisiae. Genetics 155:441 (1987).

31. A. J. Brake, C. Brenner, R. Najarian, P. Laybourn and J. Merryweather, Structure of genes encoding precursors of yeast peptide mating pheromone a-factor. in "Protein transport and secretion", M. J. Gething ed., Cold Spring Harbor Laboratory, Cold Spring Harbor, New York (1985) p103.

32. R. Betz, J. W. Crabb, H. E. Meyer, R. Wittig and W. Duntze, Amino acid sequences of a-factor mating peptides from Saccharomyces cerevisiae. J. Biol. Chem. 262:546 (1987).

33. J. Nikawa, S. Cameron, T. Toda, K. M. Ferguson and M. Wigler, Rigorous feedback control of cAMP levels in Saccharomyces cerevisiae. Genes & Develop. 1:931 (1987).

34. R. J. Resnick and E. Racker, Phosphorylation of the RAS2 gene product by protein kinase A inhibits the activation of yeast adenylate cyclase. Proc. Natl. Acad. Sci. USA, 85:2474 (1988).

35. A. Papageorge, D. Lowy and E. M. Scolnick, Comparative biochemical properties of p21 ras molecules coded for by viral and cellular ras genes. J. Virol. 44:509 (1982).

36. R. Ballester, M. E. Furth and O. M. Rosen, Phorbol ester- and protein kinase c-mediated phosphorylation of the cellular Kirsten ras gene product. J. Biol. Chem. 262:2688 (1987).

MOLECULAR INTERACTIONS OF p21

Patrick D. Bailey and G.W. Guthrie Montgomery

Department of Chemistry
University of York
York
YO1 5DD
England

INTRODUCTION

The intrinsic GTPase activity of the p21 proteins appears to be central to the control of the transduction of cellular signalling across the cell membrane. Oncogenic p21 species that have reduced GTP hydrolysis have been studied to investigate if hydrolysis correlates with an increase in transforming potential, as the GTP-bound enzyme represents the active state of the complex[1,2]. The isolation of cytoplasmic cellular extracts containing GTPase activating protein (GAP) has clarified this relationship, indicating that in vivo the level of the active GTP-bound complex will result from both the intrinsic hydrolysis of the p21 protein and that induced by the ability of the GAP protein to recognise and stimulate the p21 molecule.[3]

The presence of GAP activity has been reported in cytosolic extracts of Xenopus oocytes [4], an extract from NIH 3T3 cells [5], and also in bovine brain[6] where the protein was isolated and purified to apparent homogeneity, corresponding to a protein of 125 kDa. This protein was found to be highly labile and was denatured on freeze thawing.

The action of G proteins in coupling particular receptor/effector responses can be mediated by other factors. Acylation of membrane associated proteins such as src may be responsible for cytoskeletal attachment, and furthermore this may be responsible for the constraint of associated molecules to one common domain within the cell [7]. The importance of the covalent attachment of lipid at a C-terminal cysteine for membrane attachment has been established in p21[8,9] although the extent and nature of the incorporation of G-proteins, and in particular p21, within the bilayer has recently been actively debated (see ref 10).

Many membrane bound enzymes rely on their lipid environment for their correct functional activity [11,12]. We have previously characterised the affinity of c-Ha ras p21 for the bilayer by ESR and have identified an association which is discernable from that of integral membrane proteins [13]. We present here the results of two areas of research a) the isolation of a cold labile GTPase stimulating protein from normal rat heart and b) further data as to the affinity of the c-Ha ras p21 protein for the phospholipid bilayer.

RESULTS AND DISCUSSION

1. Membrane Affinity

Our initial work on the association of p21 for the cell membrane utilized phospholipid vesicles containing a 5-doxyl stearate spin probe. This enabled us to determine the influence of p21 on the fluidity of the "synthetic membrane", and thereby identify the nature of its interaction. Our results indicated that about 70 phospholipids are associated with each molecule of p21, and that (unlike trans-membrane proteins) no tight "annulus" of lipid exists.

As an extension of this work, we investigated the GTPase activity of the molecule in incubations containing vesicles of phospolipid (prepared as described previously [13]). GTP hydrolysis was measured over a range of temperatures which we had established covered that where the bilayers were highly fluid, regularly ordered and in the intermediate heterogeneous state. The dependence of an integral membrane protein for the bilayer is seen as departures from linearity in the increase of enzymic activity with increasing temperature[11] The data obtained (figure 1) does not appear to contain any such changes in linearity, and hence indicated the the p21 has only a tentative association with the bilayer and is independent of the membrane for enzymic activity. This observation is consistent with the interpretation that this molecule has no dependance for incorporation within the bilayer, as is characteristic with classical trans-membrane proteins. The activation energy for the hydrolysis of GTP obtained from the gradient of an Arrhenius plot derived from this data was 26 kJ mol^{-1} (corrected for non-enzymic hydrolysis; data not shown).

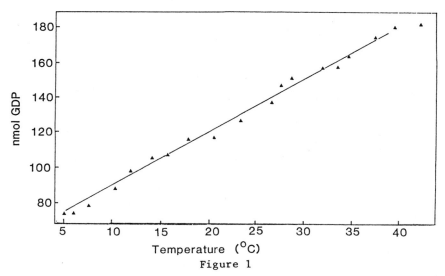

Figure 1

The increase in p21 GTPase activity with increasing
incubation temperature. The GTPase activity of c-Ha
ras p21 was assayed by incubation in 1.54 mM GTP/ 2.5
mM DDT/ 1 mM EDTA/ 10 mM $MgCl_2$ for 24 h.
The p21 was purified as described previously [13]
without the use of denaturants). Hydrolysis was
measured by the release of GDP in the incubation. GTP
and GDP concentrations were measured by reverse phase
HPLC on a C14 column run isocratically with 50 mM
phosphate pH 6.5/ 0.35% acetonitrile/ 50 mM tert-butyl
ammonium bromide (essentially the method of Tucker et
al.[14]).

Figure 1 supports the assumption that the affinity for bilayer
phospholipid in this non-enzymic form of the c-Ha ras p21 molecule is not
involved in maintenance of a functional conformation, as would be the case in
an integral trans-membrane protein; but rather to assist membrane attachment
and perhaps insertion in the post-translationally modified protein. Our model
for the extent of the bilayer perturbation by p21 is shown in figure 2. The
difference of this affinity to that of a trans-membrane protein which has a
tightly associated phospholipid annulus resulting from membrane association
is clearly depicted. The population of phospholipid influenced by p21 only
extends to about 70 proximal molecules. Trans-membrane proteins appear to re-
organise the ordering of as many as 170 adjacent phospholipids.

2. GAP activity

Figure 3 indicates the stimulation of GTP hydrolysis by Gly[12] c-Ha
p21 by protein fractions purified by affinity chromatography, isolated from
cytosolic extracts from normal rat heart.

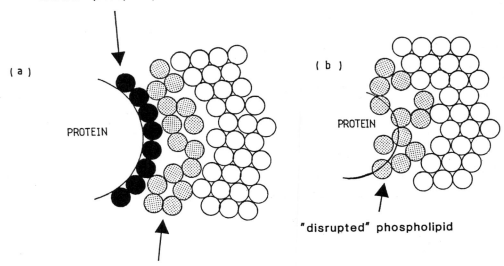

"annular" phospholipid

(a)

PROTEIN

(b)

PROTEIN

"disrupted" phospholipid

"disrupted" phospholipid

Figure 2

Diagramatic representation of (a) the assumed
interaction of a trans-membrane protein with its
surrounding phospholipid and (b) our proposed model of
the affinity of c-Ha ras p21 for the membrane[13].

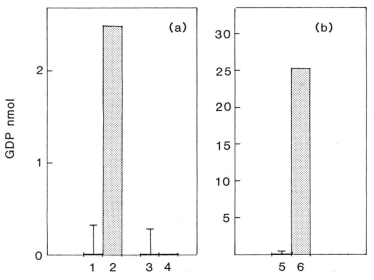

Figure 3

Figure 3

The stimulation of p21 GTP hydrolysis by proteins isolated from normal rat heart.

a) A 24 h incubation with Gly^{12} p21 (lanes 1 & 2) and Lys^{12} p21 (lanes 3 & 4). Lanes 1 & 3 represent the hydrolysis of the p21 species alone (normalised to zero), lanes 2 & 4 represent the hydrolysis with the addition of the proteins isolated by a p21 affinity column. Approximately 30 μg p21 was used in each incubation.

b) A 90 minute incubation with Gly^{12} p21 in the absence (lane 5) and presence (lane 6) of proteins isolated by gel-filtration and a subsequent separation by a p21 affinity column. Approximately 30 μg of p21 was used in each incubation, corresponding to 15 μg prepared as described in figure 1 and 15 μg solubilised by denaturants as by the method of Lacal [1]. The inclusion of p21 isolated by these two distinct methods, within one experiment, was to optimise the probability of detecting GAP activity.

A cytoplasmic extract of normal rat heart was chosen for the isolation of GAP activity as this tissue has been previously reported to contain a significant level of p21 expression [15]. Tissue homogenates were initially purified by the precipitation of particulate matter by centrifugation at 12000 g for 10 min. The chromatographic separation of this preparation was by two gel matrices; Sephacryl S-200 Superfine and an affinity column by coupling C-Ha ras p21 to epoxy activated Sepharose 4B. The buffer used to elute the gel filtration column was 50mM Tris-HCl pH 7.0/0.05% cholate/0.5M NaCl/1mM PMSF/0.02% sodium azide, and fractions having mobilities corresponding to the molecular weight range of approximately 80kDa to 160kDa were collected. The affinity matrix was eluted with several buffers containing 50mM Tris-HCl pH 7.0 with varying concentrations of NaCl (0-0.5M) and sodium cholate (0-0.1%).

The assay of GAP activity was by the stimulation of the hydrolytic rate of the p21. Compensation for the hydrolysis of GTP by proteins in the fraction under assay was only necessary in partially purified fractions and crude cytoplasmic extracts. No GTPase activity was detectable in the "GAP" fractions, but the results for the 24 h incubation are compensated for non-enzymic GTP breakdown which was detectable due to the duration of these assays.

GAP fractions contained approximately 20 fmol of each protein species, as extrapolated from their band density on silver-stained SDS gels.

This stimulation was not found for the corresonding Lys[12] form of p21. This result is consistent with the observation by Trahey & McCormick[4] of the stimulation of the cellular rather than the oncogenic form of N-<u>ras</u> p21 by the GAP protein isolated from <u>Xenopus</u> oocytes. Fractions isolated from affinity matrices which contained GAP activity were found to have a highly reproducible protein composition when assayed by SDS polyacrylamide gel electrophoresis, containing three main bands with apparent molecular weights of 60k, 130k and 140k (see figure 4, lane 1). The protein separation of the column of covalently bound p21 was notably different to the column matrix without bound p21 (data not shown), which suggests that the protein fractionation was due to affinity chromatography, and not non-specific interations.

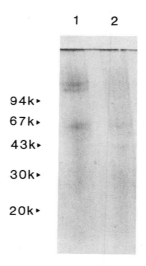

Figure 4

Silver-stained SDS polyacrylamide gel of
two fractions assayed for GAP activity.

Activity typically eluted from the column on increasing the NaCl concentration to 0.3 M and the cholate concentration of 0.1%. Higher salt and detergent concentrations resulted in only one major band (60k) to be detected in the eluate (Figure 4, land 2). Under these conditions no GAP activity was detected, suggesting either the 130k or 140k protein contained the GAP activity.

However, the isolated protein was found to be highly labile, irreversibly denatured by freezing or even mild sonication, and in addition Gibbs et al.[6] have reported the reversible inactivation of GAP activity by NaCl. Therefore we have yet to identify which of the three proteins isolated is responsible for the activation of the Gly^{12} p21 enzymic activity.

Current research indicates that GAP activity can be isolated from a wide variety of tissues, and that the protein is easily irreversibly denatured. In our experiments, incubations which displayed high GTPase levels due to GAP stimulation reverted to rates typical of in vitro GTP hydrolysis by p21 alone (typically about 4 pmol GTP/ nmol p21.minute at 1.54mM GTP), upon freezing, thawing and reincubating at $22^{\circ}C$.

Although rat heart was chosen as our initial source of protein, due to the high level of p21 expression in this tissue[15], it now appears that brain may contain higher levels of GAP protein [6]. In addition the inclusion of 10% glycerol to buffers was reported by Gibbs et al. to enable GAP activity to be recovered after storage at $4^{\circ}C$. The GAP activity was lost in samples isolated from rat heart when stored with 10% glycerol both frozen and at $4^{\circ}C$.

REFERENCES

1. Lacal, J.C., Srivastava, S.K., Anderson, P.S. & Aaronson, S.A. (1986) Cell 44, 609-617.
2. Gibbs, J.B., Sigal, I.S., Poe, M. & Scolnick, E.M. (1984) Proc. Natl. Acad. Sci. U.S.A. B1, 5704-5708.
3. Vogel, U.S., Dixon, A.F., Schaber, M.D., Diehl, R.E., Marshall, M.S., Scolnick, E.M., Sigal, I.S. & Gibbs, J.B. (1988) Nature 335, 90-93.
4. Trahey, M. & McCormick, F. (1987) Science 238, 542-545.
5. Cales, C., Hancock, J.F., Marshall, C.J. & Hall, A. (1988) Nature 332, 548-551.
6. Gibbs, J.B., Schaber, M.D., Allard, W.J., Sigal, I.S. & Scolnick, E.M. (1988) Proc. Natl. Acad. Sci. U.S.A. 85, 5026-5030.
7. Neer, E.J. & Clapham, D.E. (1988) Nature 333, 129-134.
8. Willumsen, B.M., Christensen, A., Hubbert, N.L., Papageorge, A.G. & Lowy, D.R. (1984) Nature 310, 583-586.
9. Willumsen, B.M., Norris, K., Papageorge, A.G., Hubbert, N.L. & Lowy, D.R. (1984) EMBO Journal 3, 2581-2585.
10. TIBS - 12th December 1987 - Letters pp 461-462.
11. Montgomery, G.W.G. & Gooday, G.W. (1985) FEMS Microbiol. Lett. 27, 29-33.
12. Hesketh, T.R., Smith, G.A., Houslay, M.D., McGill, K.A., Birdsall, M.J.M., Metcalfe, J.C. & Warren, G.B. (1976) Biochemistry 15, 4145-4151.
13. Montgomery, G.W.G., Jagger, B.A. & Bailey, P.D. (1988) Biochemistry 27, 4391-4395.
14. Tucker, J., Sczakiel, G., Feuerstein, J., John, J., Goody, R.S. & Wittinghofer, A. (1986) EMBO Journal 5, 1351-1358.
15. Spandidos, D.A. & Dimitrov, T. (1985) Bioscience Reports 5, 1035-1039.

RAS-TRANSFORMED CELLS CONTAIN MICROTUBULES RICH IN

MODIFIED α-TUBULIN IN THEIR TRAILING EDGES

A.R. Prescott, R. Magrath, and R.M. Warn

School of Biological Sciences
University of East Anglia
Norwich, NR4 7TJ, U.K.

SUMMARY

The distribution of microtubules rich in acetylated α-tubulin has been examined in a variety of transfected, transformed cell lines expressing, either reversibly or constitutively, ras oncoproteins. These cell lines all show large numbers of cell processes or 'tails' rich in acetylated tubulin microtubule bundles. The incidence of these bundles correlated with p21 oncoprotein expression. This microtubule distribution was very different from that seen in the untransformed varients of the cells examined where acetylated microtubules were predominently perinuclear. This difference may be of general significance in distinguishing motile cells from non-motile ones.

INTRODUCTION

A major feature of at least some ras-transformed cell lines is their increased motility. Microinjection of cells with ras-protein (Bar-Sagi & Feramisco, 1986) or transfection with inducible ras genes (McKay et al, 1986) results amongst other changes in a marked increase in cell motility (Dr. H. Patterson, personal communication). We have studied a number of fibroblast or fibroblast-like cell lines tranfected with ras oncogene constructs, some under promotional control. The cell lines were: T15, a NIH-3T3 derived line transfected with a construct consisting of multiple copies of the human N-ras proto-oncogene under control of a mouse mammary tumour virus long terminal repeat promoter (McKay et al., 1986). MR4: a Rat 1 derived line transfected with a mutant human H-ras gene under the control of a metallothionein promoter (Reynolds et al., 1987); EC816: a constitutive over expresser of normal human H-ras; and L1043: a constitutive expresser of human mutant (ETJ24) H-ras. All the lines except MR4 were developed by Dr. C. Marshall and collaborators at the Chester Beatty Institute, London.

Induction of the promoter in the T15 and MR4 cells by dexamethason
and zinc respectively results in distinct cell shape changes
(Linstead et al., 1988). Increased cell motility resulting from p2
oncoprotein expression is coupled with a change from polygonal to
polarised spindle shape. These polarised cells are characterised
by one (or ocassionally more) thin cell processed or 'tails'.
These tails are a common feature of the ras transformed cells
we have examined

A careful study of the microtubule network of the T15 cell line
(Prescott et al., 1988b) has shown that the cell processes in
transformed cells are rich in post-translationally modified
microtubules. There are three known major post-translational tubli
modifications. α-tubulin can be detyrosinated at its C-terminus by
a carboxypeptidase (Barra et al., 1974) and it can be acetylated at
lysine 40 by an acetylase (L'Hernault and Rosenbaum, 1983).
β-tubulin can be phosphorylated (Gard and Kirschner, 1985). The
majority (90%) of the microtubules in a number of interphase cell
lines have been found to turn over very rapidly with a half-life in
the region of 10 minutes (Schulze and Kirschner, 1986, Sammak and
Borisy, 1988, Prescott et al., 1988a). However, a minority of the
microtubules have a half-life of an hour or more (Schulze and
Kirschner, 1987). It is on these microtubules that the α-tubulin
modifications are generally thought to occur (detyrosination:
Gundersen, et al., 1987, acetylation: Piperno et al., 1987). Direc
evidence for this correlation has recently been obtained for one
cell type (Schulze et al., 1987).

Using monoclonal antibodies against the two modified α-tubulin
epitopes, detyrosination and acetylation, it has been demonstrate
that the tails of induced T15+ cells are filled with a bundle of
microtubules rich in the modified α-tubulins (Prescott et al.,
1988b). Such antibody staining distinguishes the more motile
transformed cells (T15+) from the static untransformed cells (T15-)
By examination of the occurrence of post-translationally modified
microtubules in other ras-transformed cell lines we have
investigated whether these changes in microtubule organisation are
general feature of ras transformation and the cell shape changes
which follow in these cell lines. The results are compared with the
data for T15 cells. In the light of these observations their
significance for cell motility and its role in metastasis is
discussed.

MATERIALS AND METHODS

Cell culture

T15 cells were cultured in Dulbecco's Modified Eagle Medium (DMEM
with 10% aseptic calf serum added in a 10% CO_2 atmosphere at 37°C.
Cells for immunofluorescence study were seeded onto glass
coverslips, placed into medium with (T15+) or without (T15-) 2μM
dexamethesone, and allowed to grow to confluence for at least 24
hours. Major changes in the phenotype of the cells were normally
not seen until three days had elapsed in dexamethasone containing
medium. L1043 and EC816 cells were grown in DMEM with 5% aseptic

calf serum seeded onto glass coverslips. MR4 cells were grown in DMEM with 10% aseptic calf serum with (MR4+) or without (MR4-) added zinc sulphate (100μM).

Antibodies

p21 oncoprotein was stained with a panreactive mouse monoclonal antibody (Cetus Corporation, Emeryville, CA). 6-11-B1 is a mouse monoclonal antibody which recognises acetylated α-tubulin (Piperno et al., 1987), YL 1/2 (Kilmartin et al., 1982), a rat monoclonal antibody which recognises tyrosinated α-tubulin, was supplied by Sera Lab., Sussex.

Immunofluorescence

Cells on coverslips were fixed in 90% methenol in MES buffer 100mM 2-(N-morpholino)-ethane sulphonic acid,(MES), 1mM EGTA, 1mM $MgSO_4$, pH 6.9) at -20°C for 5 min. The cells were then extracted with acetone, also at -20°C, and washed three times with phosphate-buffered saline/bovine serum albumin 0.02% w/v) (Weber et al., 1975). Non-specific staining was prevented by blocking with rabbit serum diluted (Sera Lab) 1:10 in P.B.S.. The microtubules were stained sequentially with 6-11-B1, Rhodamine conjugated anti-mouse IgG (Dako, High Wycombe), YL 1/2 and fluorescein conjugated anti-rat IgG (Lorne Diagnostics, Reading). Cells stained for F-actin were fixed in 4% paraformaldehyde in PBS and stained with FITC conjugated phalloidin (5μg/ml). The cells were finally mounted in Citifluor (Citifluor Ltd., London) and photographed with an Olympus OM2n camera mounted on a Zeiss standard microscope, using Kodak TMax 400 film pushed to 1600 ASA.

RESULTS

Acetylated microtubule distribution in untransformed cells

The distribution of microtubules rich in acetylated α-tubulin (acetylated microtubules) in the untransformed variants of either the reversibly transformable cell lines or the parental NIH-3T3 cell line is predominantely perinuclear. For example Fig. 1A shows NIH-3T3 cells stained with the 6-11-B1 antibody against acetylated α-tubulin. The groups of cells in the bottom right and top left of Fig 1A typify this distribution. Microtubule staining in general, at the cell periphery was fragmented in appearance and only a few complete rather curly microtubules could be seen in the mre sparse cultures.

Rarely in the parent NIH-3T3 cells and occasionally in the T15-line individual cells were seen which were crowded with acetylated microtubules. These included well spread giant cells with a larger than usual proportion of their microtubules acetylated (see for instance the cell in the bottom left hand corner of Figs. 1A and B) and also cells with long processes filled with a bundle of acetylated microtubules. The latter were more commonly seen in more sparse cultures.

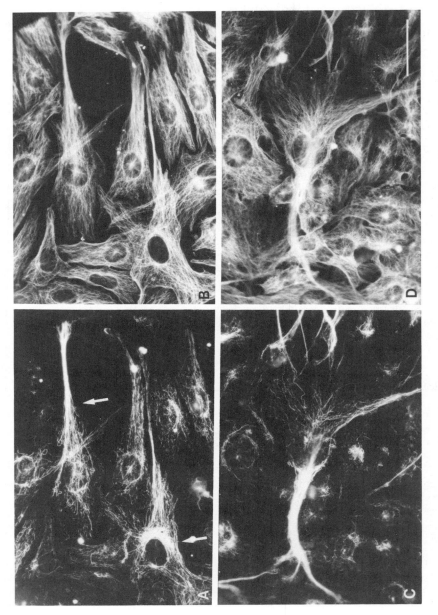

Figure 1. The distribution of acetylated (A and C) and tyrosinated (B and D) microtubules in the same NIH-3T3 cells (A and B) and L1043 cells (C and D). Arrows, giant cells filled with acetylated microtubules. Scale bar 50 µm.

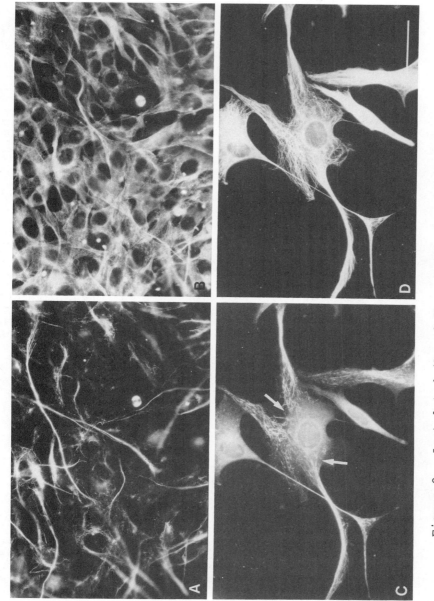

Figure 2. Acetylated (A and C) and tyrosinated (B and D) microtubules in transformed T15 cells. Arrows – origins of tail microtubule bundles in the cell body. Scale bar 50 μm.

Acetylated microtubules in ras transformed cells

We have previously demonstrated that transformation of the T15 cell line by dexamethasone induction of p21 oncoprotein results in a marked increase in the number of cell processes containing modified microtubule bundles (Prescott et al., 1988b). These bundles were a prominent feature of confluent T15+ cultures (Fig 2A and 3A) and enabled one to distinguish the cell tails from the overall microtubule staining pattern (Fig 2B and 3B). The acetylated microtubules of the tail originated in the cell body near the base of the tail or from the perinuclear array (Fig 2C). Usually only one tail is present (Fig 2A) but occassionally two or more occur (Fig 2C). Along the tail proper they formed a dense bundle where it was difficult to discern individual microtubules. Ultimately they were seen to spread out into the terminus of the tail, or 'foot' region (Fig 2C). The predominance of the acetylated microtubule bundles present in the cell tails (Fig 3A) correlated with the appearance in the T15+ cells of p21 oncoprotein (Fig 3B) and also with an increase in cell motility. Levels of the protein varied between individual cells and this heterogenieity was also seen in the variety of cell shapes seen. As the levels of p21 rose in the cells then the number of tails seen with 6-11-B1 increased. Ultimately the cells clumped into foci containing high levels of p21 (Fig 3D), these foci contain abundant cell tails and rounded cells (Prescott et al., 1988b).

Cell tails rich in modified microtubules were a common feature of other cell lines transformed with the ras oncoprotein p21. These included another line which could be reversibly transformed (MR4) and two lines which constitutively expressed p21, either by over expression of the normal protein (EC816) or lower levels of a mutant protein (L1043). The MR4 cell line when induced with zinc and stained with the 6-11-B1 antibody against acelylated tubulin (Fig 4A) showed a number of cell tails rich in acelylated microtubule bundles. Confluent MR4+ cultures of equivalent density to the induced T15+ cells generally had fewer tails than the T15's. However, the number of individual cells showing high levels of p21 (by immunofluorescence) were also lower in the MR4+ cells than the T15+ cells even though Western blot analysis showed the overall levels to be equivalent (data not shown). Once again the acetylated microtubules in the cell body were very fragmented in appearance, even more acetylated fragments were visible in the MR4+ cells than in the T15+ cells, though the counterstained tyrosinated microtubules (fig 4B) were complete. Modified microtubule tails were however a distinctive feature of the transformed MR4+ cells.

Cell lines which constitutively expressed the ras oncoprotein could obviously only be compared in their expression of modified microtubules with their parental cell line NIH-3T3. However a feature of the cell lines examined (EC816 and L1043) was their changing morphology as a result of increasing density. At low density the number of polarised cells was much less than that seen on approaching confluence. The number of modified microtubule bundles seen in the cell tails was similar to the situation in NlH-3T3 cells at low density whilst raising the density led to an increase in the number of cell tails containing acetylated or detyrosinated microtubules. As was shown in the T15 cell line these tails could be single or multiple, reflecting the changes in direction of movement by the cells (Fig 5A). Figure 4C shows EC816 cells with tails rich in acetylated microtubules, the surrounding

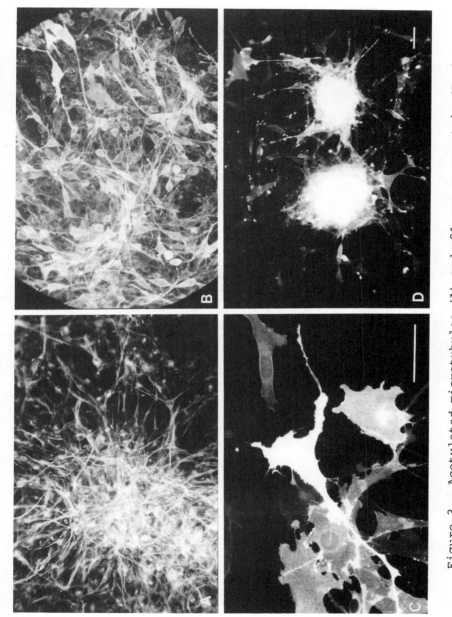

Figure 3. Acetylated microtubules (A) and p21 ras oncoprotein (B, C and D) in T15+ (A, B and D) and EC816 (C) cells. A and C are cells on different coverslips from the same culture dish. Scale bar (A, B & C on C) – 50μm.

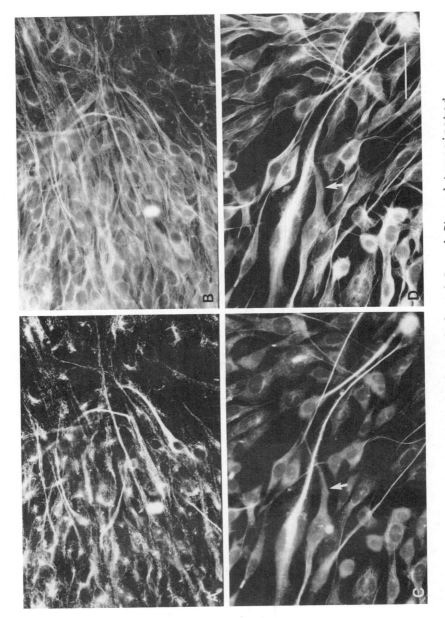

Figure 4. Acetylated microtubules (A and C) and tyrosinated tubulin microtubules (B and D) in MR4+ (A and B) and EC816 (C and D) cells. Arrows – process only weakly staining for acetylated microtubules. Scale bar 50 μm.

cells have many fewer acetylated microtubules mainly only sections of microtubules showing staining. Interestingly cell processes were apparent in this cell line which contained microtubules which were only weakly acetylated (Fig 4C) although unmodified microtubules were present (Fig 4D).

The L1043 cell line commonly consisted of a heterogeneous mixture of cells some large and well spread while others were more polarised, often overlying the more spread ones. The polarised cells were characterised by strongly staining acetylated microtubule bundles in the tails while the spread cells contained mainly perinuclear acetylated microtubules (Fig 1C). Staining for modified microtubules readily distinguished these two cell types from one another, the polarised cells were less apparent when the modified microtubules were stained (Fig 1D).

F-actin stress fibres in ras-transformed cells

The presence of F-actin stress fibres in the T15 cell line proved to be rather variable, both in the transformed and untransformed states (Prescott et al., 1988b). However, the general conclusion was that the presence of modified microtubules present in cell tails correlated with a reduction in stress fibre expression in these cells. The altered microtubule and microfilament organisation also correlated with an increase in cell motility.

Figure 5 shows the distribution of stress fibres and acetylated microtubules in EC816 cells. F-actin stress fibres were most marked in the cells in which the acetylated microtubules were restricted to the cell body (Fig 5A and B). Cells with prominent acetylated microtubules in their tails contained fewer stress fibres although they were not completely lost. However, there were instances where cells with strongly stained acelylated microtubules tails (Fig 5C) also contained F-actin stress fibres (Fig 5D). Thus in the EC816 cell line there was also a general but not absolute correlation between reduced stress fibre development and the presence of cell tails packed with modified microtubules.

DISCUSSION

In this study we demonstrate that a variety of cell lines transfected with ras oncogenes share common features in their microtubule distribution. In particular microtubules rich in acetylated α-tubulin were commonly seen in bundles in the tails of cells which were expressing the ras p21 oncoprotein. Both the reversibly transformable lines (T15 and MR4) and constitutive lines (EC816 and L1043) differed from either their parental or untransformed counterparts in this feature. The latter contained mainly acetylated microtubules only in the perinuclear region of the cell. These microtubules were usually wavy in distribution and the staining fragmented in appearance, representing intermittant acetylation along microtubules.

As the levels of ras oncoprotein (p21) increased in the reversibly transformable lines (MR4 and T15) the numbers of tails stained also increased. Cells which expressed p21 at high levels ultimately clumped together into foci. Phase microscopy, scanning electron microscopy (Linstead et al., 1988), and unmodified microtubule staining showed the foci to contain many rounded cells. However, acetylated α-tubulin staining showed that the foci contained large

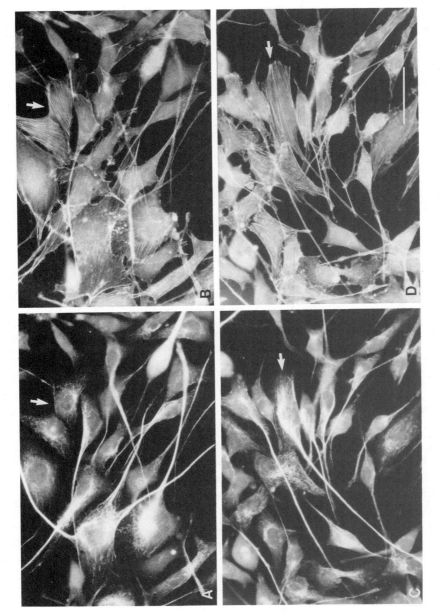

Figure 5. The distribution of acetylated microtubules (A and C) and F-actin microfilaments (B and D) in EC816 cells. Arrows: (A&B) – cells with distinct stress fibres and no tails. (C&D) – cells with tails and some stress fibres.

numbers of cell tail microtubules. EC816 cells rich in p21 also showed marked polarisation of the cell and distinctive tails. However, some tails were less heavily stained for acetylated tubulin a feature not seen in the transformed T15 or MR4. This may represent a difference in the levels of p21 expressed in the various cell lines or a difference in the protein type - N-ras versus H-ras.

We have discussed elsewhere (Prescott et al., 1988b) the possible roles of these modified microtubules in the tail of transformed cells, which may be in maintaining the integrity of the trailing edge of the cell fullfilling either a structural or organisational function. This may be a common feature of motile cells, the most widely studied of which are embryonic chick heart fibroblasts (Abercrombie, 1980).

Cell motility is likely to be important in metastasis. Metastatic migration may be influenced by factors secreted by transformed cells, such as autocrine motility factor. This factor is secreted at high levels by other NIH-3T3 derived cell lines transfected with _ras_ oncogenes but not the parent line (Liotta et al., 1986). Thus the changes in cell motility and cell ultrastructure may be due to the switching on of motility inducing factors, either directly or indirectly as the result of _ras_ activity. How these changes are induced by _ras_ remains to be elucidated.

Acknowledgements

The authors wish to thank Dr. Chris Marshall of the Chester Beatty, London for supplying cell lines for this study and his colleague Dr. Hugh Patterson for helpful discussions. We also thank Dr. G. Piperno for the 6-11-B1 antibody and Deborah Clemitshaw for typing the manuscript.

GLUTATHIONE S-TRANSFERASE/RAS FUSION PROTEIN: A TOOL FOR AFFINITY CHROMATOGRAPHY OF RAS ASSOCIATED PROTEINS

Hiroshi Maruta

Ludwig Institute for Cancer Research
PO Royal Melbourne Hospital
Victoria 3050, Australia

SUMMARY

Two proteins have previously been shown to interact directly with ras protein: Yeast adenylate cyclase comprizing 200 kDa and 70 kDa peptides is activated by ras protein (1) while a mammalian cytosolic protein of 125 kDa called GAP stimulates GTPase activity of normal ras protein but not that of its oncogenic varients (2). In this paper, using a ras fusion protein as a tool for affinity chromatography, I provide an evidence for the direct interaction of ras protein with at least three other proteins, i.e., tubulin, actin and cAMP-dependent protein kinase.

INTRODUCTION

To understand the exact role of ras protein in a growth factor-induced signal transduction which leads to either cell proliferation or differentiation, depending on the type of cell in question, one has to identify its upstream regulator as well as downstream target in the given cell and characterize their interactions with ras protein. Molecular biologists would accomplish this task by creating or selecting several distinct mutants in which a ras protein-dependent signal transduction pathway(s) are blocked and cloning the responsible genes by complementation. Biochemists would make a slightly different approach: He or she is to isolate directly a set of ras associated proteins which may affect the biological activity of ras protein or whose activity may be affected by ras protein. So far two distinct proteins have been shown to interact directly with ras protein: One is yeast ras-dependent adenylate cyclase complex (1) and the other is a mammalian GAP which stimulates GTPase activity of normal ras protein (2). It still remains to be clarified, however, whether the GAP is a target or regulator of ras protein. In addition, at least three other proteins are expected to interact with ras protein during its maturation, the process required for its translocation to the plasma membrane. They are ras carboxyl peptidase (RCP) which removes the C-terminal tripeptide to leave the Cys 186 as the new C-terminal residue (3), ras carboxyl methylase (RCM) which methylates carboxyl group of the same Cys (4) and ras acyltransferase (RAT) which transfers palmitate to thiol group of the same Cys (5). However, none of these enzymes has been isolated or characterized as yet. Therefore, to affinity-purify these enzymes and possibly several other ras associated proteins, I have created a series of ras fusion proteins which form a tight complex with glutathione (GSH)-sepharose affinity column. In these fusion proteins, the C-terminus of glutathione S-transferase (GST) is linked to the N-terminus of normal or modified human ras protein with a 13 amino acid spacer as shown in Fig. 1. A fusion protein called GST/c-HaRas 186 reported in this paper contains a truncated ras protein in which the C-terminal tripeptide of c-HaRas protein has been removed. Thus, it serves as a substrate for both RCM and RAT but not for RCP, and its GTPase activity is highly stimulated by GAP from bovine brain.

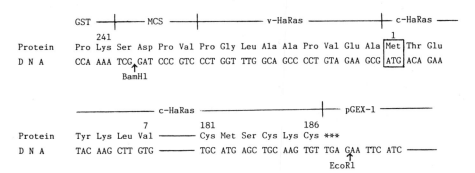

Fig. 1. Outline of the primary structure of a ras fusion protein, GST/c-HaRas 186, and the corresponding base sequence of its gene. The arrows indicate BamH1 and EcoR1 restriction sites which are used for insertion of ras sequence into pGEX-1 carrying GST gene. The number 241 corresponds to the C-terminus of GST, while the numbers 1 and 186 indicate the N- and C-termini of the truncated ras protein, respectively. For detail, see the following text under MATERIALS AND METHODS.

MATERIALS AND METHODS

Plasmid. In order to fuse the N-terminus of a truncated c-HaRas protein, deprived of the C-terminal tripeptide (-Val-Leu-Ser-COOH), to the C-terminus of GST from Schistosoma japonicum (Sj), I have prepared from E. coli plasmid pRAS1 (6) a BamH1/EcoR1 DNA fragment of 604 bp containing (5' to 3') the last codon (GTC) of MCS polylinker site, the 5' flanking sequence of 27 bp of v-HaRas, 186 codons encoding c-HaRas protein from Met 1 to Cys 186 and a termination codon (TGA) by the polymerase chain reaction (7) using a pair of priming oligonucleotides, 5' GGGATCCCGTCCCTGGTTTGGCAGCC 3' (sense) and 5' GGAATTCTCAACACTTGCAGCTCATGCAGCC 3'(antisense). The BamH1/EcoR1 fragment was then ligated with BamH1/EcoR1 fragment of E. coli plasmid pGEX-1 encoding SjGST (8) to prepare a new plasmid pGEX-1/c-HaRas186. The primary structure of the Ras fusion protein, GST/c-HaRas 186, is outlined in Fig. 1.

Ras fusion protein. The fusion protein GST/c-HaRas 186 of 440 amino acids was prepared from E. coli strain NM522 (9) carrying the plasmid pGEX-1/c-HaRas186 essentially by the procedure of Smith and Johnson (8) except that the IPTG induction of the plsmid expression was performed at 27°C instead of 37°C for 24 hrs because the fusion protein becomes poorly soluble when the cells are grown at 37°C (data not shown). The cells harvested from 1 liter culture were resuspended in 100 ml of L buffer containing 50 mM Tris-HCl, pH 7.5, 25% sucrose and 0.1 mg/ml lysozyme, incubated at 4°C for 15 min, lyzed by addition of 0.5% Nonidet P-40 and 0.03% SDS. The released DNAs were digested by 0.04 mg/ml pancreas DNase I in the presence of 5 mM $MgCl_2$ at 4°C for 15 min and the lysate was centrifuged at 20,000xg for 15 min. The supernatant was applied to a GSH-sepharose 6B column (bed volume: 15 ml) equilibrated with W buffer containing 20 mM Tris-HCl, pH 7.5 and 2 mM $MgCl_2$, and approximately 60 mg of the fusion protein was bound to the column. In order to form a stable complex of the ras fusion protein with a non-hydrolyzable GTP analogue, the column was then washed with 100 ml of A buffer containing 10 mM NaF, 0.3 mM $AlCl_3$ and 1 mM DTT in W buffer. The fusion protein could be recovered from the column with 100 ml of E buffer containing 50 mM Tris-HCl, pH 9.6, and 5 mM glutathione, dialyzed against A buffer overnight, concentrated with sucrose and stored at -20°C.

Affinity chromatography on ras/GST fusion protein bound to GSH-sepharose. GSH-sepharose 6B was prepared by coupling GSH to an epoxy-activated sepharose 6B (Pharmacia) according to the procedure of Simon and Vander Jagt (10). Bovine brain extract was prepared as previously described (2). Briefly, one fresh bovine brain (320 g) in 560 ml of B buffer was homogenized in a Waring blendor and centrifuged at 20,000xg for 40 min. The supernatant (450 ml) was applied first to a GSH-sepharose 6B column (bed volume: 30 ml) equilibrated with B buffer at a flow rate of 20 ml/hr, in order to let most of endogenous GST bind to the column. The column was washed with the same buffer and the bound GST (approximately 70 mg) was eluted with E buffer (100 ml). The flow-through fraction, essentially free of GST, was supplemented with 10 mM NaF and 0.3 mM $AlCl_3$, and then applied to a second GSH-sepharose 6B column (bed volume: 10 ml) which had been bound to approximately 40 mg of the ras/GST fusion protein in A buffer. The column was washed with 100 ml of A buffer and then the bound GAP and other proteins was eluted with 0.2 NaCl in A buffer (50 ml). Under these conditions the ras/GST fusion protein remains on the column. The eluate was dialyzed against B buffer overnight, concentrated with sucrose and stored at 0°C.

Microtubule, actin and catalytic subunit of cAMP-dependent protein kinase. Microtubules composed of tubulin and several other microtubule-associated proteins (MAPs) have been purified by two cycles of assembly and disassembly from bovine brain according to the procedure of Sloboda et al (11). G-actin was prepared from an acetone powder of rabbit skeletal muscle (Sigma Chemical M-6890) according to the procedure of Spudich and Watt (12). Catalytic subunit as well as regulatory subunit of cAMP-dependent protein kinase was purchased from Sigma Chemicals.

Normal human HaRas protein and its oncogenic or effector-site varients. Human c-HaRas protein, its oncogenic varient in which Gly 12 and Ala 59 are replaced by Val and Thr, respectively, and effector-site varient in which Thr 35 is replaced by Ala were prepared from E. coli after IPTG induction of the corresponding plasmid expression according to the procedure of Gibbs et al (13).

Assay for GAP activity and protein phopshorylation. GAP activity was measured according to Trahey and McCormick (14). Protein phosphorylation was assayed at 37°C for 10 min in the presence of (gamma-^{32}P) ATP in TMD buffer containing 50 mM Tris-HCl, pH 7.5, 5 mM $MgCl_2$ and 1 mM DTT, the phosphorylated proteins being separated by SDS-polyacrylamide gel electrophoresis, and the radioactivity of each protein band was quantitated by radioautography.

RESULTS

Major ras associated proteins in bovine brain extract. The following two precautions have been taken prior to affinity chromatography of ras associated proteins in bovine brain extract on GST/ras fusion protein bound to GSH-sepharose. First, bovine brain extract should be passed through a GSH-sepharose column to be deprived of any endogenous GST. Otherwise the GST would compete with GST/ras fusion protein in binding to GSH-sepharose. Second, since ras protein is active as a signal transducer and in binding to GAP only in the GTP-bound form and not in the GDP-bound form, GTP or GDP bound to ras/GST fusion protein should be replaced by a complex of GDP and AlF_4 which acts as a non-hydrolyzable GTP analogue thereby keeping the fusion protein in the active form. Otherwise as soon as GAP or other ras associated proteins interact with the fusion protein, the bound GTP would be hydrolyzed and the signal transducer activity of the fusion protein including the interaction with GAP would be attenuated. Two peptides of 56 kDa and 42 kDa comigrating with tubulin and actin on SDS polyacrylamide gel, respectively, appeared to be the major soluble proteins that bind the ras/GST fusion protein, when the post-microsomal supernatant (S-100) fraction prepared from bovine brain was first deprived of the endogenous GST by a GSH-sepharose column, then applied to a ras/GST/GSH-sepharose column, and the bound proteins were eluted with 0.2 M NaCl (see Fig. 2). Under these conditions only 10% of total GAP activity of S-100 fraction was bound to the ras/GST/GSH-sepharose column probably due to the low affinity of GAP for ras protein. None of these proteins was bound to a plain GST, instead of ras/GST fusion protein, on a GSH-sepharose column.

Phosphorylation of ras protein and its associated proteins. In addition to GAP, several other minor proteins also were bound to the ras/GST fusion protein or its complex with the two major ras associated proteins of 56 kDa and 42 kDa (RAP 56 and RAP 42). They include a few protein kinases which phosphorylate either ras protein, RAP56 or RAP42 in the presence of ATP. The phosphorylation of both RAP56 and RAP42 was cAMP/calcium-independent, whereas the phosphorylation of ras protein was cAMP-dependent (date not shown). In fact, ras protein was phosphorylated directly by purified catalytic subunit of cAMP-dependent protein kinase and the phopshorylation was completely inhibited by the regulatory subunit of cAMP-dependent protein kinase in the absence of cAMP. Shih and his colleagues have recently not only corfirmed the cAMP-dependent phosphorylation of ras protein but also shown that protein kinase C also phosphorylates ras protein at the same site (=Ser 177) both in vitro and in vivo (15). However, the exact role of the phosphorylation in the biological function of ras protein still remains to be clarified.

Identification of tubulin and actin with the major ras associated proteins. In order to examine whether or not RAP56 and RAP42 are tubulin and actin, respectively, tubulin and actin purified from bovine brain and rabbit skeletal muscle, respectively, were applied to the ras/GST/GSH-sepharose affinity column. Indeed tubulin or actin alone was able to bind to the ras/GST fusion protein. Since ras protein as well as tubulin and actin are acidic proteins, it is likely that the interaction of ras protein with these two major cytoskeletal proteins is rather specific.

DISCUSSION

In this paper the first evidence is provided for the direct interaction of ras protein with two major cytoskeletal proteins, i.e., tubulin and actin. Since there is a significant sequence homology between a calcium-dependent F-actin binding protein called calpactin II (or lipocortin I) and ras protein (16), and the dysfunction of a ras-related YPT1 protein in yeast leads to the complete disorganization of microtubules (17), perhaps it is not totally surprizing to find the direct interaction of ras protein with these cytoskeletal proteins. We are currently examining whether ras protein affects the assembly/disassembly kinetics of both tubulin and actin or not. In future we are to take a similar fusion protein approach to identify and affinity purify a variety of cytoplasmic and nuclear proteins or even genes which interact with many other oncoproteins including GAP.

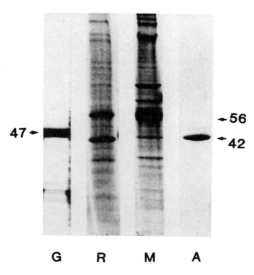

Fig. 2. 10% polyacrylamide (+SDS) gel electrophoresis. The samples are: G, GST/c–HaRas 186 fusion protein of 47 kDa; R, bovine brain soluble proteins bound to GST/c–HaRas 186 fusion protein; M, microtubules containing tubulin of 56 kDa; A, actin of 42 kDa.

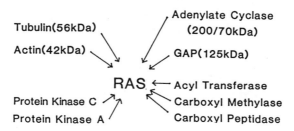

Fig. 3. A list of proteins which have been shown or expected to interact directly with ras protein.

ACKNOWLEDGEMENTS

The author is very grateful to Dr. Tony Burgess for his continuous support and great encourgement and also to Dr. Steve Ralph for his gift of E. coli plasmid pGEX-1 as well as to Dr. Richard Simpson and his colleagues for their supply of the priming oligonucleotides.

REFERENCES

1. J. Field et al., Purification of a RAS-Responsive Adenylyl Cyclase Complex from Saccharomyces cerevisiae by Use of an Epitope Addition Method, Mol. Cell. Biol. 8:2159 (1988).
2. J. B. Gibbs et al., Purification of ras GTPase activating protein from bovine brain, Proc. Nat. Acad. Sci. 85: 5026 (1988).
3. L. Goodman et al., Characterization of DPR1, a gene involved in the processing and intracellular localization of ras proteins, in: "NATO Advanced Research Workshop on Ras Oncogenes", D. A. Spandidos, ed., Plenum Press, New York (1989).
4. S. Clarke et al., Posttranslational modification of the Ha-ras oncogene protein: Evidence for a third class of protein carboxyl methyltransferase, Proc. Nat. Acad. Sci., 85: 4643 (1988).
5. J. E. Buss and B. M. Sefton, Direct identification of palmitic acid as the lipid attached to p21^{ras}, Mol. Cell. Biol. 6: 116 (1986).
6. R. B. Stein et al., Photoaffinity labeling with GTP of viral p21 ras protein expressed in E. coli, J. Virol, 50: 343 (1984).
7. R. K. Saiki et al., Primer-directed enzymatic amplification of DNA with a thermostable DNA polymerase, Science 239: 487 (1988).
8. D. B. Smith and K. S. Johnson, Single-step purification of polypeptides expressed in E. coli as fusions with glutathione S-transferase, Gene 67: 31 (1988).
9. J. A. Gough and N. E. Murray, Sequence diversity among related genes for recognition of specific targets in DNA molecules, J. Mol. Biol. 166: 1 (1983).
10. P. C. Simons and D. L. Vander Jagt, Purification of glutathione S-transferases from human liver by glutathione-affinity chromatography, Anal. Biochem. 82: 334 (1977).
11. R. D. Sloboda et al., Microtubule-associated proteins and the stimulation of tubulin assembly in vitro, Biochemistry 15: 4497 (1976).
12. J. A. Spudich and S. Watt, The regulation of rabbit skeletal muscle contraction. I. Biochemical studies of the interaction of the tropomyosin-troponin complex with actin and the proteolytic fragments of myosin, J. Biol. Chem. 246: 4866 (1971).
13. J. B. Gibbs et al., Intrinsic GTPase activity distinguishes normal and oncogenic ras p21 molecules, Proc. Nat. Acad. Sci. 81: 5704 (1984).
14. M. Trahey and F. McCormick, A cytoplasmic protein stimulates normal N-ras p21 GTPase, but does not affect oncogenic mutants, Science 238: 542 (1987).
15. T. Y. Shih et al., Novel phosphorylation and mutational studies of ras p21, in: "NATO Advanced Research Workshop on Ras Oncogenes", D. A. Spandidos, ed., Plenum Press, New York (1989).
16. C. B. Klee, Calcium-dependent phospholipid-(and membrane-) binding proteins, Biochemistry 27: 6645 (1988).
17. H. D. Schmitt et al., The ras-related YPT1 gene product in yeast: A GTP-binding protein that might be involved in microtubule organization, Cell 47: 401 (1986).

ARISTOLOCHIC ACID I INDUCED TUMORS IN WISTAR RATS CONTAIN

ACTIVATING MUTATIONS IN CODON 61 OF THE H-ras PROTOONCOGENE

H.H. Schmeiser, J. Lyons*, J.W.G. Janssen*, C.R. Bartram*, H.R. Scherf, W. Pfau and M. Wiessler

Institute of Toxicology and Chemotherapy, DKFZ, Im Neuenheimer Feld 280, D-6900 Heidelberg, FRG.
* Section of Molecular Biology, Department of Pediatrics II, University of Ulm, Prittwitzstrasse 43, D-7900 Ulm, FRG

Aristolochic acid I (AAI), a nitrophenanthrene derivative, is the main compound of the naturally occurring carcinogen aristolochic acid. Previous studies showed that AAI is directly mutagenic in the Ames test (Schmeiser et al., 1984) and leads to DNA adduct formation after metabolic activation in vitro and in vivo (Schmeiser et al., 1988). After oral administration of AAI to male Wistar rats DNA-AAI adducts in forestomach, stomach, liver, kidney, bladder and lung could be detected by use of the ^{32}P-postlabeling method.

According to the protocol of Mengs et al. (1982), male Wistar rats were treated with AAI at a dose of 10 mg/kg/d for 90 days. This treatment resulted in the induction of various tumors of the forestomach, earduct, pancreas and small intestine of the rats.

High-molecular-weight DNA isolated from 3 squamous cell carcinomas of the forestomach were assayed for their ability to induce tumors in nude mice by the tumorigenicity assay described by Fasano et al. (1984). All 3 tumor DNAs were capable of inducing tumors in nude mice. Southern blot analysis of the nude mouse tumors with specific probes for oncogenes of the ras family showed that the transforming activity was due to rat H-ras in all 3 cases.

In order to identify the mutation responsible for the transforming activity we amplified DNA sequences around codons 12 and 61 of the rat H-ras gene by PCR as described by Saiki et al. (1988). Using the dot-blot screening procedure reported by Verlaan-de Vries et al. (1986) all transfectants showed hybridization to 20mer oligomers representing the normal H-ras sequences around codon 61 but no hybridization to 20mer oligomers representing the normal H-ras sequences around codon 12. By direct sequencing of the enzymatically amplified DNA (Mc Mahon et al., 1987) from the tranfectants we could show, that the transforming activity was due to AT→TA transversions in either the second or third position of the 61st codon. In each case the presence of the activating mutation of the H-ras oncogene was confirmed in the original primary forestomach tumor DNA by means of PCR and subsequent selective oligonucleotide hybridization. Two fore-stomach tumors showed an AT TA transversion at the second position of codon 61, resulting in leucine instead of the normal glutamine, whereas one forestomach tumor showed the same mutation at the third position corresponding to a substitution of histidine for glutamine.

These mutational changes in adenine residues at codon 61 of the H-ras proto-oncogene are consistent with our results on the metabolism and DNA adduct formation of AAI. Indeed, we characterized a N^6-deoxyadenosine-AAI adduct (manuscript in preparation) as the major adduct formed by reaction of AAI and DNA after activation with Xanthine oxidase. The observation of carcinogen-specific patterns of base substitution in forestomach tumors of rats suggests that these alterations can result from the direct interaction of the electrophilic ultimate carcinogenic form of AAI with the c-H-ras gene in vivo.

References

Fasano, O., Birnbaum, D., Edlund, K., Fogh, J. and Wigler, M., 1984, New human transforming genes detected by a tumorigenicity assay, Mol. Cell. Biol., 4:1695

McMahon, G., Davis, E. and Wogan, G.N., 1987, Characterization of c-Ki-ras oncogene alleles by direct sequencing of enzymatically amplified DNA from carcinogen-induced tumors, Proc.Natl. Acad. Sci. USA, 84:4974

Mengs, U., Lang, W. and Poch, J.A., 1982, The carcinogenic action of aristolochic acid in rats, Arch. Toxicol., 51:107

Saiki, R.K., Gelfand, D.H., Stoffel, S., Scharf, S.J., Higuchi, R.,Horn, G.T., Mullis, K.B. and Ehrlich, H.A., 1988, Primer-directed enzymatic amplification of DNA with a thermostable DNA polymerase, Science, 239:487

Schmeiser, H.H., Pool, B.L. and Wiessler, M., 1984, Mutagenicity of the two main components of commercially available carcinogenic aristolochic acid in Salmonella typhimurium, Cancer Lett., 23:97

Schmeiser, H.H., Schoepe, K.-B. and Wiessler, M., 1988, DNA adduct formation of aristolochic acid I and II in vitro and in vivo, Carcinogenesis, 9:297

Verlaan-de Vries, M., Bogaard, M.E., van den Elst, H., Boom, J.H., van der Eb, A.J. and Bos, J.L., 1986, A dot-blot screening procedure for mutated ras oncogenes using synthetic oligodeoxynucleotides, Gene,, 50:313

β-ADRENOCEPTOR IS REQUIRED FOR ACTIVATION OF AVIAN ERYTHROCYTE G$_S$ BY GUANINE NUCLEOTIDES

Alan K. Keenan, Deirdre Cooney
and Rosemary Murray

Department of Pharmacology
University College
Belfield, Dublin 4
Ireland.

INTRODUCTION

During catecholamine-induced activation of turkey erythrocyte adenylate cyclase, hormone binding to the β-adrenoceptor (R) increases affinity of the latter for the stimulatory guanine nucleotide binding protein G$_S$. Exchange of GDP for GTP at the level of the G$_S\alpha$ subunit then occurs, leading to G$_S$ activation and increased conversion of ATP to cAMP via activation of the catalytic unit of adenylate cyclase C. While guanine nucleotide analogues such as Gpp(NH)p have been reported to cause marked cyclase stimulation in frog erythrocytes in the absence of hormone [1], the turkey system has been said to be refractory to such stimulation [2]. We now demonstrate that in both chicken and turkey erythrocyte membranes, Gpp(NH)p is a significant stimulant of adenylate cyclase activity and that this stimulation is abolished stereoselectivity by the β-adrenoceptor antagonist (+-) - propranolol. Furthermore we show that hormone-independent activation of G$_S$ in a reconstituted system requires the simultaneous presence of R and we propose therefore that the target species for guanine nucleotide activation in avian erythrocytes is an R-G$_S$ complex which exists in equilibrium with free R and G$_S$.

EXPERIMENTAL METHODS

Chicken and turkey erythrocyte membranes were purified according to Steer and Levitzki [3] and stored in liquid nitrogen until required. Turkey erythrocyte β-receptor and G$_S$ were purified according to Hekman et al [4]. Adenylate cyclase was assayed by the method of Salomon et al [5] and protein was determined either by the Lowry [6] or Peterson [7] method.

Reconstitution experiments: G$_S$ was reconstituted alone or in the presence of R by combination with phosphatidylethanolamine/phosphatidylserine/cholestrerol mixtures in a lipid: protein weight ratio of 100:1. Following removal of excess detergent by Sephadex G-50 chromatography, vesicles were assayed for incorporated R and G_S using [^{125}I] cyanopindolol and [^{355}] GTPγS respectively.

G_S activation: Vesticles containing G_S +- R were treated with Gpp(NH)p and Mg (0.6μM and 0.5mM respectively) at 30°C and the extent of G_S activation with time was monitored by adding aliquots of the activation mixture to a crude rabbit myocardial extract which was then assayed for adenylate cyclase activity.

RESULTS

The stimulation of adenylate cyclase activity in turkey and chicken erythrocyte membranes by a number of activators is outlined in Table 1.

Table 1. Stimulation of adenylate cyclase activity in A, chicken and B, turkey erythrocyte membranes

Stimulant	Specific Activity, A	Specific Activity, B
None (basal)	2.0 +- 0.8	2.4 +- 1.3
Isoprenaline	5.0 +- 1.5	59.1 +- 1.7
Gpp(NH)p	73.0 +- 7.5	126.9 +- 0.5
Gpp(NH)p+propranolol	5.0 +- 1.0	9.0 +- 0.7
Gpp(NH)p+isoprenaline	128.2 +- 11.6	648.2 +- 4.8
NaF (10mM)	85.5 +- 5.0	270.4 +- 10.2
Forskolin (100μM)	4.1 +- 2.0	42.0 +- 0.6

Data are mean values +- s.e.m. of three independent experiments performed in duplicate. All stimulant concentrations were 10μM unless otherwise stated. Specific activity units: pmol cAMP/min/mg protein.

In both cases maximal stimulation was obtained with isoprenaline + Gpp(NH)p while the diterpene forskolin was a weak activator, in contrast to the situation with many other cell types where forskolin is the most potent stimulant. The stimulation by Gpp(NH)p alone was 57% (chicken) or 20% (turkey) of maximal and was abolished in each case by (+-) - propranolol. The stereoselectivity of propranolol's inhibitory action was examined in chicken membranes and the concentration-dependent inhibition of Gpp(NH)p stimulation by (+) - and (-) - propranolol is compared in Figure 1. As can be seen, the potency of the (-) - isomer was almost two orders of magnitude greater than that of the (+) - isomer (IC$_{50}$ values were 13.6nM and 1.0μM respectively).

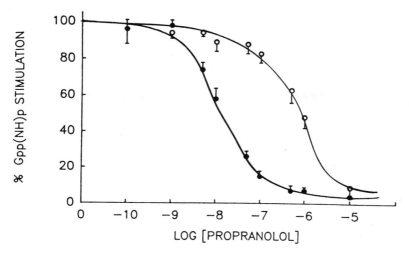

Figure 1. The steroselective inhibition of Gpp(NH)p-stimulated adenylate cyclase activity by (-) - propranolol (o) and (+) - propranolol (o). Curves were fitted using the ALLFIT programme and values shown are means +- s.e.m. of three independent experiments.

When purified R and G$_S$ wre co-reconstituted into phospholipid vesicles, Gpp(NH)p (in the presence of Mg) caused a time-dependent activation of G$_S$ which was again abolished in the presence of (+-) -propranolol, as shown in Figure 2. The extent of Gpp(NH)p -induced activation was 17% of that obtained in the presence of isoprenaline + Gpp(NH)p. When reconstituted vesicles contained only G$_S$, no time-dependent activation was obtainable with Gpp(NH)p in the absence or presence of   (+-) -propranolol (data not shown).

DISCUSSION

This study demonstrates that guanine nucleotide stimulation of adenylate cyclase activity in avian erythrocytes can occur in the absence of added hormone and is directly regulated by the β-adrenoceptor antagonist (+-)-propranolol. It has generally been stated that hormone (H) is necessary for G$_S$ activation and that it is only the H-R complex which has high affinity for G$_S$, i.e. free R would be expected to have a low affinity for G$_S$ in the absence of H. However from the present results it can be inferred that either G$_S$ activation is independent of interaction with a H-R complex, or R and G$_S$ are to some extent precoupled in the absence of H. A form of association between R and G$_S$ under "resting" conditions is indeed suggested by the observation that propranolol stereoselectively inhibited Gpp(NH)p stimulation. In support of this we have carried out a detailed study of propranolol's interaction with the chicken erythrocyte system [8] and shown a high degree of correlation between inhibition of Gpp(NH)p stimulation and displacement of [125I] cyanopindolol binding to receptors by propranolol. In the same study propranolol did not affect NaF stimulation, which is consistent with a fully receptor-independent mechanism for NaF action.

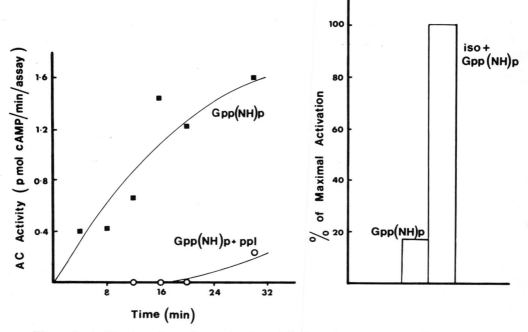

Figure 2. A. The time-dependent activation of G$_S$ in vesicles containing β-receptor
(propranolol concentration was 10μM); B. Gpp(NH)p induced activation as a
percentage of maximal.

In experiments using reconstituted vesicles the receptor-dependent G_S activation obtained with Gpp(NH)p was of the same relative magnitude (compared to maximal) as that observed in native turkey membranes, from which we infer that R-G_S coupling modes are to some extent comparable in the two systems. If this is true, the absolute requirement of R demonstrated in the reconstituted system would imply that in native membranes guanine nucleotides preferentially recognise precoupled R-G_S complexes and that maximal activation of G_S is achieved when the number of R-G_S complexes is further increased by hormone. Such complexes could in turn be dissociated by antagonists (such as propranolol), rendering the system unresponsive to stimulation by guanine nucleotides. An active role for antagonists in destabilising R-G_S complexes has been proposed for the D_2-dopamine receptor system [9] and it has further been shown that guanine nucleotides increase antagonist affinities for R in certain circumstances [10].

In conclusion therefore, we wish to propose that G_S exists in unstimulated avian erythrocyte membranes as an equilibrium mixture of free and receptor-bound forms (Figure 3) and that the relative amounts of the two forms depend on whether the system has been exposed to hormone or antagonist. This hypothesis, which has been suggested by the studies with native membranes, is much strengthened by the demonstration that R constitutes an absolute requirement for G_S activation in reconstituted vesicles.

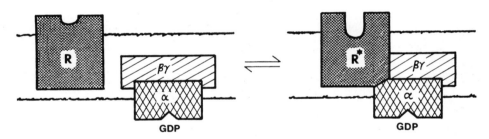

Figure 3. Proposed interconversion of free and receptor-bound G_S in avian erythrocyte membranes. Binding of R to G_S is presumed to be to the α subunit of the $\alpha\beta\gamma$ heterotrimer, which also contains bound GDP in the absence of stimulation by hormone.

ACKNOWLEDGEMENTS

We would like to acknowledge Professor E. J. M. Helmreich, Univesity of Wurzburg, in whose laboratory part of this study was carried out. We would also like to thank Dr. M.

Hekman for providing samples of purified R and G_S. This work was supported by grants from the Fritz Thyssen Foundation (to A.K.), Boehringer Ingelheim (to D.C.) and the Medical Research Council of Ireland (to R.M.). Finally we wish to thank Mr. R. Monks and Mrs. M. Doyle for their assistance in the preparation of this manuscript.

REFERENCES

1. R. J. Lefkowitz, C. Mullikin and M. G. Caron, Regulation of β-adrenergic receptors by guanyl-5'yl imidophosphate and other purine nucleotides, J. Biol. Chem. 251: 4686-4692 (1976).

2. A. Levitzki, Slow GDP dissociation from the guanyl nucleotide site of turkey erythrocyte membranes is not the rate limiting step in the activation of adenylate cyclase by β-adrenergic receptors, FEBS Lett. 115: 9-10 (1980).

3. M. L. Steer and A. Levitzki, The control of adenylate cyclase by calcium in turkey erythrocyte ghosts. J. Biol. Chem. 250: 2080-2084 (1975).

4. M. Hekman, D. Feder, A. K. Keenan, A. Gal, H. W. Klein, T. Pfeuffer, A. Levitzki and E. J. M. Helmreich, Reconstitution of β-adrenergic receptor with components of adenylate cyclase, EMBO J. 3:3339-3345 (1984).

5. Y. Salomon, C. Londos and M. Rodbell, A highly sensitive adenylate cyclase assay. Anal. Biochem. 58: 541-548 (1974).

6. O. H. Lowry, N. J. Rosebrough, A. L. Farr and R. J. Randall, Protein measurement with the Folin phenol reagent, J. Biol. Chem. 193:265-275 (1951).

7. G. L. Peterson, A simplification of the protein assay method of Lowry et al, which is more generally applicable. Anal. Biochem. 83:346-356 (1977).

8. R. Murray and A. K. Keenan, The β-adrenoceptor is precoupled to G_S in chicken erythrocyte membranes, Cellular Signalling, in press (1988).

9. K. A. Wreggett and A. DeLean, The ternary complex model: Its properties and application to ligand interactions with the D_2-dopamine receptor of the anterior pituitary gland. Mol. Pharmacol. 26:214-227 (1984).

10. P. H. Lang and B. Lemmer, Evidence for two specific affinity states of ³H-antagonist binding to cardiac β-adrenergic receptors and influence of Gpp(NH)p, J. Cyclic Nucleotide and Protein Phos. Res. 10:341-360 (1985).

EXPRESSION OF RAS ONCOGENE IN *XENOPUS LAEVIS*

Ellen Z. Baum, Geraldine A. Bebernitz, and
Philip M. Sass

Molecular Biology Section
Lederle Laboratories, American Cyanamid Company
Pearl River, N.Y. 10965

Oocytes of the South African clawed toad, *Xenopus laevis*, are an excellent model system for the study of various proto-oncogenes, including myc (1-4), src (5), and ras. The large size of *Xenopus* oocytes (~1.3 mm in diameter) permits the microinjection of biological material into these cells and subsequent biochemical analysis. In addition, the developmental processes of oogenesis and embryogenesis are easily monitored in the laboratory. In this communication, we summarize the effects of endogenous and exogenous ras protein on meiosis in oocytes. In addition, we present our preliminary characterization of the *Xenopus* homolog of mammalian Kirsten ras.

Xenopus oocytes are blocked in cell division at first meiotic prophase. To become an egg, the oocyte must resume meiosis, an event termed "maturation" (6,7). In the frog, gonadotropin released by the pituitary is transported via the blood to the ovary, where it causes the follicle cells surrounding the oocytes to produce progesterone, which is the natural inducer of maturation. In vitro, in oocytes surgically removed from the ovary, maturation can be triggered by a variety of inducing agents, including progesterone added to the culture medium (6,7). Progesterone-induced maturation apparently involves a membrane interaction since progesterone coupled to glass beads which cannot be internalized will still induce, and it will not induce if microinjected (6,7 and references therein).

A host of biochemical events in oocytes is triggered following the addition of progesterone (6,7 and references therein). The level of cAMP declines, due to decreased adenylate cyclase activity . Intracellular calcium and pH increase. There is an increase in protein phosphorylation, including ribosomal protein S6, brought about by cAMP independent protein kinase activity. Protein synthesis also increases . Maturation promoting factor (MPF), which is stored in the oocyte in an inactive form, undergoes autophosphorylation (8,9) and is capable of inducing maturation when oocyte cytoplasm is serially transferred. The most obvious physical manifestation of oocyte maturation is the breakdown of the germinal vesicle or oocyte nucleus (GVBD), which is scored in mature oocytes by the appearance of a white spot on the top of the darkly-pigmented animal pole (6,7).

In examining some of the events that accompany progesterone-induced maturation, it was possible to predict that certain agents might induce or inhibit maturation. For example, progesterone-induced maturation might be inhibited by agents which increase cAMP levels. In fact, cholera toxin (10) and phosphodiesterase inhibitors (11), agents which increase cAMP, do inhibit progesterone-induced maturation. In eukaryotic cells, cAMP works through the activation of cAMP dependent protein kinases, and it was found that microinjecton of the catalytic subunit inhibits progesterone-induced maturation (6, 12). Conversely, microinjection of the regulatory subunit will bind up the catalytic subunit in its inactive form (12). Inhibition of this kinase is sufficient to induce maturation. Consistent with this is the observation that the heat stable inhibitor of protein kinase A also induces (12). The tyrosine kinase encoded by the src oncogene has been shown to act synergistically with progesterone to induce meiosis (13), whereas the kinase encoded by v-mos can induce meiosis independently of progesterone (P.M.S., unpubl.). A variety of other agents are capable of inducing maturation, including insulin (14), calcium ionophores, metal ions, and miscellaneous surface acting agents, such as local anaesthetics and sulfhydryl reagents (15). In addition, microinjected ras protein also induces oocyte maturation (16).

In yeast, ras can stimulate the adenylate cyclase (17, 18). It might then be expected that ras microinjected into oocytes would activate *Xenopus* adenylate cyclase and raise the level of cAMP, inhibiting progesterone-induced maturation, In fact, ras protein induces oocyte maturation. Birchmeier et al. (16) demonstrated that transforming H-ras is 100x more potent as an inducer than the wild-type protein. In contrast to progesterone-induced maturation, the level of cAMP does not decline, indicating that ras can activate a pathway to maturation that is not mediated by adenylate cyclase. Similar results have been obtained by Trahey and McCormick for microinjected N-ras (19). These investigators also demonstrated that wild-type ras protein will induce maturation if bound with a non-hydrolyzable analog of GTP, GppNp.

Ras genes have been identified in a wide range of organisms, including yeast, Dictylostelium, Drosophila and mammals (20), and it was expected that *Xenopus* would also contain these genes. The investigation of ras in *Xenopus* might provide information on the function of this protein in normal cells, since there is no evidence that ras protein stimulates adenylate cyclase in organisms other than yeast.

```
              10v         20v         30v         40v         50v
CKRAS4B   MTEYKLVVVGAVGVGKSALTIQLIQNHFVDEYDPTIEDSYRKQVVIDGETC
                                                      SYRKQVVIDGETC
CXL-1                                                 SYRKQVVIDGETC
                                                          10^
              60v         70v         80v         90v        100v
CKRAS4B   LLDILDTAGQEEYSAMRDQYMRTGEGFLCVFAINNTKSFEDIHHYREQIKR
          LLDILDTAGQEEYSAMRDQYMRTGEGFLCVFAINN.KSFEDIHHYREQIKR
CXL-1     LLDILDTAGQEEYSAMRDQYMRTGEGFLCVFAINNIKSFEDIHHYREQIKR
              20^         30^         40^         50^         60^
             110v        120v        130v        140v        150v
CKRAS4B   VKDSEDVPMVLVGNKCDLPSRTVDTKQAQDLARSYGIPFIETSAKTRQGVD
          VKDSEDVPMVLVGNKC.LPSR:VDTKQAQDLARS GIPFIETSAKTRQGVD
CXL-1     VKDSEDVPMVLVGNKCALPSRSVDTKQAQDLARSSGIPFIETSAKTRQGVD
              70^         80^         90^        100^        110^
             160v        170v        180v
CKRAS4B   DAFYTLVREIRKHKEKMSKDGKKKKKKSKTKCVIM
          DAFYTLVREIRKHKEKMSK.GKKKKK SKTKC I:
CXL-1     DAFYTLVREIRKHKEKMSKAGKKKKK-SKTKCSIL
             120^        130^        140^
```

Fig.1. Amino acid sequence comparison between human Kirsten
ras and *Xenopus* ras (one letter amino acid code). CKRAS4B:
amino acid sequence of human Kirsten ras, exon 4B (24). CXL-
1: amino acid sequence of the 800 bp Xenopus cDNA clone
hybridizing to human Kirsten ras probe. The middle line
compares these two sequences. Matching amino acids are
indicated by the one letter code. For non-matching amino
acids, positive(:), zero-value (.), and negative (blank)
relatedness are indicated.

Indications that functional ras protein is found in oocytes originate from studies of microinjected ras antibodies. The microinjection of ras antibodies into oocytes inhibits insulin-induced meiosis, but not progesterone-induced meiosis (21-23). Thus, it appears that a ras-like protein is a normal component of the insulin maturation pathway.

To isolate oocyte cDNA homologues of ras oncogene, we first determined the optimal hybridization conditions for Southern blots of *Xenopus* genomic DNA probed with randomly primed human Kirsten ras cDNA (24). Optimal hybridization was at 37° in 30% formamide, 5X SSPE, 5X Denhardt's, 0.1% SDS, and 100 ug/ml sheared denatured salmon sperm DNA. Final washing was at 55° in 150 mM NaCl, 20mM Tris·HCl pH 7.8, and 1 mM EDTA. Under these conditions, two bands of approximately 4 kb and 1 kb were detected in *Xenopus* DNA which had been digested with Eco RI.

Kirsten ras cDNA was then used to probe an oocyte cDNA library in λgt10 kindly provided to us by Dr. D. Melton (25). We screened 375,000 phage and identified three positive candidates which contained inserts of 800, 820, and 1300 bp. The DNA sequence of the 800 bp clone has been determined and is 76% homologous to human Kirsten ras cDNA (data not shown). This *Xenopus* clone is a partial cDNA apparently lacking the amino terminus, and it contains one open reading frame of 447 nt. There is striking homology between the *Xenopus* and human proteins ((Fig. 1): 142 out of 150 amino acids (95%) are identical, including the lysine-rich tail encoded by exon 4B of cK-ras (24). Thus, ras protein is highly conserved between these two species.

The 800 bp clone has been used as a hybridization probe on Southern and Northern blots of *Xenopus* DNA and RNA, respectively. We find a single-copy gene hybridizing to this probe, encoding an mRNA of 2.6 kb. These cDNA clones will be used to address the function of endogenous ras oncogene in oocytes. The microinjection of sense and anti-sense RNA will allow us to increase and decrease the translatable ras RNA in oocytes, and to determine their effects on oocyte maturation under a variety of conditions. In addition, chimeric ras proteins can be constructed and tested for biological activity in this system.

Acknowledgements We thank Gail Howard for DNA sequence data

References

1. Taylor, M.V., Gusse, M., Evan, G.I., Dathan, N., and Mechali, M., *Xenopus* myc proto-oncogene during development: expression as a stable maternal mRNA uncoupled from cell division, *EMBO* **5**, 3563-3570 (1986).
2. Godeau, F., Persson, H., Gray, H.E., and Pardee, A.B., c-myc expression is dissociated from DNA synthesis and cell division in *Xenopus* oocyte and early embryonic development, *EMBO* **5**, 3571-3577 (1986).

3. King, M.W., Roberts, J.M., and Eisenman, R.N., Expression of the c-myc proto-oncogene during development of *Xenopus laevis, Mol. Cell. Biol.* **6**, 4499-4508 (1986).
4. Nishikura, K., Expression of c-myc proto-oncogene during the early development of *Xenopus laevis, Oncogene Res.* **1**, 179-191 (1987).
5. Steele, R.E., Two divergent cellular src genes are expressed in Xenopus laevis, *Nucleic Acids Res.* **13**, 1747-1761 (1985).
6. Maller, J.L., Interaction of steroids with the cyclic nucleotide system in amphibian oocytes, *Adv. in Cyclic Nucleotide Research* **15**, 295-336 (1983).
7. Wasserman, W.J., Penna, M.J, and Houle, J.G., The regulation of Xenopus laevis oocyte maturation, *in:* "Gametogenesis and the Early Embryo," J.G. Gall, ed., Alan R. Liss, Inc. N.Y. (1986).
8. Lohka, M.L., Hayes, M.K., and Maller, J.L., Purification of maturation-promoting factor, an intracellular regulator of early mitotic events, *Proc. Nat. Acad. Sci.* **85**, 3009-3013 (1988).
9. Cyert, M.S., and Kirshner, M.W., Regulation of MPF activity in vitro, *Cell* **53**, 185-195 (1988).
10. Sadler, S.E., and Maller, J.L., Progesterone inhibits adenylate cyclase in *Xenopus* oocytes, *J. Biol. Chem.* **256**, 6368-6373 (1981).
11. O'Connor, C., and Smith, L.D. Inhibition of oocyte maturation by theophylline: possible mechanism of action, *Dev. Biol.* **52**, 318-322 (1976).
12. Maller, J.L., and Krebs, E.G., Progesterone-stimulated meiotic cell division in *Xenopus* oocytes, *J. Biol. Chem.* **252**, 1712-1718 (1977).
13. Spivack, J.G., Erikson, R.L., and Maller, J.L., Microinjection of pp60 v-src into *Xenopus* oocytes increases phosphorylation of ribosomal protein S6 and accelerates the rate of progesterone-induced meiotic maturation, *Mol. Cell. Bio.* **4**, 1631-1634 (1984).
14. Maller, J.L., and Koontz, J.W., A study of the induction of cell division in amphibian oocytes by insulin, *Dev. Biol.* **85**, 309-316 (1981).
15. Baulieu, E., Schorderet-Slatkine, S., Steroidal and peptidic control mechanisms in membrane of *Xenopus laevis* oocytes resuming meiotic division, *J. Steroid Biochem.* **19**, 139-145 (1983).
16. Birchmeier, C., Broek, D., and Wigler, M., RAS Proteins can induce meiosis in *Xenopus oocytes, Cell* **43**, 615-621 (1985).
17. Toda, T., Uno, I., Ishikawa, T., Powers, S., Kataoka, T., Broek, D., Cameron, S., Broach, J., Matsumoto, K., and Wigler, M., In yeast, RAS proteins are controlling elements of adenylate cyclase, *Cell* **40**, 27-36 (1985).
18. Broek, D., Samiy, N., Fasano, O., Fujiyama, A., Tamanoi, F., Northup, J., and Wigler, M., Differential activation of yeast adenylate cyclase by wild-type and mutant RAS proteins, *Cell* **41**, 763-769 (1985).
19. Trahey, M., and McCormick, F., A cytoplasmic protein stimulates normal N-*ras* p21 GTPase but does not affect oncogenic mutants, *Science* **238**, 542-545, (1987).
20. Barbacid, M., *ras* Genes, *Ann. Rev. Biochem.* **56**, 779-827 (1987)

21. Deshpande, A.K., and Kung, H., Insulin induction of *Xenopus laevis* oocyte maturation is inhibited by monoclonal antibody against p21 *ras* proteins, *Mol. Cell. Biol.* **7**, 1285-1288 (1987).
22. Korn, L.J., Siebel, C.W., McCormick, F., and Roth, R.A., *Ras* p21 as a potential mediator of insulin action in *Xenopus* oocytes, *Science* **236**, 840-843 (1987).
23. Sadler, S.E., Schechter, A.L., Tabin, C.J., and Maller, J.L., Antibodies to the *ras* gene product inhibit adenylate cyclase and accelerate progesterone-induced cell division in *Xenopus laevis* oocytes, *Mol. Cell. Biol.* **6**, 719-722 (1986).
24. McCoy, M.S., Bargmann, C.I., and Weinberg, R.A., Human colon carcinoma Ki-*ras*2 oncogene and its corresponding proto-oncogene, *Mol. Cell. Biol.* **4**, 1577-1582 (1984).
25. Rebagliati, M.R., Weeks, D.L., Harvey, P.A., and Melton, D.A., Identification and cloning of localized maternal RNAs from *Xenopus* eggs, *Cell* **42**, 769-777 (1985).

AN APPRAISAL OF FOUR MONOCLONAL ANTIBODIES USED IN THE

IMMUNOHISTOCHEMICAL DETECTION OF *RAS* p21

J. R. Gibson[1], G. L. Johnson[2], and I. Gibson[1]

[1]School of Biological Sciences, University of East Anglia,
Norwich NR4 2PA, and [2]Department of Pathology, Norfolk and
Norwich Hospital, Norwich, NR2 2HA

INTRODUCTION

The precise role of *ras* p21 in both the normal cell and in human
carcinogenesis is far from established. It is important that reliable
immunocytochemical methods should exist for the detection of p21 enabling
the determination of a) expression at single cell level and b) its role in
benign, premalignant and malignant tissue.

There have been a number of immunohistochemical studies which have
attempted to elucidate not only the role of p21 but also its usefulness as a
diagnostic and prognostic marker. Unfortunately there is inconsistency between different reports. Increased *ras* expression has been correlated with
the depth of invasion and/or clinical stage of colorectal cancer (Horan
Hand et al., 1984) and stomach cancer (Tahara et al., 1986). No such correlation has been found by others (Williams et al., 1985; Ohuchi et al.,
1987; Noguchi et al., 1986) In mammary tumours Horan-Hand et al., (1984)
detected enhanced expression of p21 in the majority of carcinomas but Ghosh
et al., (1986) found that normal, benign and malignant cells stained with
similar intensity The authors of the latter report suggest that the monoclonal antibody RAP-5 is of limited usefulness. Robinson et al., (1986)
found that RAP-5 recognizes not only *ras* protein but also other widely
distributed, unrelated epitopes.

The purpose of this study is to determine the effectiveness of four
commonly used monoclonal antibodies in detecting *ras* p21 in fixed paraffin-
embedded specimens. Animal cells known to express p21 at high levels were
incorporated in 1% agarose to form a model tissue system, and also grown as
experimental tumours in nude mice. The effects of two fixatives were
compared. These were formol-saline and PLPD (the dichromate derivative
periodate-lysine-paraformaldehyde. Holgate et al., 1986). PLPD was recently
reported to preserve the *ras* p21 epitope detected by the monoclonal antibody
Y13-259. (Going et al., 1988)

MATERIALS AND METHODS

Cell Lines and Experimental Tumours

T15 cells were derived by transfection of NIH 3T3 cells with a steroid inducible plasmid containing normal human N-ras. (McKay et al.,1986) These cells were maintained in Dulbecco's modification of minimum Eagles medium (DMEM) containing 10% newborn calf serum and 2μM dexamethasone.

EC816 cells were derived from NIH 3T3 cells. They constitutively overexpress normal Harvey ras and were maintained in DMEM containing 5% newborn calf serum.

NIH 3T3 cells were grown in DMEM with 10% newborn calf serum.

Cell suspensions were prepared from monolayers by treatment with 2x trypsin/EDTA (GIBCO), then washed with fresh culture media. The cells were embedded in 1% ultra low gelling and melting point agarose (UKB, Sweden) and, once set, pieces of approximately 0·3 ml volume were fixed in a) formol-saline and b) PLPD.

EC816 cells innoculated subcutaneously into nude mice produced a rapidly growing tumour after an initial delay period of three weeks. Samples of tumour tissue were placed into 4% formol-saline and PLPD for 24 hours. A further sample was placed in PLPD for 5 days. All PLPD fixed material was washed for 2 to 4 hours in running tap water. Fixed tumour and cell-agarose specimens were then paraffin-embedded in the usual manner.

Antibodies

The monoclonal antibodies tested were:

1) Y13-259. (Furth et al., 1982) A rat monoclonal antibody which reacts with certain amino acids in the highly conserved domain between residues 60 and 80 of the *ras* protein. (Lacal et al., 1986)

2) Cetus pan-reactive anti-p21. This antibody was raised against a synthetic peptide consisting of amino acids 29 - 44 of normal c-ras. It is commercially available from the Cetus Corporation.

3) 142-24E5. Raised against synthetic residues 96 - 118 of v-Ha-ras p21. (Niman H, L, 1986) This antibody was obtained from the NCI Repository Facility, Bethesda, USA.

4) RAP-5. Raised against a synthetic peptide consisting of amino acids 10 to 17 of mutated human Ha-ras p21 (Horan-Hand et al., 1984)

Immunocytochemistry

Paraffin-embedded sections were stained using the avidin-biotin complex (ABC) immunoperoxidase method. Wax sections (4–5μm) were dried on a 56°C hotplate for 1 hour, dewaxed in xylene and rehydrated through a series of ethanols to water. All sections were then washed in tris buffered saline (TBS). Nonspecific binding was blocked with 10% serum for 20 minutes at room temperature. Normal rabbit serum was used prior to RAP-5, Cetus, and 142-24E5 staining and normal goat serum prior to the application of Y13-259. The sections were then incubated with primary monoclonal antibody at the optimal dilution in TBS for 60 minutes at room temperature. (Optimal dilutions: Y13-259 1/150 ; Cetus 1/100 ; 142-24E5 1/100,000 ; RAP-5 1/15,000) Biotinylated secondary antibody was applied for 30 minutes. (Biotinylated goat anti-rat, Sigma, 1/100, for Y13-259 ; biotinylated rabbit anti-mouse, Dakopatts, 1/400

- 1/500 for Cetus, 142-24E5 and RAP-5 treated cells.) The ABC complex (Dako-patts) was applied for 30 minutes. Slides were washed for 10 minutes in TBS between each incubation step. The peroxidase reaction was developed for 5 minutes using 0·6mg/ml diaminobenzidine in 50 mM Tris/HCl, pH 7·6, containing 10 mM imidazole and 0·1 ml of 3% hydrogen peroxide. This produces a brown stain in areas of antibody binding.

Sections were washed in tap water, counterstained with Mayer's haema-toxylin, dehydrated, cleared, and mounted under DPX.

Negative controls (TBS instead of primary antibody) were included for each antibody and a known positive cell-agarose section was placed on each slide.

RESULTS

Immunoperoxidase staining intensity was scored as follows: + definite but weak staining ; ++ moderate staining ; +++ intense staining.

Agarose-Cell Tissue Models

1) Y13-259. Immunostaining of PLPD-fixed paraffin embedded T15(+) and EC816 cells using Y13-259 produced strong (++-+++) cell membrane associated staining (rimming) of most cells. A subpopulation of EC816 cells did not appear to overexpress normal H-ras. The majority of NIH3T3 cells exhibited rimming of '+' intensity. Formalin fixation resulted in an inferior staining pattern. Although it was possible to distinguish the more positive cells from NIH 3T3 cells, there appeared to be nonspecific cytoplasmic staining.

2) Cetus pan-reactive anti-p21. This antibody gave similar results with both fixatives. The majority of T15(+) and EC816 cells showed a clear, ++-+++, cell membrane associated staining pattern, although negative EC816 cells were present (15 - 20 %). Ras p21 was not detected in NIH 3T3 cells.

3) 142-24E5. T15(+) and EC816 cells fixed in both PLPD and formalin gave results analagous to those achieved with the Cetus antibody. Immuno-staining of NIH 3T3 cells resulted in the majority of cells showing '+' rimming, with approximately 15% showing membrane staining of ++ intensity.

4) RAP-5. In formalin-fixed sections RAP-5 did not clearly distinguish between NIH 3T3, T15(+), and EC816 cells, all of which were positive (++-+++). The staining was cytoplasmic rather than membrane associated.

Unsatisfactory results were also obtained when using RAP-5 with PLPD fixed cell-agarose sections. Most NIH 3T3 cells showed cytoplasmic staining (+), and at higher antibody concentration (1/10,000) nuclear staining occured. Although the cytoplasm of EC816 cells stained more strongly than that of NIH 3T3 cells, T15(+) cells were negative.

EC816 Tumour

Y13-259 did not produce a definite result with formalin-fixed paraffin-embedded tumour material. Staining of PLPD-fixed tumour with this antibody resulted in most cells exhibiting nuclear staining. Cytoplasmic staining was occasionally observed and was most noticeable in mitotic cells. Very few cells showed specific membrane localization of the epitope. The nuclei of normal fibroblasts in the epidermis also appeared to stain.

Cetus pan-reactive anti-p21 failed to detect positive cells in the PLPD-fixed tumour specimens, and 142-24E5 produced barely detectable staining. Both antibodies gave a reaction with tumour sections fixed in formalin. With 142-24E5 this reaction was weak and both cytoplasmic and cell membrane staining was observed. Normal fibroblasts in the epidermis also stained. Staining was not improved by using the antibody at 1/50,000 rather than 1/100,000 and preliminary tests suggest that trypsinisation does not enhance the results.

The Cetus antibody tested gave the most satisfactory results with formalin-fixed paraffin-embedded tumour. Clear linear cell membrane associated immunostaining (++) was evident and normal fibroblasts were negative.

The use of RAP-5 on formalin-fixed paraffin-embedded EC816 tumour sections resulted in moderate cytoplasmic, but not membrane, staining. When this antibody was applied to PLPD fixed tumour sections both nuclei and cytoplasm were positively stained.

DISCUSSION

The use of monoclonal antibodies with a high specificity for ras gene products should promote the immunohistochemical study of p21 expression in both normal and tumour specimens. It is clear from this study that the epitope specifically recognized by an antibody may not survive fixation.

It is difficult to explain the different staining patterns obtained when using different monoclonal antibodies. Ras proteins are localized on the inner side of the plasma membrane. (Willingham et al., 1980) The use of RAP-5 with EC816 tumour sections fixed in formalin resulted in a predominance of cytoplasmic staining, whereas the Cetus antibody caused membrane staining of the same material.

Going et al., (1988), in their appraisal of PLPD as a fixative suitable for use with Y13-259, found that in human tissue immunostaining was largely cytoplasmic as compared with control cells and tumours. They observed occasional nuclear staining. It was their impression, using experimental ras positive tumour material, that cytoplasmic staining increased and membrane localization diminished in tissues held for 10 minutes at room temperature prior to fixation. This delay was thought to result in a staining artefact possibly due to perioperative ischaemia or hypoxia.

In our experiments equal sized pieces of EC816 tumour were placed in formalin and PLPD at the same time. The Cetus pan-reactive anti-p21 antibody produced clear cell membrane associated immunostaining but RAP-5 and Y13-259 did not. This must cast serious doubt on the ability of these two antibodies to specifically detect ras proteins in PLPD or formalin fixed tissue.

Y13-259, RAP-5 and 142-24E5 all stained normal epidermal fibroblasts as well as the ras positive tumour cells. ie they failed to discriminate between normal and abnormal cells. Cetus pan-reactive anti-p21 was the only antibody which did not stain normal fibroblasts. Work is in progress to establish the lower limits of p21 detection by this antibody.

The ability of an antibody to detect enhanced levels of ras expression in a tumour cannot be confirmed by its use with a cell-agarose model. Both the Cetus antibody and 142-24E5 were able to distinguish between NIH 3T3 and T15(+) cells in PLPD fixed 1% agarose models, and both were able to identify a subpopulation of EC816 cells which did not overexpress p21. However neither antibody was effective once applied to known ras positive PLPD-fixed EC816 tumour material. It is possible that trypsinisation of the cell monolayer prior to incorporation in agar protects the epitopes from alteration by the fixative.

In conclusion, this paper reviews the value of using genetically modified cell lines in the critical evaluation of monoclonal antibodies. It highlights the limitations of these cells when used as *in vitro* models of *in vivo* tumuors and normal tissue. The epitope specifically recognized by an antibody may not survive fixation in the manner suggested by the model. It is necessary to be cautious when interpreting results obtained with either Y13-259 or RAP-5.

Acknowledgements: We thank the SERC for a postgraduate studentship to J.R. Gibson; Dr C. Marshall for kindly providing the T15 and EC816 cell lines ; Dr A. Wyllie for his kind gift of Y13-259 antibody ; Dr J. Schlom for his generosity in providing the RAP-5 antibody.

REFERENCES

Furth, M.E., Davis, L.J., Fleurdelys, B., Scolnick, E.M., 1982, Monoclonal antibodies to the transforming gene of harvey murine sarcoma virus and of the cellular *ras* gene family. J. Virol., 43: 294-304.

Ghosh, A.K., Moore, M., Harris, M., 1986, Immunohistochemical detection of *ras* oncogene p21 product in benign and malignant mammary tissue in man. J. Clin. Pathol., 39: 428-434.

Going, J.J., WIlliams, A.R.W., Wyllie, A.H., Anderson, T.J., Piris, J., 1988, Optimal preservation of p21 *ras* immunoreactivity and morphology in paraffin-embedded tissue. J. Pathol., 155: 000-000.

Holgate, C.S., Jackson, P., Pollard K., Lunny, D., Bird, C.C., 1986, Effect fixation on T and B lymphocyte surface membrane antigen demonstration in paraffin processed tissue. J. Pathol., 149: 293-300.

Horan Hand, P., Thor, A., Wunderlich, D., Muraro, R., Caruso, A., Scholm, J, 1984, Monoclonal antibodies of predefined specificity detect activated *ras* gene expression in human mammary and colon carcinomas. Proc. Natl Acad. Sci. (USA)., 81: 5227-5231.

Lacal, J.C., Aaronson, S.A., 1986, Monoclonal antibody Y13-259 recognises an epitope of the p21-*ras* molecule not directly involved in the GTP-binding activity of the protein. Mol. Cell Biol., 6: 1002-1009.

McKay, I.A., Marshall, C.J., Calés, C., Hall, A., 1986, Transformation and stimulation of DNA syhthesis in NIH 3T3 cells are a titratable function of normal p21 N-*ras* expression. EMBO,, 5: 2617-2621.

Niman, H.L., 1986, Human oncogene-related proteins in urine during pregnancy and neoplasia. Clin. Lab. Med., 6: 181-196.

Noguchi, M., Hirohashi, S. Shimosato, Y., Thor, A., Scholm, J., Tsunokawa, Y., Terada, M., Sugimura, T., 1986, Histological demonstration of antigens reactive with anti-p21 *ras* monoclonal antibody (RAP-5) in human stomach cancers. J. Natl Ca. Inst., 77: 379-385.

Ohuchi, N., Horan Hand, P., Merlo, G., Fujita, J., Mariano-Costantini, M., Thor, A., Nose, M., Callahan, R., Scholm, J., 1987, Enhanced expression of c-Ha-*ras* p21 in human stomach adenomas defined by immunoassays using monoclonal antibodies and *in situ* hybridisation. Cancer Res., 47: 1413-1420.

Robinson, A., Williams, A.R.W., Piris, J., Spandidos, D.A., Wyllie, A.H., 1986, Evaluation of a monoclonal antibody to ras peptide, RAP-5, claimed to bind preferentially to cells of infiltrating carcinomas. Br. J. Cancer, 54: 877-883.

Tahara, E., Yasui, W., Yaniyama, K., Ochaiai, A., Yamamoto, T., Nakajo, S., Yamamoto, M., 1986, Ha-*ras* oncogene product in human gastric carcinoma: correlation with invasiveness, metastasis or prognosis. Jpn J. Cancer Res. (Gann), 77: 517-522.

Williams, A.R.W., Piris, J., Spandidos, D.A., Wyllie, A.H., 1985, Immunohistochemical detection of the *ras* oncogene product in an experimental tumour and in human colorectal neoplasms. <u>Br. J. Cancer</u>, 52: 687-693.

Willingham, M. C., Paston, I., Shih, T. I., Scolnick E. M., 1980, Localisation of the *src* gene product of the Harvey strain of MSV to plasma membrane of transformed cells by electron microscopic immunocytochemistry. <u>Cell</u>, 19: 1005-1014.

IDENTIFICATION OF A PROTEIN INTERACTING WITH *ras*-p21- BY CHEMICAL CROSS-LINKING

Jean de Gunzburg[1], Rebecca Riehl and Robert A. Weinberg

Whitehead Institute, 9 Cambridge Center, Cambridge MA 02142 USA
[1]Present address : Unité INSERM U-248. Faculté de Médecine Lariboisière Saint Louis 10, avenue de Verdun, 75010 Paris, France

INTRODUCTION

There is accumulating evidence showing that *ras* oncogenes (Ha-*ras*, K-*ras* and N-*ras*) are involved in the processes of oncogenesis and cellular transformation[1]. These genes encode highly homologous proteins of molecular weight 21,000 (p21) that are found in all mammalian tissues and are very conserved throughout evolution. *ras*-p21s are bound to the inner surface of the plasma membrane ; they bind GTP and GDP, and exhibit an intrinsic GTPase activity. A cytoplasmic protein of MW 110-120,000 has recently been identified by its capacity to enhance 100-500 fold the GTPase activity of *ras*-p21s (GAP)[2] ; it has been shown to interact with a domain of p21s necessary for their biological activity (amino acids 30-42 called "effector domain") and may therefore constitute an effector of their physiological action[3-6]. Point mutations resulting in the change of amino acids 12, 13, 59 or 61 have been shown to "activate" the oncogenic potential of *ras* oncogenes leading to cellular transformation. The GTPase activity of such mutant proteins is no longer activated by GAP, blocking them in a GTP-bound state that is thought to be the active form of p21s.

Because of certain similarities with classical transducing G proteins[7,8] (peripheral membrane localization, ability to bind and hydrolyse GTP, as well as some limited sequence homology) and their potential role in the control of cellular proliferation, *ras*-p21s have been proposed to be involved in the transduction of signals emanating from growth factor receptors[9-13]. However, there is yet no experimental evidence to sustain such an attractive hypothesis.

In order to better understand the physiological role of *ras*-p21, we whished to identify proteins that they might interact with. Many such protein-protein interactions have been unraveled by the ability of the two protagonists to form a stable complex and therefore to co-immunoprecipitate, as with the products of the Rb and Ela genes[14] or the c-fos and c-jun proteins[15-17]. Such experiments have failed in the case of ras-p21s and we turned to a method that could "freeze" short lived interactions such as chemical cross-linking, followed by immunoprecipitation of the cross-linked complex.

RESULTS AND DISCUSSION

To maximize the chances of cross-linking ras-p21 to a protein relevant to its role in the control of cellular proliferation, as well as to "freeze" what could be a short-lived interaction, we decided to perform the cross-linking experiments *in vivo* on actively growing cells and chose as biological material a cell line derived from Rat-1 fibroblasts over expressing the c-H-ras p21 100-fold relative to its basal level[18]. As depicted in fig. 1, subconfluent cultures were metabolically labeled with (^{35}S)-methionine and washed free of nutrients immediately prior to cross-linking. After incubation of the cells with the cross-linker in the cold for 30 min., the cross-linker was washed away, and the cells lysed with detergents. The lysates were precleared with non-immune serum and protein A-Sepharose, and immunoprecipitated using a polyclonal rabbit serum elicited against recombinant c-H-ras p21. We chose such a serum, that recognizes all ras-p21s over monoclonal antibodies in order not to select a particular form of ras-p21 and to preclude the possibility that the antigenic determinants in p21-ras might be inaccessible to the antibody in the cross-linked complex. After recovery of the immune complex on protein A-Sepharose leads, and extensive washing, the components are analysed on reducing SDS gels. If the cross-linkiier used is cleavable by reducing agents, additional bands observed the gel represent proteins that were cross-linked to ras-p21.

In our first experiment, we surveyed a number of different cross-linkers that were sufficiently hydrophobic to pass through the plasma membrane and penetrate inside the cells[19]. Careful comparison of samples from cross-linked and uncross-linked cells shows that when DSP was used as cross-linker, a band migrating as 60kd (p60) was immunoprecipitated with the polyclonal anti-ras-p21 serum in addition to p21. This band did not appear with preimmune serum and immunoprecipitation of both p21 and p60 could be inhibited by the addition of excess cold recombinant p21 to the cell lysate. Upon incubation of the immune complex with reducing agents prior to loading on the gel, the bulk of p60 could be eluted whereas p21 remained bound to the protein A-Sepharose beads showing that p60 is immunoprecipitated by the anti-p21 antibodies when bound to p21 through the cross-linker.

Detection of p60 was dependent on the concentration of DSP and the time for which cells were exposed to the chemical. Quenching of the reactive groups with excess ammonium ions prior to the exposure of cells to DSP abolished the immunoprecipitation of p60, whereas it had no effect at the time of cell lysis showing that cross-linking actually occured *in vivo*.

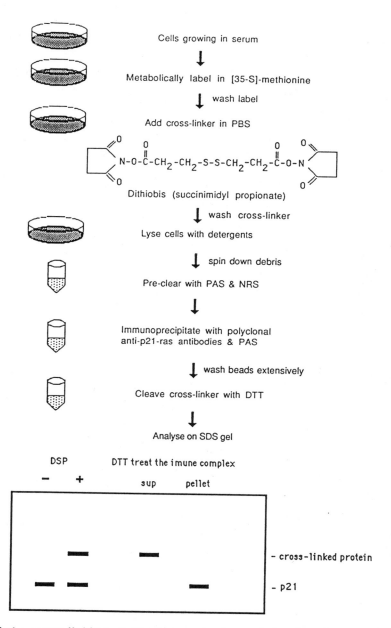

Fig1. cross-linking method. In order to maximize the chances of preserving physiological protein-protein interactions, [^{35}-S]-methionine metabolically labeled growing cells were cross-linked <u>in vivo</u> on the tissue culture dish. Polyclonal, rather than monoclonal, anti-p21-<u>ras</u> antibodies were used to immunoprecipitate detergent lysates in order to preclude the possibility that the antigenic determinant might be inaccessible to the antibody in the cross-linking.

Further evidence for the physical interaction of p21 with p60 was obtained through isolation of the cross-linked complex on sucrose gradients. It sedimented with a Mr of 200-300,000 which is too large to reflect a simple bi-molecular association. One can however not exclude that p60 is a multimeric protein and evaluation of the relative amounts of the two labeled proteins argues in favor of a p60 to p21 ratio greater than one in the complex.

A survey of various cell lines derived from Rat-1 or NIH 3T3 fibroblasts by transfection and expressing normal or activated H-ras, K-ras and -N-ras proteins at various levels was conducted in search for a relation between p21 expression and p60 detection. A similar amount of p60 was detected in all cases including the original untransfected lines suggesting that the level of p60 is limiting relative to p21, even in the non-overexpressing Rat-1 or NIH 3T3 fibroblasts. We then investigated whether the association of p60 to p21 was dependent on the mitogenic state of the cells. After extensive serum starvation of Rat-1 fibroblasts, the p21-p60 complex could still be detected ; its level was however increased 2-3 fold upon serum stimulation of the cells. The effect was only seen when serum and DSP were simultaneously added to the cells suggesting that p21-p60 association may represent an early and transient step in a serum-responsive signalling pathway.

Insight into the identity and physiological function of p60 should lead to a better understanding of the role of ras proteins in signal transduction and cell proliferation. The protein is widely distributed since it was detected by cross-linking to ras-p21 in numerous mammalian cell lines, such as mouse and rat fibroblasts, rat neuroblastoma, mouse pre-B lymphocytes and human Hela cells. It is neither pp60^{c-src} nor a related tyrosine kinase, since it is not recognised by anti-pp60^{src} or antiphosphotyrosine antibodies. Moreover, p60 was not labeled in vivo with [^{32}P]-orthphosphate nor did it display any immune-complex kinase activity. Its possible relationship with GAP, the putative effector protein of p21ras was also excluded : p60 was not recognized by anti-GAP antibodies on western blots, and the two proteins had different proteolytic patterns upon partial digestion with V8 protease.

In summary, we have identified a novel protein, p60, interacting with ras-p21 and propose that it could possibly act together with ras-p21 in the early events of a serum responsive signal transduction pathway. Purification and biochemical characterization of p60 are presently under way in an effort, through the study of this new component, to better understand the physiological role and mode of action of ras proteins.

ACKNOWLEDGEMENTS

The authors are indebted to Julian Downward, Roman Herrera, Jonathan Horowitz and Yossi Yarden for many fruitful discussions, and to Walter Carney for numerous antibody samples. They thank Meg Trahey and Frank Mc Cormick for their generous gifts of purified GAP protein and anti-GAP antibodies. This work was supported by grants from the National Institutes of Health and the American Business Foundation for Cancer Research. R.A.W. is an American Cancer Society Research Professor.

REFERENCES

1. Barbacid, M. Ann. Rev. Biochem. 56, 779-827 (1987).
2. Trahey,M. & McCormick, F. Science 238, 542-545 (1987).
3. Adari, H., Lowy, D.R., Willumsen, B.M., Der, C.J. & Mc Cormick, F. Science, 240, 518-521 (1988).
4. Calès, C., Hancock, J.F., Marshall, C.J. & Hall, A. Nature 332, 548-551.
5. Gibbs, J.B., Schaber, M.D., Allard, W.J., Sigall, I.S. & Scolnick, E.M. Proc. Natl. Acad. Sci. USA 85, 5026-5030 (1988).
6. Vogel, U.S., Dixon, R.A.F., Schaber, M.D., Dielhl, R.E., Marshall, M.S., Scolnick, E.M., Sigal, I.S. & Gibbs, J.B. Nature 303, 90-93 (1988).
7. Gilman, A.G. Ann. Rev. Biochem. 56, 615-649 (1987).
8. Levitzki, A. Science 241, 800-806 (1988).
9. Fleishman, L.F., Chahwala, S.B. & Cantley, L. Science 231, 407-410 (1986).
10. Wakelam, M.J.O., Davies, S.A., Houslay, M.D., McKay, I., Marshall, C.J. & Hall, A. Nature 323, 173-176 (1986).
11. Bar-Sagi, D. & Feramisco, J.R. Science 233, 1061-1068 (1986).
12. Hanley, M.R. & Jackson, T. Nature 328, 668-669.
13. Lacal, J.C., Moscat, J. & Aaronson, S.A. Nature 330, 269-272 (1987).
14. Whyte, P., Buchkowich, K.J., Horowitz, J.M., Friend, S.H., Raybuck, M., Weinberg, R.A. & Harlow, E. Nature 334, 124-129 (1988).
15. Courtneidge, S.A. & Smith, A.E. Nature 303, 435-439 (1983).
16. Chiu, R., Boyle, W.J., Meek, J., Smeal, T., Hunter, T. & Karin, M. Cell 54, 541-552 (1988).
17. Sassone-Corsi, P., Lamph, W.W., Kamps, M. & Verma, I.M. Cell 54, 553-560 (1988).
18. Downward, J., de Gunzburg, J., Riehl, R. & Weinberg, R.A. Proc. Natl. Acad. Sci. USA 85, 5774-5778 (1988).
19. Roth, R.A. & Pierce, S.B. Biochemistry 26, 4179-4182 (1987).

ISOLATION AND PARTIAL CHARACTERIZATION OF A NEW GENE (brl) BELONGING TO THE

SUPERFAMILY OF THE SMALL GTP-BINDING PROTEINS

Cecilia Bucci, Rodolfo Frunzio, Lorenzo Chiariotti, Alexandra
L. Brown[*], Matthew M. Rechler[*], and Carmelo B. Bruni

Dipartimento di Biologia e Patologia Cellulare e Molecolare,
Centro di Endocrinologia ed Oncologia Sperimentale del CNR,
Università di Napoli, Via S. Pansini 5, 80131 Naples, Italy
and [*]Molecular Cellular and Nutritional Endocrinology Branch,
NIDDK, NIH, Bethesda, MD, 20892 U.S.A.

INTRODUCTION

Several permanent cell lines of epithelial origin are able to grow in culture in the absence of serum components. It was originally proposed by Temin et al.(1972) that such behaviour was dependent on the production of mitogenic factors (MSA or Multiplication Stimulating Activity). The most studied system is a rat liver cell line BRL 3A isolated by Coon (1968). Subsequent studies demonstrated that MSA is the rat equivalent of the insulin like growth factor II (IGF-II) (Rechler et al., 1985), but also showed that the ability of this cell line to grow in serum-free medium is independent of the synthesis and secretion of this polypeptide (Nissley et al., 1977). We have been investigating this system mainly to understand the biological role and the expression and regulation of the IGF-II gene (Chiariotti et al., 1988). In the course of screening several cDNA libraries derived from the BRL 3A cell line we accidentally isolated some cDNA clones abundantly represented in the stable mRNA population, but not related to IGF-II. Subsequent characterization of these clones lead to the discovery that this mRNA code for a protein of molecular weight 22,800 belonging to the superfamily of ras-related genes (Bucci et al., 1988). In the present study we report some aspects of the organization, structure and expression of this novel putative GTP-binding protein.

MATERIALS AND METHODS

Chemicals and enzymes

Restriction endonucleases, DNA polymerases, T4 DNA ligase, T4 polynu-cleotide kinase, reverse transcriptase and other enzymes were purchased from commercial suppliers (Boehringer Mannheim, Promega, Amersham and Bethesda Research Laboratories). All deoxyribonucleotide triphosphates, dideoxynucle-otide triphosphates and ribonucleotide triphosphates were from Pharmacia and all radionuclides were obtained from Amersham Corp.

Cell Cultures

Rat BRL 3A and BRL 3A2 cells (Nissley et al., 1977) and mouse NIH 3T3 (kindly provided by F. S. Ambesi) were grown in modified Ham's F-12 medium containing 5% calf serum (Flow). Confluent cells were removed from the plates with 0.05% trypsin and 0.02% EDTA and harvested for DNA or RNA isolation.

Nucleic Acids Preparation and Analysis

RNA was isolated from cultured cells or from rat tissues either by ex-traction with guanidinium thiocyanate (Chirgwin et al., 1979) or by using the guanidinium thiocyanate (Fluka)/LiCl precipitation procedure (Brown et al., 1986). RNAs were electrophoresed on 1.2% agarose gels containing 2.2 M for-maldehyde (Lehrach et al., 1977). After electrophoresis the gels were blotted without further treatment and hybridized to labeled probes.

Genomic DNA was extracted from BRL-3A and BRL-3A2 cell lines and after digestion with different enzymes was electrophoresed on agarose gels (0.8%) in either TBE or TAE buffers (Maniatis et al., 1982) and then blotted accord-ing to the method of Southern (1975). Plasmid DNAs were purified according to the procedure of Clewell (1972) and were finally banded on a cesium chloride ethydium bromide equilibrium density gradient. Cell transformations were performed with the standard $CaCl_2$ technique (Mandel and Higa, 1970). Restric-tion fragments were purified by electrophoresis in 4%-6% acrylamide gels and then phenol extracted and ethanol precipitated. DNA sequencing was performed either according to Maxam and Gilbert (Maxam and Gilbert, 1980) or by the dideoxy chain terminating method (Sanger et al., 1977) after subcloning re-striction fragments into M13 vectors (Yanish-Perron et al., 1985). Data anal-ysis was simplified by the use of the computer program Microgenie (Beckman) run on an IBM AT computer.

Radioactive probes for hybridization experiments were prepared either by conventional nick translation (specific activity: 2×10^8 cpm/μg) (Rigby et al., 1977) or by the random priming DNA labeling technique (specific activity 2-3 x 10^9 cpm/μg) (Feinberg and Vogelstein, 1983). Blots were hybridized at $42°C$ in 50% formamide, 6X SSC, 10X Denhardt's, 100 μg/ml sheared denatured salmon sperm DNA, 0.5% SDS. Southern blots were washed at stringent conditions (0.1X SSC, 0.1% SDS, $65°C$). Northern blots were washed at $42°C$, 0.2X SSC, 0.1% SDS.

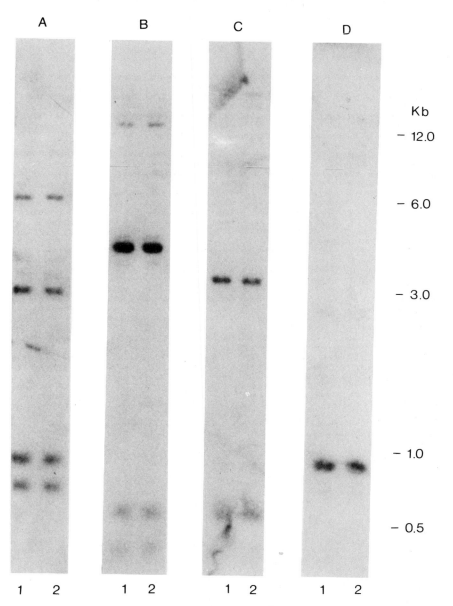

Fig. 1. Southern blots of genomic DNA (10 μg/lane) from BRL 3A (lane 1) and BRL 3A2 (lane 2) digested with: BglII (A), BamHI (B), EcoRI (C) and HindIII (D).

RESULTS

We have previously isolated from a cDNA library from Buffalo rat liver
cell line, BRL 3A, overlapping clones encoding a 201 amino acids protein
exhibiting homology with several ras-related genes. This gene was named BRL-
ras, but since it does not exhibit properties similar to those of real ras
genes we have renamed it brl. The brl cDNA recognizes two mRNA species of 2.5
and 1.5 Kb, which are abundantly expressed in several cell lines and tissues
(Bucci et al., 1988). The genomic DNA of BRL 3A and BRL 3A2 cells (Nissley et
al., 1977) was restricted with several endonucleases and hybridized to a 750
bp HincII- SacI cDNA fragment spanning the entire coding region. The same
pattern of hybridization was observed (Figure 1). The number of bands detected on Southern blots suggest the existence of a single gene. In addition
titration experiments with decreasing concentration of plasmid DNA indicate
that the gene is present in single copy (data not shown). Finally the total
size of the different fragments with the various enzymes gives an approximate
estimate of the size of the genomic region encompassing the gene (> 17 Kb).

It was shown that the steady state levels of mRNAs were not only very
high in all cells and tissues examined but also not affected during development (Bucci et al., 1988). Further studies do show some exception. In fact
the expression in lung and intestine is higher in foetuses and decreases in
adults (Figure 2 A, B). Low levels of mRNAs are present in the cortex and
are further reduced after birth (Figure 2 C).

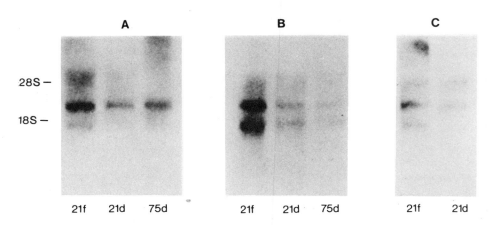

Fig. 2. Northern blots of total RNA (15 μg/lane) isolated from rat lungs (A),
        intestine (B) and cortex (C). RNAs were extracted from foetuses of 21
        days (21f) or from adult animals of 21 or 75 days (21d, 75d). The
        blots were hybridized with a cDNA PstI 650 bp fragment spanning the
        entire coding region. The relative migration of the ribosomal RNAs
        (28S and 18S) is indicated on the left.

The expression of many genes coding for essential functions in the cell is regulated during the cell cycle (Denhardt et al., 1986). To establish whether the _brl_ gene belongs to this category we performed Northern blot experiments with total RNA extracted from mouse NIH 3T3 fibroblast cell line either grown in the presence of serum or starved in medium without serum for 48 hrs. No significative difference in the levels of expression was observed (Figure 3). Moreover the same two species of mRNA are detected in the mouse cell line suggesting that the organization of the gene has been kept constant during the evolution in rodents.

DISCUSSION

We have previously compared the amino acid sequence of the _brl_ gene product with a few members of the _ras_-related genes and have found that the homology (30-35%) was mostly due to the preservation of domains involved in the binding and hydrolysis of GTP (Bucci et al., 1988). Very recently a large number of _ras_-related polypeptides has been identified in bovine (_smg_) and rat (_rab_) brain (Touchot et al., 1987; Kikuchi et al., 1988; Matsui et al., 1988; Zahraoui et al., 1988).

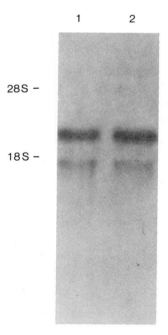

Fig. 3. Northern blot of total RNA (10 µg/lane) extracted from growing (lane 1) or starved (lane 2) NIH 3T3 cells. The blots were probed with a cDNA HincII-SacI 750 bp fragment spanning the coding region. The relative migration of the ribosomal RNAs (28S and 18S) is indicated on the left.

Table 1. Homology comparison of the GTP and COOH-terminus domains of several small Mr G proteins.

brl	---GDSGVGK^{15}	--QIWDTAGQERF^{64}	--VLGNKIDL^{123}	-IPYFETSAK^{151}	-NVEQAF^{160}	-SCSC^{201}
	*°°° **	********* °	° *** *°	° *°**	* °°*	*
rab1	---GDSGVGK^{24}	--QIWDTAGQERF^{73}	--LVGNKCDL^{128}	-IPFLETSAK^{156}	-NVEQSF^{165}	--GCC^{205}
rab2	---GDTGVGK^{19}	--QIWDTAGQESF^{68}	--LIGNKSDL^{123}	-LIFMETSAK^{151}	-NVEEAF^{160}	--GCC^{212}
rab3	---GNSSVGK^{35}	--QIWDTAGQERY^{84}	--LLGNKCDM^{139}	-FEFFEASAK^{167}	-NVKQTF^{176}	-DCAC^{220}
rab4	---GNAGTGK^{21}	--QIWDTAGQERF^{70}	--LCGNKKDL^{125}	-LMFLETSAL^{153}	-NVEEAF^{162}	-ECGC^{213}
smgB	---GNSSVGK^{35}	--QIWDTAGQERY^{84}	--LVGNKCDM^{139}	-FDFFEASAK^{167}	-SVRQAF^{176}	-NCSC^{219}
smgC	---GNSSVGK^{43}	--QIWDTAGQERY^{92}	--LVGNKCDM^{147}	-FDFFEASAK^{175}	-NVKQTF^{184}	-NCGC^{207}

The numbers at the end of each domain refer to the position of the last amino acid within each protein. Residues identical among all proteins are indicated by an asterisk and conservative ones by an open circle.

We have compared the amino acid sequence of brl with the three smg and the four rab proteins (Table 1). smg p25A is the bovine equivalent of rat rab3 (99% homology). The strong homology observed between the functional domains does not extend to other regions of these proteins. In fact the overall homology of brl with the other members, as well as that of each member compared to the others, ranges between 28 and 35%. The only exception is represented by the smg members that exhibit 77–85% homology and have therefore been considered as a separate family (Matsui et al., 1988). The structure of the COOH-terminal domain of brl, XCXC (Table 1), is similar to that of smg and also rab4 and different from that of ras and rho, CXXX, and yptl, rab2 and SEC4, CC. This domain is subject to posttranslational modifications and is required for membrane attachment (Buss and Sefton, 1986). Structural identity might therefore indicate a similar subcellular localization of the proteins, but not necessarily that they belong to a same family with similar functions.

The interest in the function and biological role played by cellular proto-oncogenes of the ras family has resulted in recent years in the discovery of an increasingly growing number of so called ras related genes. In Table 2 we have listed the different members that have been identified and have summarized their most relevant features. Despite the numerous attempts to elucidate their role, the progresses have been so far quite limited. In fact, only for a few members, mostly confined to yeast genes (YPT1, SEC4, RAS1, RAS2), the molecular function has been partially understood. All these

Table 2. Characteristics of small molecular weight G proteins.

Name [a]	Mol.wt./a.a.	GTP-binding/ hydrolysis	Identified as:	mRNAs (Kb)	Expression	Function	References
H-ras	21,000/188	+	cDNA	5.8, 2.2, 1.2	Low except rat brain and mouse heart	?	Barbacid,1987; Parada et al., 1982; Ellis et al.,
K-ras	21,000/188-9	+	cDNA	5.2, 2.0		?	
N-ras	21,000/188	+	cDNA	5.0, 2.0, 1.5		?	1982; Hall & Brown, 1985
RAS1	40,000/309	+	gene	?	?	cya	Barbacid, 1987
RAS2	41,000/322	+	gene	?	?	cya	Barbacid, 1987
R-ras	23,400/218	+	gene	1.0	Low	?	Lowe et al., 1987
rho	21,000/192	+	cDNA	4.5, 3.3	Variable	?	Madaule & Axel, 1985
YPT1	23,200/206	+	gene	0.83		Ca^{2+}	Anderson & Lacal, 1987 Schmitt et al., 1988
ypt1	23,500/205	+	cDNA	3.2, 1.6	High	?	Haubruck et al., 1987
ral	23,500/206	?	cDNA	2.8, 1.3	?	?	Chardin & Tavitian, 1986
rab 2	23,400/212	+	cDNA	?	?	?	Touchot et al., 1987
rab 4	24,000/213	+	cDNA	?	?	?	Zahraoui et al., 1988
smgA	25,000/220	+	protein/cDNA	?	High brain	?	Kikuchi et al., 1988
smgB	24,700/219	?	cDNA	?	?	?	Matsui et al., 1988
smgC	25,900/227	?	cDNA	?	?	?	Matsui et al., 1988
SEC4	23,500/215	?	gene	?	?	secretion	Salminen & Novick, 1987
ARF	21,000	+	protein	?	?	ADP rib.	Kahn & Gilman, 1986
Gp.	21,000	+	protein	?	?	?	Evans et al., 1988
brl	22,800/201	?	cDNA	2.5, 1.5	Ubiquitous	?	Bucci et al., 1988

[a] When the same gene has been found in more than one species the best characterized one has been included.

proteins contain GTP binding and hydrolysis domains and these functions have
been experimentally verified for most of them. Another common characteristic
is the relatively small molecular weight (Mr 21-25,000). It seems therefore
more appropriate to identify this superfamily of proteins as: Small molecula
weight G proteins, according to a recent proposal (Matsui et al., 1988).
Although it seems quite reasonable to postulate a role in signal transductio
for all these factors the unraveling of their precise function will represen
a great challenge in the near future.

REFERENCES

Anderson, P. S. and Lacal, J. C., 1987, Expression of the Aplysia californic
    rho in Escherichia coli: purification and characterization of its
    encoded p21 product, Mol. Cell. Biol. 7:3620-3628.
Barbacid, M., 1987, ras genes, Annu. Rev. Biochem. 56:779-827.
Brown, A. L., Graham, D. E., Nissley, S. P., Hill, D. J., Strain, A. J. and
    Rechler, M. M., 1986, Developmental regulation of insulin-like growth
    factor II mRNA in different rat tissues, J. Biol. Chem. 261:13144-
    13150.
Bucci, C., Frunzio, R., Chiariotti, L., Brown, A. L., Rechler, M. M. and
    Bruni, C. B., 1988, A new member of the ras gene superfamily identi-
    fied in a rat liver cell line, Nucleic Acids Res. 16:9979-9993.
Buss, J. A. and Sefton, B. M., 1986, Direct identification of palmitic acid
    as the lipid attached to p21 ras, Mol. Cell. Biol. 6:116-122.
Chardin, P. and Tavitian, A., 1986, The ral gene: a new ras related gene
    isolated by the use of a synthetic probe, EMBO J. 5:2203-2208.
Chiariotti, L., Brown, A. L., Rechler, M. M., Frunzio, R. and Bruni, C. B.,
    1988, Structure and function of the developmentally regulated rat
    growth factor II, in: "Gene expression and regulation: The legacy of
    Luigi Gorini", X., Bissell, G. Dehò, G. Sironi and M. L. Torriani
    eds., Elsevier Science Publishers B.V. (Biomedical Division),
    Amsterdam.
Chirgwin, J. M., Przybyla, A. E., Mc Donald, R. J. and Rutter W. J., 1979,
    Isolation of biologically active ribonucleic acid form sources en-
    riched in ribonuclease, Biochemistry 18:5294-5299.
Clewell, D. B., 1972, Effect of growth conditions on the formation of the
    relaxation complex of supercoiled ColEl deoxiribonucleic acid and
    protein in Escherichia coli, J. Bacteriol. 110:667-676.
Coon, H. G., 1968, Clonal culture of differentiated cells from mammals: rat
    liver cell culture, Carnegie Inst. Washington Yearbook, 67:419-421.
Denhardt, D. T., Edwards, D. R. and Parfett, C. L. J., 1986, Gene expressio
    during the mammalian cell cycle, Biochim. Biophys. Acta 865:83-125.
Ellis, R. W., De Feo, D., Furth, M. E. and Scolnick, E. M., 1982, Mouse cel
    contain two distinct ras gene mRNA species that can be translated in
    a p21 onc protein, Mol. Cell. Biol. 2:1339-1345.
Evans, T., Brown, M. L., Fraser, E. D. and Northup, J. K., 1988, Purificati
    of the major GTP-binding proteins from human placental membranes, J.
    Biol. Chem. 261:7052-7059.
Feinberg, A. P. and Vogelstein, B., 1983, A technique for radiolabeling
    DNA restriction endonuclease fragments to high specific activity,
    Anal. Biochem. 132:6-13.
Hall, A. and Brown, R., 1985, Human N-ras: cDNA cloning and gene structure,
    Nucleic Acids Res. 13:5255-5268.
Haubruck, H., Disela, C., Wagner, P. and Gallwitz, D., 1987, The ras-relate
    ypt protein is an ubiquitous eukaryotic protein: isolation and se-
    quence analysis of mouse cDNA clones highly homologous to the yeast
    YPT1 gene, EMBO J. 6:4049-4053.

Kahn, R. A. and Gilman, A. G., 1986, The protein cofactor necessary for ADP-
    ribosylation of Gs by cholera toxin is itself a GTP binding protein,
    J. Biol. Chem. 261:7906-7911.
Kikuchi, A., Yamashita, T., Kawata, M., Yamamoto, K., Ikeda, K., Tanimoto, T.
    and Takai, Y., 1988, Purification and characterization of a novel
    GTP-binding protein with a molecular weight of 24,000 from bovine
    brain membranes, J. Biol. Chem. 263:2897-2904.
Lehrach, H., Diamond, D., Wozney, J. M. and Boedtker, H., 1977, RNA molecu-
    lar weight determination by gel electrophoresis under denaturing
    conditions, a critical reexamination, Biochemistry 16:4743-4751.
Lowe, D. G., Capon, D. J., Delwart, E., Sakaguchi, A. Y., Naylor, S. L. and
    Goeddel, D. V., 1987, Structure of the human and murine R-ras genes,
    novel genes closely related to ras proto-oncogenes, Cell 48:137-146.
Madaule, P. and Axel, R., 1985, A novel ras-related gene family, Cell
    41:31-40.
Mandel, M. and Higa, A., 1970, Calcium dependent bacteriophage DNA infection,
    J. Mol. Biol. 53:154-162.
Maniatis, T., Fritsch, E. F. and Sambrook, J., 1982, Molecular cloning: A
    laboratory manual, Cold Spring Harbor Laboratory Press, New York.
Matsui, Y., Kikuchi, A., Kondo, J., Hishida, T., Teranishi, Y. and Takai, Y.,
    1988, Nucleotide and deduced amino acid sequences of a GTP-binding
    protein family with molecular weights of 25,000 from bovine brain, J.
    Biol. Chem. 263:11071-11074.
Maxam, A. M. and Gilbert, W., 1980, Sequencing end-labeled DNA with base-
    specific chemical cleavages, Meth. Enzymol. 65:497-559.
Nissley, S. P., Short, P. A., Rechler, M. M., Podskalny, J. M. and Coon,
    H. G., 1977, Proliferation of Buffalo rat liver cells in serum-free
    medium does not depend upon multiplication-stimulating activity (MSA),
    Cell 11:441-446.
Parada, L. F., Tabin, C. J., Shih, C. and Weinberg, R. A., 1982, Human EJ
    bladder carcinoma oncogene is homologue of Harvey sarcoma virus ras
    gene, Nature 297:474-478.
Rechler, M. M., Bruni, C. B., Whitfield, H. J., Yang, Y. W.-H., Frunzio, R.,
    Graham, D. E., Coligan, J. E., Terrell, J. E., Acquaviva, A. M. and
    Nissley, S. P.,1985, Characterization of the biosynthetic precursor
    for rat insulin-like growth factor II by biosynthetic labeling, radio-
    sequencing, and nucleotide sequence analysis of a cDNA clone, in:
    "Cancer Cells 3 / Growth factors and trasformation" J.Feramisco, B.
    Ozanne and C. Stiles, eds., Cold Spring Harbor Laboratory, Cold Spring
    Harbor, New York.
Rigby, P. W. J., Dieckmann, M., Rhodes, C. and Berg, P., 1977, Labeling de-
    oxyribonucleic acid to high specific activity in vitro by nick trans-
    lation with DNA polymerase I, J. Mol. Biol. 113:237-251.
Salminen, A. and Novick, P. J., 1987, A ras-like protein is required for a
    post-Golgi event in yeast secretion, Cell 49:527-538.
Sanger, F., Nicklen, S. and Coulson, A. R., 1977, DNA sequencing with chain-
    terminating inhibitors, Proc. Natl. Acad. Sci. USA 74:5463-5467.
Schmitt, H. D., Puzicha, M. and Gallwitz, D., 1988, Study of a temperature-
    sensitive mutant of the ras-related YPT1 gene product in yeast suggest
    a role in the regulation of intracellular calcium, Cell 53:635-647.
Southern, E. M., 1975, Detection of specific sequences among DNA fragments
    separated by gel electrophoresis, J. Mol. Biol. 98:503-518.
Temin, H. M., Pierson, R. W., Jr. and Dulak, N. C., 1972, The role of serum
    in the control of multiplication of avian and mammalian cells in
    culture, in: "Growth, nutrition and metabolism of cells in culture",
    G. H. Rothblatt and V. J. Cristofalo, eds., Academic Press, New York.
Touchot, N., Chardin, P. and Tavitian, A., 1987, Four additional Members of
    the ras gene superfamily isolated by an oligonucleotide strategy:
    Molecular cloning of YPT-related cDNAs from a rat brain library, Proc.
    Natl. Acad. Sci USA 84:8210-8214.

Yanisch-Perron, C., Vieira, J. and Messing, J., 1985, Improved M13 phage cloning vectors and host strains: nucleotide sequences of the M13mp18 and pUC19 vectors, Gene 33:103–119.

Zahraoui, A., Touchot, N., Chardin, P. and Tavitian, A., 1988, Complete coding sequences of the ras related rab 3 and 4 cDNAs, Nucleic Acids Res. 16:1204.

STRUCTURAL INVESTIGATIONS ON THE G-BINDING DOMAIN OF p21

Fred Wittinghofer, Jacob John, Emil Pai,
Ilme Schlichting and Paul Rösch

Max-Planck-Institut für medizinische Forschung
Abteilung Biophysik
6900 Heidelberg

INTRODUCTION

Ever since the ras gene product, p21, was discovered, it became clear, that it is a guanine nucleotide binding protein with significant homology to other G-binding proteins such as protein biosynthesis factors and signal-transducing G-proteins (Barbacid, 1987). Also, a lot of circumstantial evidence suggests that p21 is involved in a growth promoting signal-transduction pathway. Fig.1 shows the scheme for the involvement of p21 in such a process:

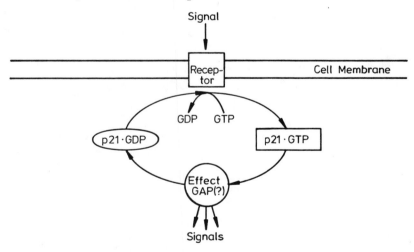

Fig.1 Schematic diagram of the cycling of p21 between an "active" and "inactive" conformation

p21 is cycling between two different conformations, an "inactive" GDP containing and an "active" GTP containing conformation. In its GDP conformation the protein, presumably, receives a signal that catalyzes the exchange of GDP for GTP.

The p21.GTP complex is then able to interact with the effector to transmit a signal. In the process of this interaction, GTP on the protein is hydrolyzed to GDP and the protein returns to the inactive resting state.Although the precise role of p21 is unknown it seems possible that the effector protein has already been discovered. GAP, a protein that has recently been described, is able to interact with p21 only in its GTP conformation and accelerates the GTPase reaction rate of normal cellular p21 (Trahey,M. and McCormick, 1987).

RESULTS
We are interested in describing in structural terms the transition between the two conformational states that are important far all G-binding proteins. For this we use NMR and X-Ray crystallography. Since the structure of the p21xGDP complex has recently been described by Kim and coworkers (deVos et al., 1988) we have concentrated our efforts on the GTP form of p21. We have expressed the G-binding domain of p21 as a C-terminally truncated 1-166 amino acid polypeptide chain called $p21_C{}'$. We have crystallized this protein as a complex with GppNp and GppCp and one of the crystals is shown in Fig.2.

Fig.2 Crystals of the G-binding domain of p21 ($p21_C{}'$) complexed to the GTP analogue GppNp. The crystal is approximately 300µm along its longest axis.

They can easily be grown to large sizes, they diffract to high resolution (below o.2nm) and are stable in the X-ray beam (Scherer et al., 1989). They have been used, together with heavy atom derivatives to prepare an electron density map of the protein-guanosine triphosphate complex which is currently being fitted to the polypeptide chain.
We have also undertaken biochemical and NMR experiments to investigate the effect of the C-terminal deletion on the properties of p21. We have shown before that the measurement of the GDP off-rate can be a precise monitor of the GDP binding site of p21 (John et al., 1988). In Table1 we show the GDP

Table1. Rate constants of p21 proteins measured in standard buffer (10mM Mg^{2+}) at $37^{O}C$

Protein	GDP off-rate k_{-1}, $x10^3$, [min^{-1}]	GTPase rate k_2, $x10^3$, [min^{-1}]
$p21_C$	7.9	28
$p21_C'$	7.8	37
$p21_T$	2.3	3.8
$p21_T'$	3.3	3.1

dissociation rate constants k_{-1} and the GTPase rate-constants k_2 of truncated p21 as compared to normal protein.

One observes that the deletion of 23 amino acids from the C-terminus has no appreciable effect on these reaction rates. Table 1 also shows that the single point mutation Gly12→Val reduces the GDP off-rate and the GTPase rate truncated proteins, although the effect on the dissociation rate is somewhat smaller for p21(1-166). This indicates that the effect of the activating mutation Gly12→Val is preserved in the truncated protein.

We also used NMR spectroscopy to investigate the structural effects of the truncation of the protein. We have shown before that the ß-phoshorous atom of GDP undergoes a large 4ppm downfield shift on binding to p21, which is the same magnitude as with EF-Tu. Fig.3 shows the phosphorous NMR spectra of the $p21_C xGDPxMg^{2+}$ complex compared to the $p21_C'xGDPxMg^{2+}$ complex. One can see that α- and ß-P-resonances have the same chemical shift in the two proteins.

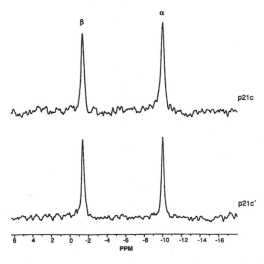

Fig.3 Phosphorous-NMR spectra of the MgxGDP complexes of truncated and normal p21.

Proton-NMR measurements of p21 have been performed in our laboratory to study the solution structure of the protein. Although the protein is too large for a complete 3D structure determination, the aromatic portion of the spectrum is nevertheless resolved well enough to assign the majority of aromatic protons. In Fig.4 the NOESY-spectrum of the aromatic portion of the proton NMR spectrum between 5.95 and 7.00ppm is shown. With such type of spectrum one can identify through-space interactions between protons that are in close proximity of each other (within 0.5nm).

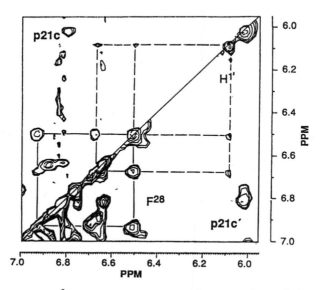

Fig.4 NOESY ^{1}H-NMR spectrum of normal and truncated p21 in theMgxGDP complex. The spectra of p21$_C$ and p21$_C$' were both cut in half along the diagonal and each half recombined along the diagonal.

One can see that the two spectra look identical. We have assigned earlier the protons of the ortho-, meta- and para-position of a phenylalanine residue (Schlichting et al., 1988) that we can now identify as F28 (Schlichting, unpublished). Fig.4 shows an NOE,i.e.a proton-proton interaction, of the aromatic protons of F28 to the 1'H-proton of the GDP-ribose, which means that they are less than 0.5nm away from the 1'H-proton of GDP. This is in agreement with the finding of the crystal structure determination by Kim and coworkers (deVos et al., 1988) where F28 has been found to be situated on top of the guanine base of GDP. Fig.4 also shows that the position of F28 relative to the ribose ring is preserved in the truncated protein p21$_C$'. Fig.5 shows another 2D-NMR comparison of the two proteins in form of the COSY-spectra, and here again the two half-spectra are aligned along the diagonal middle line. With a COSY spectrum one is able to identify interactions between protons that are near neighbours through covalent bonds.
The two spectra show one characteristic difference, namely the appearance of an additional peak which can be identified as a tyrosine. The reason for the appearance of this new tyrosine residue is not known but it indicates that this tyrosine changes its environment when the last 23 amino acids are deleted, possibly by becoming more mobile.

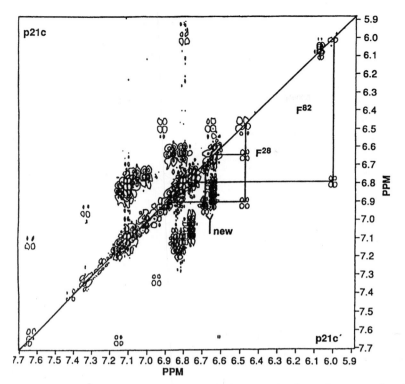

Fig.5 COSY ¹H-NMR spectra of truncated and normal p21 both complexed to MgxGDP The two spectra are aligned along the diagonal as in Fig.4.

We have also tested the biological function of the truncated protein by microinjecting the p21T' into PC12 cells (data not shown). Not surprisingly we find that the C-terminal truncation abolishes the biological effect of the protein. In conclusion we can say that although the p21 1-166 polypeptide is biologically inactive its biochemical and structural properties resemble very much those of the intact protein.

REFERENCES

Barbacid,H. (1987), Ann.Rev.Biochem. 56, 779-827

deVos,A., Tong,L., Milburn,M.V., Matias,P.M., Jancarik,J., Noguchi,S., Nishimura, S., Miura,K., Ohtsuka,E. and Kim,S.-H. (1988), Science 239, 888-893

John,J., Frech,M. and Wittinghofer,F. (1988), J.Biol.Chem. 263, 11792-11799

Rösch,P. (1988), Biochem. Biophys. Res. Commun. 150, 444-448

Scherer,A., John,J., Linke,R., Goody,R.S., Wittinghofer,A., Pai, E. and Holmes,K.C.(1988) , J.Mol.Biol., in press

Schlichting, I., Wittinghofer,A. and Rösch,P. (1988) Biochem. Biophys. Res. Commun. 150, 444-448

Willumsen,B.M., Christensen, A., Hubbert,N.L. Papageorge,A.G. and Lowy,D.R. (1984), Nature (London) 310, 583-586

MOLECULAR BIOLOGY OF GAP AND ITS INTERACTION WITH ONCOGENIC ras p21

U.S. Vogel, R.E. Diehl, M.S. Marshall, M.D. Schaber, R.B.
Register, W.S. Hill, A. Ng, E.M. Scolnick, R.A.F. Dixon, I.S.
Sigal, and J.B. Gibbs

Department of Molecular Biology, Merck Sharp and Dohme Research
Laboratories, West Point, PA 19486, USA

INTRODUCTION

The interaction of ras p21 with guanine nucleotides is central to
its biological function. Mutations that impair the intrinsic GTPase or
that enhance GTP for GDP exchange, activate ras cell-transforming
activity. Both of these types of mutations result in an increased level
of ras p21-GTP complex. By analogy with G-protein regulatory cycles
therefore, the ras p21-GTP complex would be predicted to be the
biologically active species. This prediction appears to be true for both
the mammalian and yeast systems. Discrepancies between the in vitro
GTPase activities and biological potencies which were observed, prompted
the testing of the model through the quantitation of the GTP and GDP
nucleotides bound to both normal and oncogenic ras proteins in vivo.
These measurements were first made for ras proteins expressed in yeast
cells where the ras proteins could be overproduced. Yeast RAS1 and RAS2
proteins, overexpressed in [32P] orthophosphate-labelled cultures of S.
cerevisiae cells, were found to be bound almost entirely to GDP, whereas
increased amounts of GTP were bound to RAS proteins containing oncogenic
mutations that impair GTPase activity. Unexpectedly, normal mammalian
ras p21 was bound to near equimolar proportions of both GTP and GDP
nucleotides, whereas the oncogenic forms were bound almost exclusively to
GTP (Gibbs et al., 1987). In contrast, analysis of nucleotides bound to
ras proteins in Xenopus oocytes revealed that normal ras proteins were
bound to GDP, whereas the respective oncogenic forms were bound to GTP
(Trahey and McCormick 1987, Sigal et al., 1988a). The different
proportions of GDP and GTP nucleotides bound to mammalian ras p21 in S.
cerevisiae and the higher eukaryote, Xenopus, was explained by the
discovery of an activity designated as GAP (GTPase Activating Protein,
Trahey and McCormick, 1987). This activity stimulated GTP hydrolysis of
normal ras p21 by more than 200-fold in vitro, but had no effect on the
oncogenic variants. In a model of ras p21 action, GAP could be imagined
to be either an upstream negative regulator or the downstream target of
ras where it acts similar to the ribosome in the EF-Tu system. In both
of these models, the biologically potent oncogenic ras p21 mutant
proteins would remain bound to GTP resulting in constitutive ras p21
activity.

For the characterization of GAP at a molecular level, the GAP protein was purified to homogeneity from bovine brain cerebra as previously described (Gibbs et al., 1988a). Purified GAP, a monomeric protein of 125 kDa, was able to act catalytically in stimulating the GTPase activity of normal Harvey (Ha) ras p21, but not that of the oncogenic form, [Val12]Ha ras p21. GAP also acted upon S. cerevisiae RAS2 protein and the N-terminal domain of yeast RAS1 protein, but was unable to stimulate the GTPase activity of the ras-related YPT1 protein from yeast. Both [Pro12]Ha ras p21 and [Pro11]Ha ras p21 have a twofold increased GTPase, but whereas [Pro11]Ha ras p21 is sensitive to GAP, [Pro12]Ha ras p21 is not (Gibbs et al., 1988a). This difference in sensitivity to GAP might be a conformational effect where the proline at position 12 prevents the adaptation of an active conformation required for GAP-stimualted GTPase activity. In vivo, one would predict that [Pro12]Ha ras p21 having a GAP-insensitive GTPase would be bound to more GTP than normal Ha ras p21, but less GTP than [Val12]Ha ras p21 because of the higher intrinsic GTPase of the former. Since [Pro12]Ha ras p21 does not transform cells unless it is overexpressed, the approximate 20-fold difference in the intrinsic GTPase rates between [Pro12]Ha ras p21 and [Val12]Ha ras p21 reflects the levels of GTP complex required to constitutively activate cellular proliferation. In fact, [Pro12]Ha ras p21 is biologically more potent than normal Ha ras p21 (Ricketts and Levinson, 1988). Apparently the intrinsic GTPase activity of ras p21, independent of GAP, also determines biological potency.

Another group of mutations that were investigated with respect to GAP responsiveness, were those that impaired effector function (Sigal et al., 1986; Willumsen, et al., 1987). These mutations comprise the putative "effector region" at ras p21 residues 32-40, which is a candidate for a site on ras p21 that might interact with the as yet unidentified ras target. Mutations in the 32-40 region reduced the ability of mammalian ras p21 to function in both yeast (as assayed by the complementation of the ras2⁻ mutant defect) and mammalian cells, but did not affect biochemical properties such as GTP binding and membrane localization (Sigal et al., 1988b). This region is conserved among various evolutionarily diverse ras proteins. Additional support for the importance of the 32-40 region being involved in ras p21/target interaction was obtained by genetic analysis in the yeast S. cerevisiae. The Ser mutation at residue 42 of yeast RAS2 protein (analogous to residue 35 of mammalian ras p21) reduced the ability of the protein to function in yeast. This mutant was used in a genetic screen to find a second site suppressor of the RAS mutant. The dominant suppressor isolated, SSR2, was found to have a single amino acid substitution in the CYR gene that encodes yeast adenylyl cyclase (Marshall et al., 1988). Other groups have observed that p21 having substitutions in the 32-40 region are not GAP-responsive, a result implying that GAP might be the ras p21 target (Adari et al., 1988; Cales et al., 1988). A difficulty in interpreting some of the data obtained by using some of these mutations, is that several of these amino acid substitutions such as an aspartate to alanine at either position 35 and 38 also give a decreased intrinsic GTPase activity as we have determined using pure protein preparations (Sigal et al., 1988a). A more serious problem to address is that the lack of GAP-responsiveness does not necessarily imply the lack of protein/protein interaction. For example, the oncogenic forms of the ras proteins are biologically active and not stimulated by GAP. The 32-40 region mutant proteins are biologically inactive, yet they are also not stimulated by GAP. The pertinent question remains as to whether these mutant proteins interact with GAP.

A kinetic competition assay was developed to look at GAP binding, where competition was assayed by the ability of nonradioactive ras p21-GTP or p21-GDP complexes to inhibit GAP stimulated Ha ras p21 [32P]GTP hydrolysis. The Ha ras p21-GTP complex was found to compete with an IC_{50} value of 110 uM, whereas the GDP complex of Ha ras p21 did not compete when tested up to 1.5 mM. Similarly, the GTP, but not GDP complexes of oncogenic ras p21 variants, competed for GAP binding (Vogel et al., 1988). These results were the first demonstration that the oncogenic ras p21-GTP complex binds to GAP, a result that is consistent with GAP being the ras p21 target. The oncogenic mutations seem to affect the conformation of the protein that is necessary for the intrinsic and GAP-induced GTPase, and do not affect ras p21/GAP interaction. Similarly, the 32-40 region mutant proteins that were examined did not compete for GAP binding, which coincides with their reduced biology as assayed by microinjection into Xenopus oocytes and mammalian cells (Sigal et al., 1986; Sigal et al., 1988a). In summary, the above-mentioned results are consistent with, but not proof of, GAP being the ras p21 target.

MOLECULAR CLONING OF GAP

Recently we have reported the cloning of the cDNA for GAP (Vogel et al., 1988). Peptides were isolated from enzymatic digests of GAP and the derived amino acid sequence used to construct synthetic oligonucleotide probes. Several overlapping cDNA clones were isolated from bovine brain cDNA libraries. An open reading frame of 3135 base pairs encoding a protein of 115,742 daltons was obtained. The predicted amino acid sequence was identical to that determined for GAP peptides using protein purified from bovine brain. Furthermore, authenticity of the GAP cDNA isolated was established by functional expression in E. coli. E. coli-expressed GAP was also able to stimulate the GTPase activity of normal but not that of oncogenic ras p21.

We have recently obtained GAP cDNAs from mouse brain and human placenta cDNA libraries. Comparison of the human and bovine sequences reveals an extremely high level of sequence identity (approximately 96%) with a somewhat higher degree of sequence divergence observed in the N-terminal region of the protein (Fig. 1).

On Northern blot analysis of RNAs derived from a variety of species and cell lines, a single message was observed when probed with the bovine brain GAP cDNA (Vogel et al., 1988). Similar levels of GAP mRNA were seen by blot hybridization analysis of poly(A)⁺ RNA prepared from NIH3T3 cells and cell lines expressing oncogenic [Val12]Ha ras p21 or overexpressing normal ras p21 (Fig. 2). This result would suggest that GAP transcription in these cell lines is not altered by ras transformation.

Secondary structure prediction using Chou-Fasman analysis portrays an extremely hydrophobic N-terminal domain (rich in glycine, alanine and proline) proceded by a region with a high periodicity of turns, and a C-terminal domain with a high probability of α-helices. The hydrophobic domain may be important for GAP, which is localized in the cytoplasm, to interact with membrane-bound ras p21.

Homology searches using the current databases available, showed that GAP is a novel protein. A fairly low degree of sequence homology was observed with the noncatalytic domain of S. cerevisiae adenylyl cyclase.

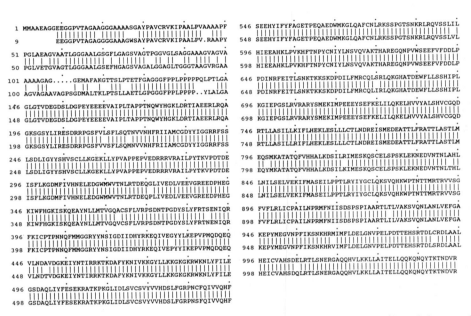

```
  1 MMAAEAGGEEGGPVTAGAAGGGAAAASGAYPAVCRVKIPAALPVAAAAPF      546 SEEHYIFYFAGETPEQAEDWMKGLQAFCNLRKSSPGTSNKRLRQVSSLIL
    |||||||||| ||||||  ||||| |||||||||||||||| |||             |||||||||||||||||||||||||||||||||||||||||||||| |||
  9 EEGGPVTAGAGGGGAAAGWSAYPAVCRVKIPAALPV.RAAPY            548 SEEHYIFYFAGETPEQAEDWMKGLQAFCNLRKSSPGTSNKRLRQVSSLVL

 51 PGLAEAGVAATLGGGAALGSGFLGAGSVAGTPGGVGLSAGGAAAGVAGVA      596 HIEEAHKLPVKHFTNPYCNIYLNSVQVAKTHAREGQNPVWSEEFVFDDLP
    ||| ||| ||| |||||| || ||||||| |||| || || |||||| |        |||||||||||||||||||||||||||||||||||||||||||||||||
 50 PGLVETGVAGTLGGGAALGSEFHGAGSVAGALGGAGLTGGGTAAGVRGAA      598 HIEEAHKLPVKHFTNPYCNIYLNSVQVAKTHAREGQNPVWSEEFVFDDLP

101 AAAAGAG.....GEMAFAKGTTSLPTETFGAGGGFPPLPPPPPQLPTLGA      646 PDINRFEITLSNKTKKSKDPDILFMRCQLSRLQKGHATDEWFLLSSHIPL
    | |||| |  | |||    |  |||| |||||||||||| |                |||||||||||||||||||||||||||| |||||||||||||||||||||
100 AGVAGAAVAGPSGDMALTKLPTSLLAETLGPGGGFPPLPPPP..YLALGA      648 PDINRFEITLSNKTKKSKDPDILFMRCQLIRLQKGHATDEWFLLSSHIPL

146 GLGTVDEGDSLDGPEYEEEEVAIPLTAPPTNQWYHGKLDRTIAEERLRQA      696 KGIEPGSLRVRARYSMEKIMPEEEYSEFKELILQKELHVVYALSHVCGQD
    ||||||||||||||||||||||||||||||||||||||||||||||||||        ||||||||||||||||||||||||||||||||||||||||||||||||||
148 GLGTVDEGDSLDGPEYEEEEVAIPLTAPPTNQWYHGKLDRTIAEERLRQA      698 KGIEPGSLRVRARYSMEKIMPEEEYSEFKELILQKELHVVYALSHVCGQD

196 GKSGSYLIRESDRRPGSFVLSFLSQTNVVNHFRIIAMCGDYYIGGRRFSS      746 RTLLASILLKIFLHEKLESLLLCTLNDREISMEDEATTLFRATTLASTLM
    |||||||||||||||| | ||||| ||||||||||||||||||||||||||       ||||||||| ||||||||||||||||||||||||||||||||||||||||
198 GKSGSYLIRESDRRPGSFVVSFLSQMNVVNHFRIIAMCGDYYIGGRRFSS      748 RTLLASILLRIFLHEKLESLLLCTLNDREISMEDEATTLFRATTLASTLM

246 LSDLIGYYSHVSCLLKGEKLLYPVAPPEPVEDRRRVRAILPYTKVPDTDE      796 EQSMKATATQFVHHALKDSILRIMESKQSCELSPSKLEKNEDVNTNLAHL
    ||||||||||||||||||||||||||||||||||||||||||||||||||        || |||||||||||||||| ||||||||||||||||||||||||| | |
248 LSDLIGYYSHVSCLLKGEKLLYPVAPPEPVEDRRRVRAILPYTKVPDTDE      798 EQYMKATATQFVHHALKDSILKIMESKQSCELSPSKLEKNEDVNTNLTHL

296 ISFLKGDMFIVHNELEDGWMWVTNLRTDEQGLIVEDLVEEVGREEDPHEG      846 LNILSELVEKIFMASEILPPTLRYIYGCLQKSVQHKWPTNTTMRTRVVSG
    || |||||||||||||||||||||||||||||||||||||||||||||||        ||||||||||||||||||||||||||||||||||||||||||||||||||
298 ISFLKGDMFIVHNELEDGWMWVTNLRTDEQPLIVEDLVEEVGREEDPHEG      848 LNILSELVEKIFMASEILPPTLRYIYGCLQKSVQHKWPTNTTMRTRVVSG

346 KIWFHGKISKQEAYNLLMTVGQACSFLVRPSDNTPGDYSLYFRTSENIQR      896 FVFLRLICPAILNPRMFNIISDSPSPIAARTLTLVAKSVQNLANLVEFGA
    |||||||||||||||||||||| ||||||||||||||||||||| |||||        ||||||||||||||||||||||||||||| |||||||||||||||||||
348 KIWFHGKISKQEAYNLLMTVGQVCSFLVRPSDNTPGDYSLYFRTNENIQR      898 FVFLRLICPAILNPRMFNIISDSPSPIAARTLILVAKSVQNLANLVEFGA

396 FKICPTPNNQFMMGGRYYNSIGDIIDHYRKEQIVEGYYLKEPVPMQDQEQ      946 KEPYMEGVNPFIKSNKHRMIMFLDELGNVPELPDTTEHSRTDLCRDLAAL
    |||||||||||||||||||| |||||||||||| || |||||||||||||        |||||||||||||| || ||||||||||||||||||||| |||| |||
398 FKICPTPNNQFMMGGRYYNSIGDIIDHYRKEQIVEPYYIKEPVPMQDQEQ      948 KEPYMEGVNPFIKSNKHRVIMFLDELGNVPELPDTTEHSRTDLSRDLAAL

446 VLNDAVDGKEIYNTIRRKTKDAFYKNIVKKGYLLKKGKGKRWKNLYFILE      996 HEICVAHSDELRTLSNERGAQQHVLKKLLAITELLQQKQNQYTKTNDVR
    |||| |||||||||||||||||||||||| |||||||||||||||||||||       |||||||| |||||||||||||||||||||||||||||||||||||||
448 VLNDTVDGKEIYNTIRRKTKDAFYKNIVKKGYLLKKGKGKRWKNLYFILE      998 HEICVAHSDQLRTLSNERGAQQHVLKKLLAITELLQQKQNQYTKTNDVR

496 GSDAQLIYFESEKRATKPKGLIDLSVCSVYVVHDSLFGRPNCFQIVVQHF
    |||||||||||||||||||||||||||||| |||||||||| | |||||
498 GSDAQLIYFESEKRATKPKGLIDLSVCSVYVVHDSLFGRPNSFQIVVQHF
```

Fig. 1. Sequence comparison between bovine GAP (top line) and human GAP (bottom line). Lines between amino acids designate identities.

A more impressive homology, due to the larger number of sequences involved, was observed between GAP and regions conserved among phospholipase C-148, the crk oncogene product, and the noncatalytic domain of the non-receptor tyrosine kinases (Fig. 3).

GAP DELETION ANALYSIS

As had been previously observed (Vogel et al., 1988), the first 128 amino acid residues of GAP are not necessary for GAP-stimulated ras p21 GTP hydrolysis. In a recent study we have extended this analysis further by generating a series of GAP mutants that had progressively larger deletions at the N- and C-termini. These mutant proteins were expressed as lacZ fusions in E. coli (M.S. Marshall et al., EMBO J., in press). A surprisingly large number of N-terminal residues could be deleted without affecting activity. A polypeptide consisting of the C-terminal 344 amino acids, which amount to approximately one-third of the protein, was found to have GAP activity in stimulating the GTPase of ras p21. The observed activity for this truncated protein was comparable to that of the full-length wild-type protein isolated from bovine brain. The purified deletion mutant was also able to interact equally well with oncogenic [Leu61]Ha ras p21 as did wild-type GAP when assayed by competition analysis. This C-terminal active domain had the same affinity and specific activity as did wild-type GAP. In contrast to the results observed for the N-terminal deletions, deletion of only 61 amino acid residues from the C-terminus rendered the protein insoluble in E. coli, which suggests that these C-terminal residues are necessary for the structural integrity of GAP.

DISCUSSION

GAP is the first mammalian protein to have been shown to interact with ras p21. We have now been able to demonstrate binding of the GTP but not GDP complexes of both normal and oncogenic forms of ras p21 to GAP. These results are consistent with, yet are not proof of a role for GAP as the ras target. Furthermore, we have shown that the oncogenic ras p21 variants are unable to adopt a conformation that is required for GAP-stimulated GTPase activity. The inability of biologically impaired ras p21 mutant proteins to interact with GAP provides us with further indication that GAP may indeed be the target of ras p21. However, we cannot exclude the possibility that GAP interacts at a distinct site that is conformationally affected by mutations in the effector region or that both GAP and a target protein interact at the same site.

The identification of a C-terminal catalytic domain is particularly interesting when taken together with the observed homologies in the N-terminal portion of GAP with regions conserved among PLC-148, CrK, and the non-receptor tyrosine kinases. The regions of homology (B and C domains, see Fig. 3) have been implicated in regulatory functions. An insertional mutation into the B box region in fps rendered the Fujinami avian sarcoma virus transformation defective, even though when expressed in E. coli the protein product was still able to display wild type enzymatic activity (Sadowski et al., 1986). Similarly, we may propose the homologous domain in GAP to interact with the C-terminal catalytic domain even though this does not seem to be the case in vitro. The putative interaction may be modified by phosphorylation (phosphorylation of GAP in vivo by the insulin receptor tyrosine kinase has been observed; Vogel et al., 1988, Gibbs et al., in press) or as yet unidentified

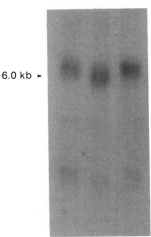

6.0 kb ►

Fig. 2. Blot hybridization analysis of GAP mRNA. Total RNA (20 μg) from [Val-12]Has ras p21 transformed cells (lane 1), Ha ras p21 transformed cells (lane 2), or NIH3T3 cells (lane 3) was electrophoresed on a formaldehyde gel, blotted onto nitrocellulose, and probed with [32P] labelled full-length GAP cDNA. Filter hybridization was carried out at 60°C in 5X SSC/0.1% SDS/1X Denhardt's solution/0.1% NaPPi. The filter was washed 3X in 2X SSC/0.1% SDS at 60°C.

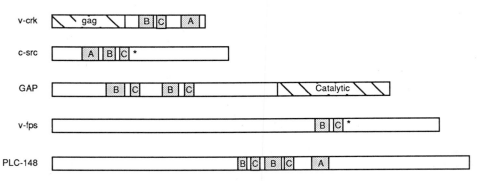

Fig. 3. Linear representations of the regions of homology between GAP and regions conserved among Crk, phospholipase C-148 (PLC-148), and the non-catalytic domain of the nonreceptor tyrosine kinases as exemplified by Src. Boxes B and C denote regions of sequence similarity and an asterix the start of the proposed catalytic domain.

proteins which could associate with this region. This modification could in turn involve a translocation to the plasma membrane, bringing GAP in close proximity to ras.

REFERENCES

Adari, H., Lowy, D.R., Willumsen, B.M., Der, C.J., and McCormick, F., 1988, Guanine triphosphatase activating protein (GAP) interacts with the p21 ras effector binding domain, Science, 240:518.

Cales, C., Hancock, J.F., Marshall, C.J., 1988, The cytoplasmic protein GAP is implicated as the target for regulation by the ras gene product, Nature, 332:548.

Gibbs, J.B., Schaber, M.D., Marshall, M.S., Scolnick, E.M., and Sigal, I.S., 1987, Identification of guanine nucleotides bound to ras-encoded proteins in growing yeast cells, J. Biol. Chem., 262-10426.

Gibbs, J.B., Schaber, M.D., Allard, W.J., Sigal, I.S., and Scolnick, E.M., 1988a, Purification of ras GTPase activating protein (GAP) from bovine brain, Proc. Nat. Acad. Sci. U.S.A., 85:5026.

Gibbs, J.B., Vogel, U.S., Schaber, M.D., Marshall, M.S., Diehl, R.E., Scolnick, E.M., Dixon, R.A.F., and Sigal, I.S., 1988b, Purification and molecular cloning of GAP, Proceeding of the EMBO G-Protein Workshop, in press.

Marshall, M.S., Gibbs, J.B., Scolnick, E.M., and Sigal, I.S., 1987, An adenylate cyclase from S. cerevisiae that is stimulated by ras proteins with effector mutations, Mol. Cell. Biol., 8:52.

Ricketts, M.H., and Levinson, A.D., 1988, High-level expression of c-H-ras1 fails to fully transform Rat-1 cells, Mol. Cell. Biol., 8:1460.

Sigal, I.S., D'Alonzo, J.S., Ahern, J.D., Marshall, M.S., Smith, G.M., Scolnick, E.M., and Gibbs, J.B., 1988a, The ras oncogene as a G-protein, "Advances in Second Messenger and Phosphoprotein Research" Vol. 21, Raven Press, New York, P.193.

Sigal, I.S., Marshall, M.S., Schaber, M.D., Vogel, U.S., Scolnick, E.M., and Gibbs, J.B., 1988b, Structure-function studies of the ras protein, LIII Cold Spring Harbor Symposium on Quantitative Biology, in press.

Trahey, M., and McCormick, F., 1987, A cytoplasmic protein stimulates normal N- ras p21 GTPase, but does not affect oncogenic mutants, Science, 238:542.

Willumsen, B.M., Papageorge, A.G., Kung, H., Bekesi, E., Robins, T., Johnsen, M., Vass, W.C., and Lowy, D.R., 1986, Mutational analysis of a ras catalytic domain, Mol. Cell. Biol., 6:2646.

Vogel, U.S., Dixon, R.A.F., Schaber, M.D., Diehl, R.E., Marshall, M.S., Scolnick, E.M., Sigal, I.S., and Gibbs, J.B., 1988, Cloning of bovine GAP and its interaction with oncogenic ras p21, Nature, 335:90.

THE COOPERATION BETWEEN VIRAL RAS GENES AND DIFFERENT IMMORTALIZING
GENES INDUCES THE TRANSFORMATION OF RAT THYROID EPITHELIAL CELLS.

A. Fusco[1,2], M.T. Berlingieri[1], M. Grieco[1], M. Santoro[1] and G. Vecchio[1].

1) Centro di Endocrinologia ed Oncologia del C.N.R. e/o Dipartimento di
   Biologia e Patologia Cellulare e Molecolare, II Facoltà di Medicina e
   Chirurgia, Università di Napoli, Napoli-Italia.
2) Istituto di Oncologia Sperimentale e Clinica, Facoltà di Medicina e
   Chirurgia di Catanzaro, Università di Reggio Calabria, via T.
   Campanella, Catanzaro.

Oncogenes of the ras family have been isolated from different human and
experimental carcinomas by transfecting high molecular weight DNA onto
NIH 3T3 fibroblasts. The activated ras genes differ from normal ones
by point mutations in specific regions of the genes (1, 2). This finding
has allowed to establish new more sophisticated and more sensitive
techniques to evidenciate point mutations of the ras genes in human
tumors. In fact, using a combinations of techniques, including specific
in vitro gene amplification by the polymerase chain reaction (PCR) and
mutation detection by cleavage at single base mismatches by RNAase A in
DNA:RNA and RNA:RNA heteroduplexes, it has been possible to evidentiate
point mutations of the ras-Ki oncogene in 21 out of 22 human carcinomas
of the exocrine pancreas and in the 40% of the human colorectal cancers
(3, 4). However, a problem that could be raised is whether or not these
mutations in the ras genes are the unique events responsible for the
carcinogenesis process. In fact, studies of chemical carcinogenesis, as
well as epidemiological analysis of malignancies in humans (5, 6, 7),
strongly suggest that the neoplastic transformation is a multistage
process and the observation that two different oncogenes are required in
concert for malignant conversion of nonestablished rat cells confirms
this point of view (8, 9). These last experiments have been performed on
fibroblast primary cells, therefore it would be extremely interesting to
know, by using an epithelial cell system, the steps necessary for

epithelial transformation. A permanent epithelial differentiated cell system could provide new insights about the role and modalities of cooperation of these genes in transformation and differentiation.

In our laboratory, a rat thyroid epithelial cell line, PC Cl 3, previously described, is available. This cell line retains in vitro the typical markers of thyroid differentiation, i.e. thyroglobulin (TG) synthesis and secretion, iodide uptake and dependence on TSH for the growth. Following infection with several murine acute retroviruses the PC Cl 3 cell line undergoes morphological changes, becomes no longer dependent on TSH for the growth, and shows modifications of the expression of the typical thyroid differentiation markers. However, no malignant phenotype is displayed by this cell line, such as growth in semisolid medium and tumorigenicity after injection into syngeneic animals or in athymic mice) (10).

For such reason this cell line can be considered a very useful model to study the steps of the carcinogenetic process in vitro. With such aim, we transfected this cell line with three immortalizing genes: Adenovirus E1A gene, the Polyoma large T gene and the viral and cellular myc. The plasmids used are represented by the plasmid pR15, carrying the Adenovirus E1A gene (11), plasmid pLT214 carrying the large T gene (12), the plasmid neo-v-myc G2 carrying the viral myc (13), the plasmid pMCGM1 carrying the human c-myc (14). The plasmids were co-transfected, following an already published procedure (15), with the pSV2neo plasmid, which confers the resistance to the geneticine, in a ratio of 10:1. After transfection, the cells were selected in G418 and the resistant cell clones were isolated. They have been firstly analyzed for the integration of the transfected gene. Fig 1 shows a Southern blot analysis obtained after digesting the DNAs extracted from PC Cl 3 and PC myc cells with the restriction enzymes Hind III, Eco RI and BamHI and hybridized with the pRyc 7.4 probe specific for the myc sequences. The lanes 4, 5 and 6 of the DNA from PC myc cells show bands other than the endogenous myc oncogenes detected in the PC Cl 3 cells (lanes 1, 2, 3). These bands presumably correspond to the human c-myc integrated sequences. Fig. 2 shows a Southern blot analysis of PC Cl 3 and PC large T hybridized with the plasmid pLT214. The bands corresponding to the integrated large T Polyoma sequences are present only in the transfected cells. The expression of the large T has been evaluated by RNA dot blot hybridization, as shown in fig. 3. The expression of the human c-myc transfected gene has been evaluated by S1 assay since Northern blot alone did not permit discriminating between the rat endogenous and the human exogenous c-myc gene expression in the PC myc cells. The S1

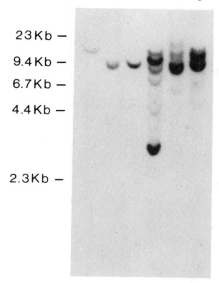

Fig. 1. Southern blot hybridization of DNA from PC Cl 3 and PC c-myc cell lines with the pRyc 7.4 probe specific for the human myc sequences (16). Lane 1, 2 and 3: PC Cl 3; Lane 3, 4, 6: PC c-myc. 20 g of DNA were digested with Eco RI (lanes 1 and 4), BamHI (lanes 2 and 5) and with HindIII (lanes 3 and 6).

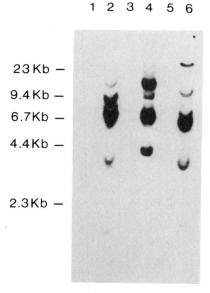

Fig. 2. Southern blot hybridization of DNA from PC Cl 3 and PC Large T cell lines with the pLT214 probe specific for the Large T gene of Polyoma virus. Lane 1, 3 and 5: PC Cl 3; Lane 2, 4, 6: PC Large T. 20 g of DNA were digested with Eco RI (lanes 1 and 2), BamHI (lanes 3 and 4) and with HindIII (lanes 5 and 6).

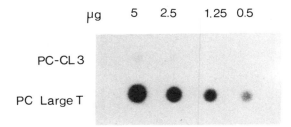

μg   5   2.5   1.25   0.5

PC-CL 3

PC Large T

Fig. 3. Dot blot hybridization of total RNA from PC Cl 3 and PC Large T cell lines with the pLT214 probe. RNA was spotted at the indicated quantities onto nitrocellulose filters and hybridized with 4x 10^6 cpm of _in vitro_ nick-translated pLT214.

     1   2   3   4   5   6   7

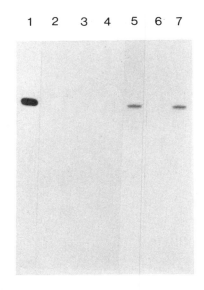

Fig. 4. Nuclease S1 Analysis of Human c-myc Transcript in Transfected Cells. Nuclease S1 analysis was carried out according to Berk and Sharp (17) by using 5' end labelled human c-myc cDNA clones. The $5'-^{32}P$-end-labelled DNA probes were heat denatured, hybridized in 80% deionized formamide to total RNA (5 and 20 g) at 55°C for 10 hr., digested with 80 units of nuclease S1 (P-L Biochemicals), and analyzed by electrophoresis on a 7 M urea 4% polyacrylamide gel. The DNA probe was 5'end-labelled by the method of Maxam and Gilbert (18). RNAs derive from the following cell lines: lanes 2 and 3: PC Cl 3; lanes 4 and 5: PC c-myc Cl 1; lanes 6 and 7: PC c-myc Cl 2. In lane 1 no nuclease S1 enzyme was added to the probe.

analysis, shown in fig. 4 and performed according to a previously published procedure (17) demonstrates that no expression of human myc gene is detectable in PC Cl 3 cells, whereas it is clearly expressed in PC myc cells.

All the transfected cell lines were analysed for the expression of the differentiation and transformation markers. The results, shown in table 1, indicate that there is no appearance of the markers typical of neoplastic transformation in all the transfected cells, whereas contrasting effects are achieved by the immortalizing genes on thyroid differentiation markers. In fact, no effect can be observed after transfection with the large T gene of Polyoma virus. Only a reduced capabilty to trap iodide appears in the PC Cl 3 cells transfected with c-myc and v-myc, whereas the differentiation is completely abolished in the PC Cl 3 cells transfected with the adenovirus E1A oncogene.

The transfected cell lines have subseguently been infected with the Harvey sarcoma virus carrying the v-ras-Ha oncogene. The infected cells have been then selected by growing them in the absence of the six growth factors, and then investigated for the acquisition of the markers typical for the fully neoplastic transformation. The results are shown in table 1, and clearly indicate that the cooperation between the, so

TABLE 1

Differentiation and transformation markers in PC Cl 3 cells transfected with the adenovirus E1A gene, large T of Polyoma virus and human and viral myc or/and infected with the Harvey murine sarcoma virus.

CELL TYPE	IODIDE UPTAKE	TG PRODUCTION	GROWTH IN AGAR	TUMOR INCIDENCE
PC Cl 3	+++	+	-	-
PC E1A	---	-	-	-
PC Large T	+++	+	-	-
PC v-myc	+--	+	-	-
PC c-myc	+--	+	-	-
PC v-ras-Ha	---	-	-	-
PC E1A + v-ras-Ha	---	-	+	+
PC Large T + v-ras-Ha	---	-	+	+
PC v-myc + v-ras-Ha	---	-	+	+
PC c-myc + v-ras-Ha	---	-	+	+
FRT-Fibro	---	-	-	-

called immortalizing genes, and the v-ras-Ha oncogene lead the cells to the neoplastic phenotype. No significant differences in the expression of the malignant phenotype is observed with the different immortalizing genes. It is interesting to notice that, since PC Cl 3 is a permanent cell line, the requirements for large T, E1A and human or viral myc for cooperation with ras indicate that the so-called immortalizing gene products may have on epithelial cells some other effect, different from that already known with primary embryo fibroblasts and therefore a step, different from immortalization, must be considered for epithelial carcinogenesis.

Our experiments have allowed us to establish a cellular system in which the steps leading to a fully malignant phenotype may be clearly identified and analyzed. Moreover, this cellular system has the great advantage of consisting of epithelial cells, which are known to represent the target cells for more than 90% of human neoplasia. On the other hand, cooperation between genes other than ras genes might be investigated in this cellular system, thus helping the comprehension of the possible role of other oncogenes belonging to a family different from the ras in the development of epithelial cancers.

Acknowledgements

This work was supported by the Progetto Finalizzato delle Ricerche of the C.N.R. and by the Associazione Italiana per la Ricerca sul Cancro.

REFERENCES

1. M. Barbacid, ras genes, Annu. Rev. Biochem. 56:779 (1987).

2. H.E. Varmus, The molecular genetics of cellular oncogenes, Annu. Rev. Genet. 18:553 (1984).

3. C. Almoguera, D. Shibata, K. Forrester, J. Martin, N. Arnheim, and M. Perucho, Most human carcinomas of the exocrine pancreas contain mutant c-K-ras genes, Cell 53:549 (1988).

4. J.L. Bos, E.R. Fearon, S.R. Hamilton, M. Verlaan-de Vries, J.H. Van Boom, A.J. van der Eb and B. Vogelstein, Prevalence of ras gene mutations in human colorectal cancers, Nature 327:293 (1987).

5. I. Beremblum, The mechanism of carcinogenesis, Cancer Res. 1:807 (1941).

6. P. Armitage and R. Doll, The age distribution of cancer and a multistage theory of carcinogenesis Br. J. Cancer 8:1 (1954).

7. J. Cairns, Mutation selection and the natural history of cancer, Nature, 255:197 (1975).

8. H. Land, L.F. Parada and R.A. Weimberg, Tumorigenic conversion of primary embryo fibroblasts requires at least two cooperating oncogenes, Nature, 304:596 (1983).

9. H.E. Ruley, Adenovirus early region E1A enables viral and cellular transforming genes to transform primary cells in culture, Nature 304:602 (1983).

10. A. Fusco, M.T. Berlingieri, P.P. Di Fiore, G. Portella, M. Grieco, and G. Vecchio, One- and two-step transformation of rat thyroid epithelial cells by retroviral oncogenes, Mol. Cell. Biol. 2:3365(1987).

11. N. Stow, The infectivity of Adenovirus genomes lacking DNA sequences from their left-hand termini, Nucleic Acid Res. 5105 (1982).

12. M. Rassoulzadegan, A. Cowie, A. Carr, N. Gleichenhaus, R. Kamen and F. Cuzin, The roles of individual Polyoma virus early protein in oncogenic transformation, Nature 300:713 (1982).

13. G. Falcone, I.C. Summerhayes, H. Peterson, C.J. Marshall and A. Hall, Partial transformation of mouse fibroblats and epithelial cell lines with the v-myc oncogene. Exp. Cell Res. 168:273 (1987).

14. D. Spandidos, Mechanisms of carcinogenesis: the role of oncogenes, transcriptional enhancers and growth factors, Anticancer Res., 5:485 (1985).

15. F.L. Graham and A.J. van der Eb, A new technique for the assay of the infectivity of human adenovirus 5 DNA, Virology, 52: 456 (1973).

16. K. Nishikura, A.A. Rushdi, J. Erikson, R. Watt, G. Rovera and C.M. Croce. Differential expression of the normal and the translocated human c-myc oncogenes in B cells, Proc. Natl. Acad. Sci. U.S.A., 80:4822 (1983).

17. A.J. Berk and P.A. Sharp, Sizing and mapping of early adenovirus mRNAs by gel electrophoresis of S1 endonuclease-digested hybrid, Cell, 12:721 (1977).

18. A.M. Maxam and W. Gilbert, Sequencing labeled DNA with base specific chemical changes cleavages, Methods Enzymol., 65:499 (1980).

NATO ADVANCED RESEARCH WORKSHOP ON "RAS ONCOGENES"
November 10-15, Vouliagmeni, Athens, Greece

SEATED (left of right)

1.	Pizon, V.	9.	Bucci, C.
2.	Tsiziyotis, C.	10.	Green, S.
3.	Marshall, C.	11.	Spandidos, D.
4.	Pintzas, A.	12.	Gallwitz, D.
5.	Dautry, F.	13.	Perucho, M.
6.	Borrello. M.G.	14.	Yiagnisis, M.
7.	Gunzburg, W.H.	15.	Lemoine, N.
8.	Lowe, P.N.	16.	Tamanoi, F.

STANDING (left to right)

1.	Fasano, O	20.	Maruta, H.
2.	Wyllie, A.H.	21.	Willumsen, B.
3.	Bailey. P.D.	22.	Denner, J.
4.	Chardin, P.	23.	Bos, H.
5.	Housiaux, P.J.	24.	Hall, Al
6.	Burgering, B.	25.	Vogel, U.
7.	Kinsella, A.R.	26.	Gaum, E.Z.
9.	Bade, E.G.	27.	Pierotti, M.A.
10.	Balmain, A.	28.	Kumar, R.
11.	Boukamp, P.	29.	Fusco, A.
12.	Kakkanas, A.	30.	Lacal, J.C.
13.	Rutherford, T.	31.	Lacal, (Mrs.)
14.	Connolly, J.A.	32.	Jacobs, A
15.	Lord, P.G.	33.	Stephenson, P.
16.	Colicelli, J.	34.	Shih, T.Y.
17.	Newbould, S.	35.	de Gunzburg, J.
18.	Wittinghoffer, A.	36.	Doppler, W.
19.	Maruta, (Mrs.)		

PARTICIPANTS

AUSTRALIA

Housiaux, Philip J.
  Ludwig Institute for Cancer Research
  Post Office Royal Melbourne Hospital
  Victoria 3050, Australia

Maruta, Hiroshi
  Ludwig Institute for Cancer Research
  Post Office Royal Melbourne Hospital
  Victoria 3050, Australia

CANADA

Connolly, Joe A.
  Dept. of Anatomy, Faculty of
  Medicine, University of Toronto
  Medical Science Building
  Toronto, Ontaria M5S 1A8, Canada

DENMARK

Willumsen, Berthe
  University Institute of Microbiology
  Oster Faraniagsgade 2A
  DK-1353 Copenhagen K, Denmark

FRANCE

Chardin, P.
  Unite N 248, INSERM
  Laboratoire de Genetique et
  expression Des Oncogenes
  Faculte De Medicine Lariboisiere
  Saint-Louis, 10 Avenue de Verdun
  75010 Paris, France

Dautry, F.
  Institut Gustave-Roussy
  Laboratoire D'Oncologie Moleculaire
  Rue Camille Desmoulins
  94805 Villejuif Cedex, France

Haliassos, A.
  Laboratoire de Biochimie Genetoque
  Inserm U 15 et 129, CNRS UA1147
  Institut de Pathologie Moleculaire
  24 Rue du Faubourg Saint-Jacques
  75104 Paris, France

Riou, Guy
  Laboratoire de Pharmacologie
  Clinique et Moleculaire
  Institut Gustave Roussy
  94800 Villejuif, France

Pizon, Veronique
  Laboratoire de Genetique et
  Expressiondes Oncogenes
  Institut National de la Sante
  et de la Resherche Medicale
  Unite NO 248 - INSERM
  Faculte de Medecine Lariboisiere
  St. Louis, 10 Avenue de Verdun
  75010 Paris, France

F.R. GERMANY

Bade, E.G.
  Universitat Konstanz
  Fakultat fur Biologie, Postfach 5560
  D-7750 Konstanz, F.R. Germany

Boukamp, Petra
  Institut fur Biochemie
  Deutsches Krebsforschungszentrum
  Im Neuenheimer Feld 280
  D-6900 Heidelberg 1, F.R. Germany

Buchmann, Albrecht
  German Cancer Research Center
  Institut for Biochemistry
  6900 Heidelberg, F.R. Germany

Fasano, Ottavio
  Europen Molecular Biology Laboratory
  Postfach 10.2209, Meyerhofstrasse 1
  6900 Heidelberg, F.R. Germany

F.R. GERMANY (continued)

Gallwitz, Dieter
  Abteilung Molekulare Genetic
  Karl-Friedrich-Nohhoeffer-Institute
  Max-Planck-Institute fur
  Buophysikalische Chemie
  D-3400 Gottingen-Nikolausberg
  F.R. Germany

Gunzburg, Walter H.
  Abt. fur Molekulare Zellpathologie
  GSF-Munchen,
  Ingolstadterlandstrassel
  D-8042 Neuherberg, F.R. Germany

Schmeiser, H.H.
  Institut fur Toxikologie and
  Chemotherapie
  Deutsches Krebsforschungszentrum
  Im Neuenheimer Feld 280
  D-6900 Heidelberg 1, F.R. Germany

Wittinghofer, A.
  Abteilung Biophysik
  Max Planck Institute fur
  Medizinische Forschung
  Johnstrasse 29
  D-6900 Heidelberg 1, F.R. Germany

GREECE

Kakkanas, Athanasios
  Hellenic Pasteur Institute
  127 vas Sofias Ave., Athens, Greece

Pintzas, Alex
  Hellenic Pasteur Institute
  127 vas Sofias Ave., Athens, Greece

Spandidos, Demetrios
  Biological Research Center
  National Hellenic Research
  Foundation
  48 Vas Constantinou Avenue
  Athens 11635, Greece

Tsiziyotis, Christos
  Biological Research Center
  National Hellenic Research
  Foundation
  48 vas Constantinous Ave.
  Athens 11635, Greece

Yiagnisis, Mary
  Biological Research Center
  National Hellenic Research
  Foundation
  48 vas Constantinous Ave.
  Athens 11635, Greece

IRELAND

Keenan, Alan K.
  Department of Pharmacology
  University College, Foster Ave.
  Blackrock, Dublin, Ireland

ISRAEL

Zeev, Lev
  Department of Biology
  The Technion - Israel Institute
  for Technology
  Haifa 32000, Israel

ITALY

Borrello, Maria G.
  Division of Experimental Oncology A
  Instituto Nazionale per la studio
  e la Cura die Tumori, 20133 Milano
  via g Venezian 1, Italy

Bucci Cecilia
  Dipartmento Di Biologia e Patologia
  Cellulare e Moleculaire, "L. Cifano"
  Via Pinsini 5, 80131 Napoli, Italy

Fusco, Alfredo
  Departmento di Biologia e Patologia
  Cellulare e Moleculare Centro
  di Endocrinologia ed Oncologia
  Sperimentale del C.N.r.
  Universita degli studi di Napoli
  via S Pansini 5, 80131 Napoli, Italy

Pierotti, Marco A.
  Division of Experimental Oncology A
  Instituto Nazionale per lo Studio
  e la cura dei Tumori
  20133 Milano, via g Venezian 1, Italy

NETHERLANDS

Bos, Hans
  Department of Medical Biochemistry
  University of Leiden
  Sylvius Laboratories, P.O. Box 9503
  2300 RA Leiden, The Netherlands

Burgering, Boudewijn
  Dept. of Medical Biochemistry
  Sylvius Laboratories
  Unversity of Leiden, P.O. Box 9503
  2300 RA Leiden, The Netherlands

320

SWITZERLAND

Doppler, Wolfgang
 Ludwig Institute for Cancer Research
 Inselspital, CH-3010 Bern
 Switzerland

UNITED KINGDOM

Balmain, Allan
 The Beatson Institute for
 Cancer Research
 Garscube Estate, Bearsden
 Glasgow G61 1BD, United Kingdom

Bailey, P.D.
 Dept. of Chemistry
 University of York
 Heslington, York YO1 5DD
 United Kingdom

Carter, Graham
 University of wales College of
 Medicine, Dept. of Haematology
 University Hospital of Wales
 Heath Park, Cardiff CF4 4XN, Wales

Denner, Joachim
 Marie Curie Research Institute
 The Chart, Oxted, Surrey RH8 OTL
 United Kingdom

Green, Stephen
 Imperial Chemical Industries plc
 Control Toxicology Lab.
 Alderley Park, Macclesfield
 Cheshire SK1U 4TJ
 United Kingdom

Hall, Alan
 The Institute of Cancer Research
 Chester Beatty Laboratories
 Fulham Road, London SW3 6JB
 United Kingdom

Hawley, T.
 School of Biological Sciences
 Univesity of East Anglia
 Norwich NR4 7TJ, United Kingdom

Jacobs, Allan
 University of wales College of
 Medicine, Dept. of Haematology
 University Hospital of Wales
 Heath Park, Cardiff CF4 4XN
 United Kingdom

Kinsella, Anne R.
 Paterson Institute for Cancer
 Research, Wilmslow road
 Manchester M20 9Bx, United Kingsom

Lemoine, Nick
 CRC Thyroid Tumor Biology Research
 Group, Dept. of Pathology
 University of Wales, College of
 Medicine, Cardiff CF4 4XN
 United Kingdom

Lowe, Peter N.
 Department of Molecular Scinces
 Wellcome Research Laboratories
 Langley Court, Beckenham
 Kent BR3 3BS, United Kingdom

Lord, P.G.
 Pharmaceuticals Division
 Imperial chemical Industries Plc
 Mereside Alderley Park
 Macclesfield, Cheshire SK10 4TG
 United Kingdom

Marhall, Chris
 Institute of Cancer Research
 Chester Beatty Laboratories
 Fulham Road, London SW3 6JB
 United Kingdom

Montgomery, G.W.G.
 Department of Chemistry
 Univesity of York, Heslington
 York YO1 500, United Kingdom

Newbould, Susan A.
 School of Biological Sciences
 University of East Anglia
 Norwich NR4 7TJ, United Kingdom

Stephenson, P.
 School of Biological Sciences
 University of East Anglia
 Norwich NR4 7TJ, United Kingdom

Rutherford, Tim
 Department of Haematology
 St. George's Hospital Medical School
 University of London, Crammer Terrace
 Tooting, London SW17 ORE
 United Kingdom

Wyllie, Andrew H.
 Dept. of Pathology, University
 Medical School, Teviot Place
 Edinburgh E48 9AG, London England

Prescott, A.R.
 School of Biological Sciences
 University of East Anglia
 Norwich, NRA 7TJ, United Kingdom

UNTIED STATES OF AMERICA

Baum, Ellen Z.
  Lederle Labs, Bldg. 96, Room 217
  Pearl River, New York 10965

Colicelli, John
  Cold Spring Harbor Laboratory
  Cold Spring Harbor, New York 11724

de Gunzburg, Jean
  Whitehead Institute for Biomedical
  Research, 8 Cambridge Center
  Cambridge, Mass. 02142

Kumar, R.
  Developmental Oncology Section
  NCI-Frederick Cancer Research
  Facility, P.O. Box B
  Frederick, Maryland 21701

Lacal, Juan Carlos
  Laboratory of Cellular and
  Molecular Biology
  National Institutes of Health
  Bldg 37, Room 1E24
  Bethesda, Maryland 20892

McCormick, Frank
  Cetus Corporation
  1400 53rd Street
  Emeryville, Calif. 94608

Perucho, Manuel
  Department of Biochemistry
  State University of New York at
  Stony Brook, N.Y. 11794-5215

Shih, Thomas Y.
  Laboratory of Molecular Oncology
  National Cancer Institute NIH
  NCI-Frederick Cancer Research
  Facility, P.O. Box B
  Frederick, Maryland 21701

Stacey, Dennis W.
  Dept. of Molecular Biology
  Building NN06, The Cleveland Clinic
  Foundation, 9500 Euclid Avenue
  Cleveland, Ohio 44106

Tamanoi, Fuyuhiko
  Dept. Biochemistry and Molecular
  Biology, University of Chicago
  920 East 57th Street
  Chicago, Illinois 60637

Vogel, Ursula
  Dept of Molecular Biology
  Merck, Sharp, and Dohme
  Research Laboratories
  West Point, Pa. 19486

INDEX